Photochemical Water Splitting

Materials and Applications

ELECTROCHEMICAL ENERGY STORAGE AND CONVERSION

Series Editor: Jiujun Zhang
National Research Council Institute for Fuel Cell Innovation
Vancouver, British Columbia, Canada

Published Titles

Electrochemical Supercapacitors for Energy Storage and Delivery: Fundamentals and Applications
Aiping Yu, Victor Chabot, and Jiujun Zhang

Proton Exchange Membrane Fuel Cells
Zhigang Qi

Graphene: Energy Storage and Conversion Applications
Zhaoping Liu and Xufeng Zhou

Electrochemical Polymer Electrolyte Membranes
Jianhua Fang, Jinli Qiao, David P. Wilkinson, and Jiujun Zhang

Lithium-Ion Batteries: Fundamentals and Applications
Yuping Wu

Lead-Acid Battery Technologies: Fundamentals, Materials, and Applications
Joey Jung, Lei Zhang, and Jiujun Zhang

Solar Energy Conversion and Storage: Photochemical Modes
Suresh C. Ameta and Rakshit Ameta

Electrochemical Energy: Advanced Materials and Technologies
Pei Kang Shen, Chao-Yang Wang, San Ping Jiang, Xueliang Sun, and Jiujun Zhang

Electrolytes for Electrochemical Supercapacitors
Cheng Zhong, Yida Deng, Wenbin Hu, Daoming Sun, Xiaopeng Han, Jinli Qiao, and Jiujun Zhang

Electrochemical Reduction of Carbon Dioxide: Fundamentals and Technologies
Jinli Qiao, Yuyu Liu, and Jiujun Zhang

Metal–Air and Metal–Sulfur Batteries: Fundamentals and Applications
Vladimir Neburchilov and Jiujun Zhang

Photochemical Water Splitting: Materials and Applications
VNeelu Chouhan, Ru-Shi Liu, and Jiujun Zhang

Photochemical Water Splitting
Materials and Applications

Neelu Chouhan
Ru-Shi Liu
Jiujun Zhang

CRC Press
Taylor & Francis Group
Boca Raton London New York

CRC Press is an imprint of the
Taylor & Francis Group, an **informa** business

CRC Press
Taylor & Francis Group
6000 Broken Sound Parkway NW, Suite 300
Boca Raton, FL 33487-2742

First issued in paperback 2019

© 2017 by Taylor & Francis Group, LLC
CRC Press is an imprint of Taylor & Francis Group, an Informa business

No claim to original U.S. Government works

ISBN-13: 978-1-4822-3759-7 (hbk)
ISBN-13: 978-0-367-86991-5 (pbk)

Library of Congress Cataloging-in-Publication Data

Names: Chouhan, Neelu. | Liu, Ru-Shi. | Zhang, Jiujun.
Title: Photochemical water splitting : materials and applications / Neelu
Chouhan, Ru-Shi Liu, Jiujun Zhang.
Description: Boca Raton : CRC Press, 2017. | Series: Electrochemical energy
storage and conversion
Identifiers: LCCN 2016032394| ISBN 9781482237597 (hardback : alk. paper) |
ISBN 9781315279657 (ebook)
Subjects: LCSH: Photoelectrochemistry. | Water--Electrolysis.
Classification: LCC QD578 .C46 2017 | DDC 546/.225--dc23
LC record available at https://lccn.loc.gov/2016032394

Visit the Taylor & Francis Web site at
http://www.taylorandfrancis.com

and the CRC Press Web site at
http://www.crcpress.com

Contents

Series Preface

The goal of the Electrochemical Energy Storage and Conversion series is to provide comprehensive coverage of the field, with titles focusing on fundamentals, technologies, applications, and the latest developments, including secondary (or rechargeable) batteries, fuel cells, supercapacitors, CO_2 electroreduction to produce low-carbon fuels, water electrolysis for hydrogen generation/storage, and photoelectrochemistry for water splitting to produce hydrogen, among others. Each book in this series is self-contained, written by scientists and engineers with strong academic and industrial expertise who are at the top of their fields and on the cutting edge of technology. With a broad view of various electrochemical energy conversion and storage devices, this unique book series provides essential reads for university students, scientists, and engineers and allows them to easily locate the latest information on electrochemical technology, fundamentals, and applications.

Jiujun Zhang
National Research Council of Canada

Preface

On January 26, 2011, during his State of the Union address, U.S. President Barack Obama stated: "We're issuing a challenge. We're telling America's scientists and engineers that if they assemble teams of the best minds in their fields and focus on the hardest problems in clean energy, we'll fund the Apollo projects of our time.... At the California Institute of Technology, they're developing a way to turn sunlight and water into fuel for our cars.... We need to get behind this innovation." This reflects the importance and relevancy of solar water splitting in fuel generation. We like to underline the word fuel that is to be a sustainable and renewable fuel, which can produce energy/power without releasing any additional carbon dioxide to the atmosphere, is the biggest challenge for the mankind. As the current convention fuels are gradually running out, sooner or later they will be completely exhausted. We have to be prepared for this situation with a better fuel substitute, otherwise a big energy mess will be created that will end life due to scarcity of food, water, and energy. Therefore, it's high time to think and act in the direction of ultimate energy sources. There is an extreme need to line up the industrialists, socialists, intellectuals, and political leaders to meet the current energy challenges, while taking great care of our environment. Hydrogen, as a product of water splitting, is a clean and green solution to the aforementioned problem. Nature provides us a clean and renewable source of hydrogen in the form of water. Unfortunately, cleavage of the water to its constituents, that is, hydrogen and oxygen, at an industrial scale represents one of the "Holy Grails" of materials sciences. To facilitate this reaction, the combination of catalytic material and solar energy has been recognized as a feasible approach to break water. Although renewable energy sources such as sunlight and water are available almost free of cost, developing stable, efficient, and cost-effective photocatalytic materials to split water is a big challenge. As devoted efforts to develop effective materials have continued over the last few decades, various materials with size and structures from nano to giant have been explored. Plentiful materials such as metal oxides, metal chalcogenides, carbides, nitrides, and phosphides of various composition like heterogeneous, homogeneous, plasmonic, mesomorphic, metamaterial, and new graphene-based materials have been tested. There have been critical discussions on the merits and demerits of the studied systems. However, some real technological breakthroughs in material development are definitely necessary for practical applications and commercialization of the technology. To accelerate the research and development activities in the area of water splitting, this book may act as a catalyst. Moreover, this book gives a comprehensive overview and description on both fundamentals and applications of photocatalytic water splitting focused on the recent advances in materials. It also highlights the need for common parameters for studying solar water-splitting phenomena. In addition, it provides insight into the various current and past practices and available databases by emphasizing the pros and cons of the existing and future technologies that are and will be used in water splitting. The book as a whole is our humble effort to give a panoramic view of the developments made in photocatalytic water splitting since the process was discovered.

Neelu Chouhan
University of Kota

Ru-Shi Liu
National Taiwan University

Jinjun Zhang
National Research Council of Canada

Introduction

This book comprises seven chapters. Chapter 1 introduces hydrogen as a green and efficient fuel to satisfy the energy needs of future generations. Relevant issues such as hydrogen fuel efficiency, production, application, safety, the hydrogen economy, environmental effects, and so on are covered in this chapter. Chapter 2 discusses the basic concepts of photochemical water splitting in order to equip readers with basic terminology and fundamental concepts such as electrochemistry of the water splitting phenomena, selection criteria of photocatalytic material, excitation binding energy, overpotential, diffusion length, carrier mobility and penetration in a photocatalyst, electrode overpotential, band gap and band edge position, band edge bending, efficiency, and so on. Chapter 3 discusses the different practical methods of hydrogen generation from water splitting using techniques such as electrolysis, thermochemical water splitting, biocatalytic water splitting, mechanoocatalytic water splitting, plasmolysis, electrolysis, magnetolysis, radiolysis, and photocatalytic and photoelectrocatalytic water splitting. This chapter gives a better understanding of how photochemical methods work and their benefits compared to other methods in water splitting. Chapter 4 describes different aspects of photoelectrochemical (PEC) water splitting, including factors affecting efficiency of PEC; semiconducting photoelectrode materials (electron transfer phenomenon, material and energetic requirements); models of the water splitting process; reactor design and operation, gradient/bias-based reactors; and reactors based on suspension and electrode type. The chapter emphasizes the electrochemistry involved in the water splitting process and various electron transfer reactions at different interfaces of electrodes/cocatalysts, electrode/electrolyte, electrode/sensitizers in the presence of the sacrificial electrolyte at active sites. This is a very important chapter that provides information about the materials involved in different stages of the photocatalytic processes, which is valuable for rational design and optimization of the PEC reactor's efficiency. It also focuses on the challenges and future perspectives of the field. Chapter 5 deals with oxide semiconductors such as ZnO, TiO_2, Fe_2O_3, and WO_3, as well as graphene oxide, which are used as photocatalytic materials for water splitting. This chapter includes the innovative ways to improve the efficiency of the devices such as band gap engineering of the metal oxide, doping, making a solid solution, and addition of quantum dots (QDs)/dyes or plasmonic materials for visible light sensitization, as well as incorporating a Z-scheme to the system. Moreover, a photocatalyst designed at nanoscale can be synthesized and unique aspects of the nanotechnology are discussed in detail. A special attention is given to some metal ion–doped metal oxide photocatalysts. Chapter 6 concentrates explaining the mechanism of the photocatalytic cleavage of water in the presence of scavenger electrolytes (electron scavenger and hole scavenger), photocorrosion, methods for photocorrosion prevention, the mechanism of heterogeneous electrocatalysis, and the mechanism of homogeneous molecular catalysis. The techniques to bridge the gap between heterogeneous electrocatalysis and homogeneous molecular catalysis are also illustrated with suitable examples. This chapter also describes the role of metallic/metallic hydroxide cocatalyst in the hydrogen evolution reaction (HER)/oxygen evolution reaction (OER) and the nature/role of the active sites on catalyst's surface. Some conceptual advancements of the active materials for hydrogen generation through water splitting are explained in brief. Chapter 7 is devoted to describing the most significant technological advances and vivid aspects of the nanostructured semiconducting materials that are used for water splitting, including their structural properties, energetic transport dynamics, and the material design and strategies to enhance the photoresponse of nanomolecular devices. Different nanoforms of the materials like nanocrystalline, thin films, mesoporous, plasmon resonant, metamaterials are discussed with advancement schemes. Current state-of-the-art key challenges with future approaches in the development of efficient PEC cells for water splitting are also discussed in this chapter.

Authors

Neelu Chouhan, PhD, is an associate professor and the head of the Department of Pure and Applied Chemistry at the University of Kota in India. She earned her BSc in 1989 from MDS University, Ajmer, India, a MSc in 1991, a MPhil in 1993, and a BEd in 1996. She earned her PhD on *Organic Conducting Materials* from Monanlal Sukhadia University, Udaipur, India, in 2006. Dr. Chouhan worked as a lecturer in chemistry at SRD Modi College, Kota (1996–1998), and at Govt. PG College, Bundi (1998–2012). She carried out her 2 years of postdoctoral research fellowship from 2008 to 2009 at the Department of Chemistry, National Taiwan University, Taiwan, and worked on photocatalytic nanomaterials for water splitting. Her research interests are organic superconductors, functional materials, nanomolecular devices, and the photochemistry of water splitting. She is the author or coauthor of more than 30 publications in international scientific journals of high impact factor with a good number of citations, contributed to seven chapters and three books of national/international publications, and has also been granted one international patent.

Ru-Shi Liu, PhD, is currently a professor at the Department of Chemistry, National Taiwan University, Taipei, Taiwan. He earned his BSc in chemistry from Soochow University, Taiwan, in 1981 and the MSc in nuclear science from National Tsing Hua University, Taiwan, in 1983. From 1983 to 1985, he worked at the Materials Research Laboratories, the Industrial Technology Research Institute, Taiwan. He earned two PhDs in chemistry—one from National Tsing Hua University in 1990 and the other from the University of Cambridge in 1992. Dr. Liu was an associate professor at the Department of Chemistry in National Taiwan University from 1995 to 1999 before he was promoted to a full professor in 1999. He has also served as an adjunct Pearl Chair professor at the National Taipei University of Technology, Taiwan, since August 2014. His research is focused on the field of materials chemistry. He is the author or coauthor of more than 550 publications in international scientific journals. He has also been granted more than 100 patents.

Jiujun Zhang, PhD, earned his BSc and MSc from Peking University in 1982 and 1985, respectively, and his PhD in electrochemistry from Wuhan University in 1988. Dr. Zhang is now a principal research officer and core competency leader at Energy, Mining, and Environment Portfolio (NRC-EME) of the National Research Council of Canada, Montreal, Canada. Zhang holds several adjunct professorships, including one at the University of Waterloo and the other at the University of British

Columbia. He is the author or coauthor of more than 400 publications with more than 20,000 citations (h-index: 62; i10-index: 142), including 230 peer-reviewed journal papers, 18 books, and 41 book chapters, and has been granted 16 U.S./European/Canada patents. His research is mainly based on electrochemical energy storage and conversion. He has been elected as a fellow of the Electrochemical Society of Electrochemistry (FISE), fellow of the Royal Society of Chemistry (FRSC), fellow of the Engineering Institute of Canada (FEIC), and fellow of the Canadian Academy of Engineering (FCAE).

1 Introduction to Hydrogen as a Green Fuel

1.1 INTRODUCTION

Hydrogen gas can be seen as a future renewable energy (RE) carrier/fuel by virtue of the fact that it gives water as a combustion product without evolving the "greenhouse gases" such as CO_2. Hydrogen is considered the most clean and storable energy carrier of the future if it can be produced from a renewable energy source via a CO_2-neutral and efficient route. Solar water splitting is a renewable and sustainable fuel production method because it can utilize sunlight, the most abundant energy source on Earth, and water, the most abundant natural resource available on Earth. Water splitting can be carried out using coupled solar cell–water electrolysis systems, but efficiency loss between the two systems and high installation cost make them a less attractive option. As an alternative route, the direct photoelectrochemical (PEC) cell is potentially more economical because it combines the functions of a solar cell and an electrolyzer in a single device.

1.2 CURRENT ENERGY SCENARIO

Energy is a lever to trigger the speed of development in all segments of life (economical, social, and political) for safer, affordable, cleaner, and more habitable environmental conditions, that required for better standard of living. A secured, uninterrupted, affordable, and adequate energy supply is required to sustain global economic growth and stability. Worldwide our current energy storage contains 1047.7 billion barrels of oil, 5501.5 trillion standard cubic foot (scf) of natural gas (NG), and 984 billion tons of coal as conventional energy sources that might be sufficient to satisfy our energy needs for 40.2, 53.8, and 205 years, respectively. In addition, these energy sources are not proportionately distributed throughout the world. For example, the United States contains about 25% of world coal reserves while Middle East countries account for about 60% of oil reserves. This results in energy insecurity in countries that have inadequate energy assets/supply that are the most probable grounds for political disturbances. According to the estimated record, nearly one-quarter of world's population (1.6 billion) still does not have electricity today. However, the continuous increase in energy requirements has been putting a lot of pressure on conventional energy sources. But the limited availability of the fossil fuels and corresponding environmental threats compel us to explore an uninterrupted supply of energy by utilizing alternative sources. All conventional sources of

1

energy are carbon rich and so their combustion leads to CO_2 emission (main greenhouse gas) that adds to the extra burden on its naturally occurring amount. CO_2 as a greenhouse gas absorbs the infrared part of the sun's radiation and reradiates it back to Earth's surface, which traps the heat and keeps Earth 30° warmer than it would be otherwise—but without greenhouse gases, Earth would be too cold to live. But the additional CO_2 leads to an extra rise in temperature (van Ruijven et al. 2011). As a consequence, Earth's average temperature increases, which will result in unpredictable changes in weather patterns in the form of floods, droughts, and submerging of low-lying areas due to melting of ice at the poles. The current concentration level of CO_2 in the atmosphere is around 390 ppm (in January 2011) and scientists suggest that this value should drop to 350 ppm, otherwise it should lead to irreversible catastrophic effects. CO_2 emission was found to be 27 gigatons in 2005 and is expected to boost up to 42 gigatons in 2030 and 62 gigatons in 2050. The countrywise contribution to CO_2 emission is shown in Figure 1.1 (U.S. Environmental Protection Agency 2013). Most of this emission comes from power, industrial, and transportation sectors.

Our energy consumption history (1990–2011) and futuristic energy projections (2011–2035) are shown in Figure 1.1b, which will be increased at a rate of 1.4% per year till 2035. This energy consumption profile reflected that the world average capacity

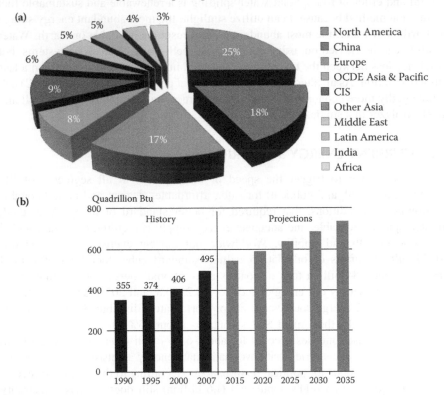

FIGURE 1.1 (a) Countrywise per capita CO_2 emission in percentage. (b) World energy consumption history (1990–2011) and futuristic energy projections (2011–2035); 1 Quadrillion BTU = 1015 BTU = 1015 x 1054 Joules. (From U.S.Environmental Protection Agency, Light-Duty Automotive Technology, Carbon Dioxide Emissions, and Fuel Economy Trends: 1975 Through 2012, March 2013.)

of energy utilization rates have continued to rise over time, from about 65% in 1990 to about 80% today, with some increases still anticipated in the future. This increasing demand for energy, imposed challenges in front of us as the threat of disruptive climate and huge capital investments in energy segment. Both are becoming problematic issues for developed and developing countries. According to a report, there is an irrational ratio of the population and energy consumption rate found among the developed (20% and 60%) and developing countries (80% and 40%) (Nezhad 2009). Meeting this demand without further damage to the environment is a great challenge. There are two main approaches to achieve a long-range energy scenario: the first scenario involves the replacement of the long-term development process with advanced energy-producing technologies and/or implementation of hybrid processes instead of conventional fuels to reduce fuel consumption and to reduce the climate change effect via new advanced technologies used for gas conversion and, so, more fixations of the gases to value-added products will be achieved. The second scenario includes the development of alternative energy resources. The six available renewable resources in nature, biomass, hydropower, wind, solar, geothermal, and biofuels, are economically, socially, and environmentally sustainable. But no single approach is able to achieve the goal. Therefore, a number of energy scenarios are given by different agencies, including the Energy Information Administration (EIA), World Energy Council (WEC), International Energy Agency (IEA), and many more, using different proportions of both approaches. The most comprehensive and authentic analysis on the world energy scenario, based on the world's facts and perceptions, was given by the IEA, which has constituted a committee of 5000 experts from 39 countries on Energy Research and Technology to develop a strategy for the world energy scenario for 2050. In their report they concluded that world energy consumption will be doubled by 2050 and carbon emission rate will increase by a factor of 2.5. Their recommendations focused on alternative resources of energy. To shape the world energy future, IEA projected strategic energy scenario planning for 2050, which is represented in Figure 1.2 (Nezhad 2007). This includes a step-by-step process to achieve the desired scenario. The process is initiated by identifying the scope of the scenario and then defining the main driving forces behind it. For developing the scenario model, system dynamics of the future energy market will be analyzed by identification of the interrelationship among the driving forces. Continuation of the status quo is not sustainable due to the rising demand for energy, particularly for fossil fuel, and unacceptable level of the CO_2 emission. Therefore, by employing the existing technologies or those that are under development, a path toward clean, competitive energy will be established by utilizing sustainable energy solutions. Subsequently, possible future energy scenario (a few models suggested by the IEA are the baseline for the accelerated technologies and blue map) models would be developed. They assume the future energy demand and level of CO_2 emissions in light of the above-defined driving forces and accordingly set periodical goals to meet the most probable energy assessment and reduced CO_2 emission level for 2050 by utilizing decision support software. Finally, strategies will be developed to accomplish the goal. No single strategy is enough to reach the desired level of energy production, consumption, and CO_2 reduction. Therefore, a fusion of the following strategies might be used for achieving the desired goal:

1. Research development, demonstration, and deployment of new technology
2. Investment strategies

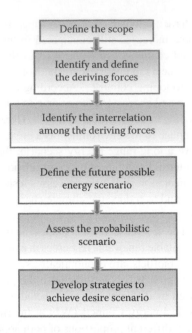

FIGURE 1.2 Stepwise energy scenario planning. (From Nezhad, H., *Software Tools for Managing Project Risk: Management Concepts,* Management Concepts, pp. 1037–55, Vienna, 2007.)

3. CO_2 emission reduction
4. Intensity reduction
5. International collaborations
6. Governmental engagement

By employing existing technologies or those that are under development, the world could be brought to a much more sustainable energy path.

1.3 FUEL: PAST, PRESENT, AND FUTURE

Energy is the main driving force behind all types of development/progress in each and every facet of life through advancement in different sectors: power, transportation, industrial, agricultural, commercial, public services, living standard growth, and so on. Its consumption is a basic criteria for determination of the living standard and prosperity of a community or country, which depend upon different factors, namely, access to energy sources, prices, climate, income, and urbanization level (Jiang and O'Neill 2004). Fuel is the axis of development that provides us energy. The major driving forces behind the world fuel supply and demand are fuel growth rate, energy consumption rate, investment requirements, demographic changes, oil resource prices, global energy density, alternative energy sources, CO_2 emission, and technology innovation and improvement (Figure 1.3a). The famous geophysicist H.K. Hubbert (1971) predicted that the fossil fuel era would be of very short duration and that the world's conventional energy resources will run out by 2100 and in the United States by 2050 (Figure 1.3b) (Hubbert 1956). His predictions are proving correct and we cannot deny the current fuel status or avoid the future scenario.

The cost paid (by means of energy crisis, climate change, ecological imbalance, pollution, human health, etc.) for this economical prosperity and development is too high. Almost three-quarters of the world's energy is provided by burning of fossil fuels (Yüksel 2008). The Energy Information Administration estimated in 2007 that the primary sources of energy consisted of petroleum 36.0%, coal 27.4%, and

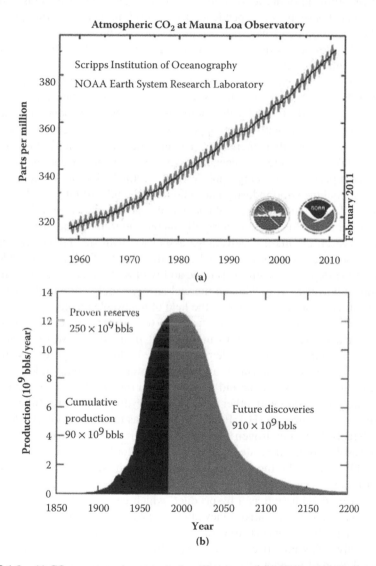

FIGURE 1.3   (a) CO_2 concentration over the last 50 years and (b) Hubbert's bell-shaped curve plot for time versus production of any exhaustible resources for the time interval 1850–2200 AD. (From Hubbert, M. K., Nuclear Energy and Fossil Fuels, *Paper presented before the Spring Meeting of the Southern District Division of Production*, American Petroleum Institute Plaza Hotel, March 7–9, San Antonio, TX, 1956.)

NG 23.0%, amounting for 86.4% share of fossil fuels in the primary energy consumption of the world. Fossil fuels are nonrenewable resources as they take millions of years to form and they are being depleted with much faster rate than new reserves are being created. However, looking beyond the scarcity of conventional oil, saving the climate of our planet is the most important reason for switching from fossil fuels to renewable resources. Now, it is high time to think seriously and act accordingly on the choice of the sources of energy, and renewable/nonconventional/ alternate energy is no doubt a good choice over conventional fuels. This can offer substantial benefits such as independence from world market fossil fuel prices and the creation of millions of new green jobs. It can also provide energy to the two billion people currently without access to energy services (Hinrich-Rahlwes et al. 2013). The common feature of all renewable energy sources, such as wind, Sun, Earth's crust (geothermal), hydropower, biomass, biofuels, and hydrogen derived from renewable resources, is that they produce little or no greenhouse gases and are virtually an unlimited "fuel." Furthermore, the attempts to introduce regenerative energy resources are met with varied opposition. Because of the serious depletion in available oil stocks and their increasing demand, the political and industrial lobby have started taking an interest in regenerative resources of energy and included them as an essential part of their energy policy. Some technologies are already competitive: the solar and the wind industry have maintained double-digit growth rates over 10 years now, leading to faster technological deployment worldwide (Hairston 1996). The use of RE technology has been rapidly increasing to meet the growing energy demand. However, the main disadvantage associated with standalone RES is their inability to provide energy security and reliability due to their unpredictable, seasonal, and time-dependent natures. But the ability of RES to face energy challenges to maintain a universal available commercial energy supply without damaging the environment will sustain and progress the field of renewable technology.

After looking at the merits and demerits of the different renewable materials, available technologies and ways to make profit from them, hydrogen has been selected study. Being the lightest element and the highest energy density material (142 kJ/kg), hydrogen has enormous and diverse potential for renewable power, and can be used as an energy carrier and an eco-benign fuel. Hydrogen in nature exists primarily in combination with other elements. For hydrogen to be useful as a fuel, it must exist as free hydrogen (H_2). Furthermore, it has the highest energy-to-weight ratio of all fuels: 1 kg of hydrogen has the same amount of energy as 2.1 kg of NG or 2.8 kg of gasoline (Global Strategy Institute, Center for Strategic and International Studies 2005). Being the most abundant element of the universe, hydrogen holds the great promise of providing a clean and renewable source of energy that produces water on combustion as a by-product, which can be recycled. Hydrogen is the cleanest and ideal alternative/renewable available fuel. Besides the wide public acceptance and great concern toward the good ecological health of the planet, there are only a few countries (e.g., Turkey, the Netherlands, etc.) with a countable contribution of regenerative resources of energy in energy/power generation. With regard to hydrogen, there are many benefits and two drawbacks for using it as a fuel with existing technology. First, storage of liquid hydrogen requires four times the storage space of conventional petroleum-based fuels. Second, the mode of hydrogen production depends on the availability of a nonrenewable resource (petroleum). Hydrogen fuel technologies can be used in areas from prototypes (to optimize power outputs from internal combustion engines [ICEs], gas turbine engines, and fuel cells [FCs]) to usable military hardware. But its broad adaptation has stalled due to its low density and because its storage and transportation are difficult. Here, advancement in technology can help to overcome these problems. One way to solve this problem is to use metal hydrides, metallic compounds that incorporate hydrogen atoms, as a storage medium for hydrogen. In this technique, the metal binds to hydrogen to produce a solid metal hydride 1000 times or much smaller than the original hydrogen gas cylinder storage ability. The hydrogen can then later be released from the solid by heating it to a given temperature. Now the question arises as to how to produce hydrogen, which is well answered in Section 1.7. After the faster expansion of research and experiments, solar and geothermal heating systems became prominent renewable segments, which can empower the enormous and diverse potential of hydrogen as a renewable power resource. After 2025, hydrogen generated by electrolysis and renewable electricity will be introduced as a third renewable fuel in the transport sector, complementing the biofuels and direct use of renewable electricity. Hydrogen is also applied as a chemical storage medium for electricity from renewable resources and used in industrial combustion processes and cogeneration for provision of heat and electricity, and also for short periods of reconversion into electricity. Hydrogen generation can have high energy losses; however, the limited potential of biofuels and probably also the battery electric mobility make it necessary to have a third renewable option. Alternatively, this renewable hydrogen could be converted into synthetic methane or liquid fuels depending on economic benefits (storage costs vs. additional losses) as well as technology and market development.

1.4 HYDROGEN AS A CHEMICAL FUEL

Hydrogen is the lightest and smallest element in the universe and possesses exclusive physical and chemical properties that offer benefits as well as challenges to its successful widespread adoption as a fuel. Moreover, the abundance of hydrogen is seen in almost 95% of the visible matter (mass) on Earth. It appears naturally on Earth's crust in very small portions and has a tendency to bind with other elements such as carbon and oxygen instead of existing free in its molecular "H_2" form. Almost all fuels are chemical fuels but the unique properties of H_2 such as high diffusivity, low viscosity, inimitable chemical nature, combustibility, and electrochemical properties are the characteristics that make hydrogen a different or better fuel than other gases. Apart from conventional fuels it falls into the category of alternative, nonconventional, or advanced fuels. This class includes well-known alternative fuels such as biodiesel, bioalcohol (methanol, ethanol, butanol), chemically stored electricity (batteries and FCs), nonfossil methane, nonfossil NG, vegetable oil, propane, and other biomass sources. The hydrogen molecule exists in two forms (*ortho* and *para*), distinguished by the relative rotation of the nuclear spin of the individual atoms in the molecule. Molecules with spins in the same direction (parallel) are called *ortho*-hydrogen and those with spins in the opposite direction (antiparallel) are called *para*-hydrogen. Figure 1.4 shows that the *ortho*-hydrogen will convert to *para*-hydrogen as the temperature of hydrogen is lowered.

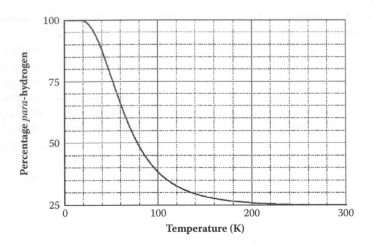

FIGURE 1.4 Equilibrium percentage of *para*-hydrogen and *ortho*-hydrogen versus temperature. (From McCarty, R.D. et al., *Selected Properties of Hydrogen (Engineering Design Data)*, NBS Monograph 168, National Bureau of Standards, Boulder, CO, 1981.)

The equilibrium mixture of *ortho*- and *para*-hydrogen at any temperature is referred to as equilibrium hydrogen. The equilibrium *ortho–para*-hydrogen mixture with a content of 75% *ortho*-hydrogen and 25% *para*-hydrogen at room temperature is called normal hydrogen. The *ortho*- to *para*-hydrogen conversion is accompanied by a release of heat, 703 J/g (302.4 British thermal units per pound mass [Btu/lb]) at 20 K (−423°F), with 527 J/g (226.7 Btu/lb) heat required for normal to *para*-hydrogen conversion. Catalysts are used to accelerate this conversion, which produces almost pure *para*-hydrogen liquid (≥95%), unless it is a slow process that occurs at a finite rate (taking several days to complete) and continues even in the solid state.

It is an odorless, invisible, nontoxic, noncarcinogenic, nonpoisonous, smoke-free, and highly buoyant (lighter than air) gas that rises and diffuses most easily when leaked. Mixtures of hydrogen with air, oxygen, or other oxidizers are highly flammable over a wide range of compositions. Addition of He, CO_2, N_2, H_2O, and argon are used to reduce the flammable range for hydrogen in air (Coward and Jones 1952). Water is the most effective substance in reducing the flammability range, and helium/Ar is the least effective (Coley and Field 1973). H_2 can be commonly identified by the thermal wave that produces low radiant heat and has the widest range of flammability among ($H_2 \approx 4\%–75\%$) the class (Table 1.1). Moreover, hydrogen has a low autoignition temperatures, higher octane rating than that of conventional gasoline, and the highest heating value, that is, 52,000 Btu/lb, among all the available fuels. Flammability range allows for lean mixtures with better fuel economy and lower combustion temperature than gasoline. Hence, hydrogen engines perform more efficiently than gasoline engines. It possesses a compression ratio higher than gasoline and lower than diesel that can be beneficial when used as a fuel for ICEs because lean

TABLE 1.1

Comparative Qualities of Hydrogen with Other Gasoline

Properties/Fuel	H$_2$	CH$_4$ (CNG)	Gasoline	Diesel
Molecular weight	2 (no-carbon)	16 (75% carbon)	100–105 (C$_4$–C$_{12}$, 88%)	200–300 (C$_9$–C$_{25}$, 87%)
Density K(g/L)	0.0899 × 10^{-3}(g), 70.990 (L)	1.8160 × 10^{-3}(g), 422.36 (L)	0.745	0.832
Autoignition temperature (°C)	385	540–630	260–460	180–320
Air/fuel ratio	34.3	17.2	14.6	14.5
Ignition energy (mJ)	0.002	0.28	0.24	–
Diffusion coefficient (cm³/s)	0.61	1.90	21.34	–
Energy density (kJ/kg)	142.00	45.30	48.6	33.8
Combustion temperature (°C)	2318 (O$_2$)	1914 (O$_2$)	2307	2327
Combustion range in air (%)	4–75	5.3–15	1.4–7.6	0.6–5.5
Explosive range in air (%)	13–79	19	–	–
Octane rating	130	87–93	91–99	–
Fuel efficiency in combustion internal engine (%)	60	–	22	45

Source: College of the Desert, *Hydrogen Fuel Cell Engines and Related Technologies, Revision 0,* College of the Desert, Palm Desert, CA, 2001.

Note: CNG, compressed natural gas.

mixtures that raise efficiency (diesel engines > hydrogen engines > gasoline [petrol] engines), especially at low power, engine idle, and lower combustion temperatures, can help to suppress the amount of emissions of nitrogen oxides, which can still occur in ICEs fueled by hydrogen. Moreover, in terms of efficiency 57 million metric tons of hydrogen is equal to about 170 million tons of oil equivalent. The growth rate of hydrogen production is around 10% per year. The aforementioned properties along with the availability of hydrogen strongly support the candidature of hydrogen as a green and clean fuel. The main purpose of fuel is to store energy, which should be in a stable form and can be easily transported to the place of production. The user employs this fuel to generate heat or perform mechanical work, such as powering an engine. It may also be used to generate electricity, which is then used for heating, lighting, or other purposes to perform work.

1.5 THE HYDROGEN ECONOMY

In the current hydrocarbon economy, gasoline is the primary fuel for transportation and on the other hand coal rules the power industries. Combustion of these hydrocarbon fuels emits CO_2 and other pollutants that is directly or indirectly hazardous to Earth. Moreover, the supply of economically usable hydrocarbon resources is also limited worldwide and has to compete with the day-by-day increasing demand for hydrocarbon fuels, particularly in China, India, and other developing countries. The solution to this situation lies in the gradual shuffling of the conventional energy sources to renewable alternatives. Some of the renewable alternative sources of energy are biofuel/biomass, hydrothermal, solar, wind, geothermal, and so on. Out of these sources, solar energy is the natural choice for clean energy as the sun gives us an inexhaustible energy source and a guaranteed powerhouse of energy during our existence on Earth that may provide an ample amount of energy free of cost (1 second, 100% exposure of the sun to total surface area of the Earth. We get the astonishingly huge amount of 400 trillion trillion watts. produces enough energy to meet the current needs of Earth for 500,000 years). If we use this solar energy for the production of hydrogen from carbon-free chemicals (H_2O, H_2S, etc.), then it will be a win–win situation from both perspectives, that is, energy and the environment. Therefore, hydrogen is considered as the most clean and eco-benign fuel, and has a dual character, that is, as an energy carrier and a storage medium. Therefore, hydrogen acts as a good fuel. In looking to the great possibilities in hydrogen as a fuel for transportation and process industries, Lawrence W. Jones of the University of Michigan coined the term hydrogen economy (Jones 1970) to combat the negative effects of hydrocarbon fuels. The sketches of modern interest in the hydrogen economy were first revealed by a technical report of Sorensen (1975). He underlined the need of clean energy in a society in his study and that it could be derived from renewable resources (Sørensen 1975) and constituted the first energy scenario, illustrating the technical usage of hydrogen. World-scale hydrogen economy promoters argued that hydrogen can be an environmentally cleaner source of energy to end users, particularly in transportation applications, without releasing major pollutants (such as particulate matter) or CO_2 at the point of end use. Other technical obstacles faced include hydrogen storage issues and the purity requirement of hydrogen used in FCs—with current technology, an operating FC requires the purity of hydrogen to be as high as 99.999%. Metal-organic frameworks (MOFs), single-wall carbon nanotubes (2.5%–3.0% w/w hydrogen; Dillon et al. 1997), fullerenes (Chen and Wu 2004), activated carbon at low temperature (Hirano et al. 1993), organic chemical hydrides (7.3% w/w hydrogen) (Newton et al. 1998), silica microspheres (Schmitt et al. 2006), and metal hydrides (5.0%–7.0% w/w hydrogen) (Schlapbach and Zuttel 2001) are often used for storage purposes. In metal hydrides, hydrogen storage is possible at very low pressure that might be lower than the pressure in a car tire; this means we can store an enormous amount of hydrogen in a very small space, for example, magnesium can store a maximum of around 8% w/w of hydrogen. Therefore, hydrogen appears to be an attractive substitute to the current fossil fuel–based energy economy for the future, unless the adverse global environmental impacts are observed in production or usage of this alternative energy. A 2004 analysis affirmed that "most of the hydrogen supply chain pathways would release significantly less CO_2 into the atmosphere than would gasoline used in hybrid electric vehicles" and that significant reductions in CO_2 emissions would be possible if carbon capture or carbon sequestration methods were utilized at the site of energy or hydrogen production. An Otto cycle ICE running on hydrogen is said to have a maximum efficiency of about 38% (8% higher than a gasoline ICE.) The combination of the FC and electric motor is two to three times more efficient than an ICE. On the other hand, hydrogen engine conversion technology is more economical than FCs.

The scientists have also estimated the potential effects of hydrogen-based energy systems on the environment/climate, which would be much lower than those from fossil fuel–based energy systems. However, such impacts will depend on the rate of hydrogen leakage during its synthesis, storage, and use. The researchers have calculated that a global hydrogen economy with a leakage rate of 1% of the produced hydrogen would produce a climate impact of 0.6% of the fossil fuel system that it replaces (Derwent et al. 2006). If the leakage rate was 10%, then the climate impact would be 6% of that of the fossil fuel system. The current study suggests that a future hydrogen-based economy would not be free from climate disturbance, although this may be considerably less pronounced than that caused by the current fossil fuel energy systems. Care should be taken to reduce hydrogen leakage to a minimum, if the potential climate benefits of a future global hydrogen economy are to be realized (Derwent et al. 2006). At this time, the hydrogen economy faces many technical challenges and high startup costs that are mainly focused on infrastructure and storage (GM 2007 [http://www.gm.com/company/gmability/adv_tech/400_fcv/fc_challenges.ht]).

For the practical and efficient implementation of hydrogen vehicles on the road, the government has to strategically construct an appropriate number of hydrogen gas fueling stations along the roadways and in the urban city centers. GM proposed that for establishing a network of 11,700 fueling stations to cover 130,000 miles of highway to connect the important transit zones in the most crowded 100 urban areas in the United States, it would cost US$10–US$15 billion. Shell Hydrogen has also estimated that the cost of setting up a functioning network of hydrogen gas filling stations in the United States and Europe will be US$20 billion and US$6 billion for Japan. Notwithstanding these estimated costs, Shell, in partnership with GM, strongly believes in the apt development and the quick establishment of futuristic technology to advocate the hydrogen economy (Shell 2004).

1.6 HYDROGEN PRODUCTION

In contrast to conventional fossil fuels, which can be mined or extracted, the uniqueness of the hydrogen lies in that it can be produced from the variety of feedstocks, including oil, coal, NG, biomass, water, and so on. Presently, the main feedstock for hydrogen is NG because the efficiency is high and the production cost is relatively low. But hydrogen has also been introduced in this scenario as the third renewable fuel that might be playing the most important role in transport sector by 2025. Currently, most of the hydrogen is generated from NG (48%), raw petroleum products (30%), coal (18%), and electrolysis of water (4%) for industrial use, but petroleum supplies may become limited in the near future (Figure 1.5a through c) (Rühl et al. 2012).

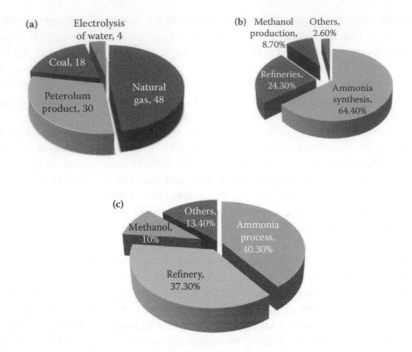

FIGURE 1.5 (a) Percentage of production of hydrogen from petroleum products, natural gases, coal, and electrolysis of water, (b) worldwide percentage consumption of hydrogen in various process industries, and (c) percentage hydrogen usages in process industries in the United States.

(From Rühl, C. et al., *Energy Policy*, 50, 109–116, 2012.)

Hydrocarbons are the most common feedstock for hydrogen production and the technologies used in this process are divided into two major categories, that is, the oxidative and nonoxidative processes, which can be further classified into following categories.

1.6.1 OXIDATIVE PROCESS

Oxidative hydrogen production from hydrocarbon occurs at high temperatures (>700°C) in the presence of the oxidant (O_2, CO_2, steam, or combination thereof). It is represented by the following generic equation:

$$C_n H_m + [O_x] \rightarrow xH_2 + yCO + zCO_2 \qquad (1.1)$$

where $C_n H_m$ is a hydrocarbon ($n \geq 1$; $m \geq n$), and $[O_x]$ is an oxidant, which can decide the nature of oxidative process could be endothermic (when $[O_x]$ is O_2) or exothermic (when $[O_x]$ is H_2O, CO_2, or $H_2O–CO_2$) or thermoneutral (when $[O_x]$ is $O_2–H_2O$, $O_2–CO_2$).

1.6.1.1 Steam Methane Reforming

Steam methane reforming (SMR) is a well-developed and widely available approach for manufacturing industrial hydrogen. Forty percent of the World's hydrogen is synthesized by using this technique. It is an endothermic reaction that is carried out in the presence of catalyst at temperatures of 800°C–1000°C and pressures of 10–50 atm.

$$CH_4 + H_2O \rightarrow 3H_2 + CO \quad \Delta H = 206 \text{ kJ/mol} \tag{1.2}$$

Figure 1.6a and b depicts the simplified block diagram of SMR of two major technologies that differs by the final treatment of the product: (1) included solvent CO_2 removal and a methanator and (2) equipped with a pressure-swing adsorption system. Hydrogen production from conventional fuel (NG, naphtha petroleum products, etc.) is started by a desulfurization unit (DSU), where organic sulfur compounds are first converted into H_2S by catalytic hydrogenation; afterward H_2S is converted into ZnS through scrubbing over a ZnO bed at 340°C–390°C (Figure 1.6a and b) (Armor 1999). Desulfurization is required to enhance the sensitivity of the SMR process and to avoid the sulfur poisoning of the catalysts used in different units of the process. Traces of halides present in feedstock (NG, naphtha, etc.) were removed by using a alumina guard bed. In the SMR unit (reactor unit with 15-m long and 12-cm inner diameter Ni alloy glass tube), hydrocarbons are transformed into hydrogen under steam flowed over the catalytic bed at 700°C–950°C and 2–2.6 MPa, as shown in Figure 1.6a and b. High-temperature (340°C–360°C, Fe–Cr catalyst) and low-temperature (200°C–300°C, Cu–Zn–Al) reactors are commonly used for CO conversion to H_2, which is an exothermic reaction. Finally, synthesized H_2 is separated from CO, CO_2, and purified. An older version of SMR used solvents (monoethanolamine, water, ammonia solutions, potassium carbonate, methanol, etc.) to remove acid gas (CO_2) and remaining residual CO_2 and CO is converted into CH_4 in the methanation reactor. The modern version incorporates a pressure swing adsorption (PSA) unit that operates under 20-atm pressure and contains adsorption beds filled with molecular sieves of suitable pore size for H_2 purification from CO_2, CO, and CH_4 impurities. PSA off-gas consists of 55% CO_2, 27% H_2, 14% CH_4, 03% CO, 0.4% N_2, and water, which is burnt as fuel in the primary reformer furnace.

1.6.1.2 Autothermal Reforming

Sorption-enhanced reformer: In the sorption-enhanced reforming technique gaseous products of catalytic reforming like CO_2 (separated by CaO) would be separated out from the reaction zone by a sorption process. Three main reactions proceed simultaneously in this method: (1) a fluidized bed reactor (FBR) contains a mixture of reforming catalyst and CO_2 acceptor at 750°C, (2) Carbonation reaction for balancing the heat input required by the endothermic reforming reaction, and (3) FBR regenerating reaction that rejuvenates the reforming catalyst and spent acceptor ($CaCO_3$) is accomplished in an adiabatic FBR regenerator at about 975°C. The beauty of this procedure lies in having fewer processing steps and improved efficiency, elimination of the need for shift catalysts, and reduction of temperature in the primary reforming reactor by 150°C–200°C.

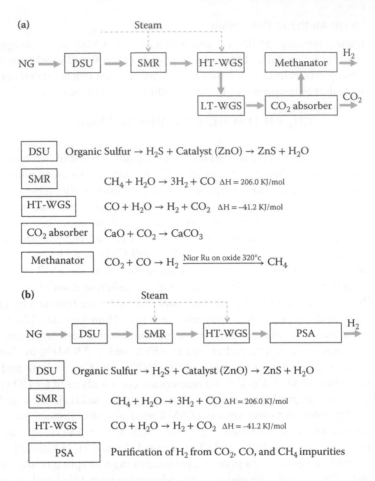

FIGURE 1.6 Schematic hydrogen production by steam methane reforming (SMR) using desulfurization unit (DSU), catalytic reforming, high-temperature/low-temperature water–gas shift (WGS) unit for CO conversion to H_2: (a) older version of SMR that includes special feature for CO_2 removal unit from solvent and methanation unit; (b) modern version of SMR with special feature, that is, the pressure swing adsorption system. (From Muradov, N. and Nazim, Z.,, *Hydrogen Fuel: Production, Transport and Storage,* CRC Press, Taylor and Francis, Washington, DC, 2009.)

H_2 membrane reformer: Hydrogen produced in the reforming reaction selectively permeates through a membrane (Pd or Pd/Ag or other Pd-based alloys of several micron thick) to exit the reaction zone. The main advantages of the H_2 membrane reformer (HMR) are mentioned in the following:

1. H_2 producing reactions are free from the chemical equilibrium
2. High methane conversion efficiency at lower temperature
3. System produces separate H_2 and CO_2
4. No need for additional CO shift converter

5. More simple and compact configuration
6. High purity hydrogen (99.999%) production
7. Higher overall efficiency (H_2 yield 93% and energy efficiency 70%)

Use of a membrane increases the efficiency of methane conversion by 7% over its thermodynamic equilibrium value.

1.6.1.3 Partial Oxidation

Partial oxidation (POx) of hydrocarbons is another way to produce hydrogen on a commercial scale. This is an exothermic process, where fuel and oxygen are combined in such a proportion that will produce a mixture of H_2 and CO. If the pure oxygen is used in POx process then the cost of the hydrogen production would be enhanced over the POx process, which uses air as oxidizer but in this case the effluent would be rich in nitrogen that needs large water-gas shift (WGS) reactors and gas purification system. The process can be carried out in the presence (low temperature 600°C–900°C) and absence (high temperature 1100°C–1500°C) of a catalyst. The catalyst POx method does not utilize the high carbonaceous feedstock that includes heavy residual oils (HROs) and coal. The catalytic process uses light hydrocarbon (naphtha, NG, etc.) feedstock.

Thermal POx: Pressurized oxygen is used for high-temperature (1250°C–1500°C) combustion at 3–12-MPa pressure. At these operational conditions, the reaction kinetics is very fast; therefore, no catalyst is used in the process. The major advantage of this method is that it can utilize all kinds of feedstock lighter than HROs, including petroleum coke and high sulfur content and high heavy metal content that is restricted to be used as fuel due to environmental hazards. The major reaction during POx of sulfurous heavy oil fractions can be presented by the following chemical reactions:

$$C_mH_nS_p + m/2O_2 \rightarrow mCO + (n/2 - p)H_2 + pH_2S \text{ (exothermic)} \tag{1.3}$$

$$C_mH_nS_p \rightarrow (m - x)C + (n - y)H_2 + (m - z)CH_4 + C_kH_l + pH_2S \text{ (endothermic)} \tag{1.4}$$

$$C + \tfrac{1}{2}O_2 \rightarrow CO \text{ (exothermic)} \tag{1.5}$$

$$C + H_2O \rightarrow H_2 + CO \text{ (endothermic)} \tag{1.6}$$

Catalytic POx: Catalytic POx (CPO) reactions are performed in the presence of a heterogeneous catalyst (refractory supported Ni and noble metals such as Rh, Pt, Pd, Ir, Ru, Re, perovskites [$LaNi_{1-x}Fe_xO_3$], etc.), and O_2(air) as an oxidant at 800°C, as follows:

$$CH_4 + \frac{1}{2}O_2 \rightarrow CO + 2H_2 \quad (\Delta H° = -38 \text{ kJ/mol}) \tag{1.7}$$

Catalytic partial oxidation (CPO) of hydrocarbon is an adiabatic reaction and a mechanistically challenged reaction that involves adsorption, desorption, and surface steps. Complexity arises due to the high catalyst surface temperature for an extremely short contact time, resulting in the lack of thermal equilibrium between solid and gaseous phases. A wide range of hydrocarbon feedstock, including heavy hydrocarbons, can be used for hydrogen production.

O_2-membrane POx: Although the oxidation of hydrocarbon seems economical by using air as an oxidant over the process of using pure O_2, air-blown POx becomes heavily diluted with nitrogen, which requires a more complex and expensive gas separation unit. The benefit of using O_2-permeable membrane (OPM) POx is that it can use air but does not involve the dilution of syngases with nitrogen, which reduces the cost of production by 25%–40%. The process requires the chemically and mechanically stable membrane (dense ceramic membrane made of a mixture of ionic and electronic conductors) that can handle a high temperature (1000°C). A few of the notable OPMs are $Ba_{0.5}Sr_{0.5}Co_{0.8}Fe_{0.2}O_{3-x}$ (Wang et al. 2002), La–Sr–Fe–Co–O (Yang et al. 2005), Sr–Fe–Co, and Ba–Sr–Co–Fe-based mixed oxide systems, and so on. One side of the membrane is exposed to the air and the other side to the reducing agents (H_2, CH_4, and CO), as shown in Figure 1.7a. Large-scale commercial use of this technique is difficult. Because of the manufacturing of the OPM without creating a crack, voids and factures are not an easy task.

1.6.1.4 Combined Reforming

The combined reforming (secondary reforming) process is based on the principle of controlling the H_2/CO ratio in synthesis gas and reduction in consumption of O_2, as shown in Figure 1.7b. This method is commonly used for the synthesis gas production that utilized in ammonia manufacturing from NG and naphtha. The main advantage of a combined reformer is that the pressure of the secondary reformer can be increased by 3.5–4.5 MPa due to the low exit temperature of the tabular reformer, which will reduce the size of the compressor by half of that used in the conventional SMR process (Solbken 1991).

1.6.1.5 Steam Iron Reforming

This is the oldest (practiced from early 1900s) commercial method, used for pure hydrogen production; it separates hydrogen from feedstock through oxidation steps by using iron oxide. Thus, this method does not require WGS and CO_2 removal stages in hydrogen manufacturing and that is the beauty of the process. Direct and indirect involvement of hydrocarbon feedstock is used in steam iron processes (SIPs) for hydrogen production.

SIP methane: In the direct method, iron oxide straightforwardly reacts with methane or hydrocarbon, which reduces iron oxide and oxidized methane, as follows:

$$Fe_3O_4 + CH_4 \rightarrow 3(1-y)Fe_{1-y}O(3Fe) + CO_y \cdot H_2O \qquad (1.8)$$

where y (= 0.05–0.17) is related to the cation vacancy in wustite. Fe-catalyzed dissociation of methane into carbon and hydrogen is thermodynamically and kinetically favorable at temperatures above 600°C. Carbon deposited at iron oxide can act

(a)

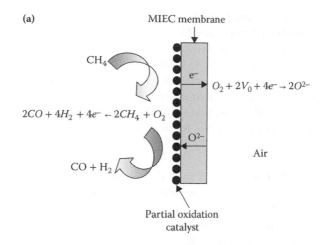

(b)

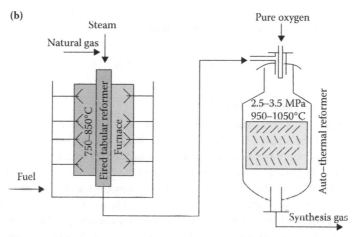

FIGURE 1.7 (a) Ceramic membrane reactor for partial oxidation of methane to syngas. (b) Schematic diagram of the combined reforming of natural gas.

(From Yang, W. et al., *TopicsCatal.*, 35, 155–167, 2005.)

as the reducing agent or can directly react with iron oxide to form Fe_3C. During the indirect SIP, hydrocarbon is initially converted into syngas followed by the reduction of the iron oxides with the major components of the syngas, that is, hydrogen (activation energy = 95 kJ/mol) and CO (activation energy = 98 kJ/mol) into iron oxide/Fe and H_2O and iron oxide/Fe and CO_2, respectively. Subsequently in a reduced iron oxide atmosphere, steam regenerates pure hydrogen as follows:

$$0.95\,Fe + H_2O(g) \rightarrow Fe_{0.95}O + H_2 \quad \Delta H = -16.3 \text{ kJ/mol} \qquad (1.9)$$

$$3.17\,Fe_{0.95}O + 0.38\,H_2O(g) \rightarrow Fe_3O_4 + 0.83\,H_2 \quad \Delta H = -46.1 \text{ kJ/mol} \qquad (1.10)$$

Recently, SIP was modified for hydrogen storage and FC applications by using sponge iron (Lehrhofer 1996; Hacker 1998).

SIP HRO: High feedstock flexibility is the major advantage of the SIP; it can be applied not only to gaseous but also to liquid and solid (biomass, coal) fuels. In this technique, by cracking of HRO, hydrogen gas is produced along with sulfur-free distillate fuel and recycling of catalyst is possible (magnetite to wustite form of iron, and vice versa), which is the main benefit of this method over the hydrogen produced by the noncatalytic POx method. The following steps are involved in this process in different parts of the reactor:

1. In the cracking reactor, HRO over the catalytic bed of magnetite-based catalyst at 450°C–600°C, is converted into cracking products (hydrogen, liquid, and coke)
2. In the regenerator column, partial combustion of coke to CO occurs, which reduces the magnetite form of the catalyst to the wustite at 800°C–850°C.
3. In the hydrogen generator, under the flow of steam hydrogen generated at 650°C, FeS is converted into Fe_3O_4 as follows:

$$FeS + H_2O(g) \rightarrow Fe_3O_4 + H_2S + H_2 \qquad (1.11)$$

1.6.1.6 Dry (CO_2) Reforming of CH_4

The growing concerns about the negative environmental impact of CO_2 (i.e., global warming) attract the sincere attention of researchers and academicians toward the CO_2 reforming of methane technology for production of H_2. Dry (CO_2) reforming of methane (DRM), also known as stoichiometric reforming, is a well-studied reaction that is of both scientific and industrial importance. It is a useful way to consume two abundant greenhouse gases, that is, CO_2 and CH_4, for the production of synthetic gas (a mixture of CO and H_2) that can be used to produce a wide range of products, such as higher alkanes and oxygenates by means of Fischer–Tropsch (FT) synthesis. CO_2 plays the role of an oxidant during reforming of methane, which is an alternative to SMR and POx processes; water (steam) and oxygen are used as oxidants, respectively. Due to its highly endothermic nature, this reaction requires quite high operating temperatures of 800°C–1000°C along with a catalyst (generally noble metals) that is able to attain high equilibrium conversion of CH_4 and CO_2 to H_2 and CO and to minimize the thermodynamic driving force for carbon deposition.

$$CH_4 + CO_2 \rightarrow 2CO + 2H_2 \quad \Delta H° = 247 \text{ kJ/mol} \qquad (1.12)$$

CO_2 is one of the major components of this equilibrium mix. At higher temperatures (>800°C), the CO_2 molar fraction in the mix dramatically drops, and the H_2 and CO molar fraction becomes predominant and reaches the plateau. Although noble metals (Rh, Ru, Pt, and Pd metals) (Pakhare and Spivey 2014) have been found to be much more resistant to carbon deposition than the most widely used Ni catalysts, they are generally uneconomical. Therefore, noble metals are used to promote the Ni catalysts to increase their resistance toward carbon deactivation. The major challenges in commercialization of this process are mentioned as follows:

1. Hydrogen produced during the reaction tends to react with the CO_2 and proceeds with a lower activation energy than the dry reforming reaction itself. For example, the typical reaction is as follows:

$$CO_2 + H_2 \rightarrow H_2O + CO \tag{1.13}$$

2. Thermodynamic limitations of noncomplete conversion of methane during CO_2 reforming at the operating pressure of the syngas plant (Rostrup-Nielsen 2000).
3. The process economy strongly depends on the pressure and the cost of CO_2 available.

From a practical viewpoint, it is preferable to operate CO_2 reforming of methane at moderate temperatures and with the CH_4:CO_2 ratio close to unity, which would require a catalyst that kinetically inhibits carbon formation under conditions that are thermodynamically favorable for carbon deposition (Hu and Ruckenstein 2002).

The CO_2 plays the role of an oxidant during reforming of methane and is an alternative to SMR and POx processes, where CO_2 plays the role of an oxidant. Sometimes, the process is also called stoichiometric reforming but more often it is referred to as dry reforming. Like SMR, it is a highly endothermic process requiring high operational temperatures of 800°C–1000°C. Owing to the presence of CO_2 in the feedstock, the process produces synthesis gas with a high CO/H_2 ratio (1:1) according to Equation 1.12.

As noted above, with the growing concerns about the negative environment impact of CO_2 (i.e., global warming), the CO_2 reforming technology is getting more attention increasingly. It should be noted, however, that the objective was to produce H_2 only. However, if the process targets the production of syngas with relatively high content of CO (e.g., for FT synthesis), then this approach can be conducive to increasing the CO/H_2 ratio in the syngas. In this case, CO_2 from the feedstock will be sequestered in the form of synthetic fuels (e.g., FT gasoline or diesel) or oxygenated compounds (alcohols, esters, etc.) that is produced as a coproduct of the methane reforming process (Armor 1999).

Practical implementation of CO_2 reforming of methane faces several key challenges, technically and economically. At the preferred pressure of the syngas plant, CO_2 reforming will result in the incomplete conversion of methane due to thermodynamic limitations (Rostrup-Nielsen 2000). Furthermore, the process economy strongly depends on the pressure and the cost of CO_2 available. From a practical view point, it is preferable to operate CO_2 reforming of methane at moderate temperatures and with the CH_4:CO_2 ratio close to unity, which would require a catalyst that kinetically inhibits the carbon formation under conditions that are thermodynamically favorable for carbon deposition (Hu and Ruckenstein 2002). Iron-, cobalt-, and nickel-based catalysts are particularly active in methane decomposition and in the CO disproportionation reaction, and noticeable deposition of carbon on the surface of these catalysts would occur at temperatures as low as 350°C. The form of carbon deposited on metal surfaces is controlled by the reaction temperature: in the lower temperature range 350°C–600°C, amorphous and filamentous carbons are the predominant form of carbon, whereas an ordered graphitic structure dominates at temperatures above 700°C (Hu and Ruckenstein 2002). Most of the reported research on CO_2 reforming of methane relates to Ni-based catalysts, because Ni exhibits high catalytic activity at lesser cost. However, Ni catalysts are prone to carbon deposition and deactivation as discussed earlier. Therefore, much research has been conducted to improve the resistance of Ni catalysts to deactivation and eliminate carbon deposition during the process. Comprehensive reviews on the topic of CO_2 reforming of methane using Ni-based catalysts and other nonprecious metal catalysts were recently published by Hu and Ruckenstein (2002) and Bradford and Vannice (2013).

The commercial Ni catalyst can be minimized by selectively passivating the catalytically active site. The suppression of carbon deposition by sulfur passivation is attributed to the strong adsorption of sulfur that controls the size of active metal ensembles. H.Topsoe has developed a Sulfur PAssivated ReforminG (SPARG) process by using a partially sulfided Ni catalyst (Rostrup-Nielsen and Bak Hansen 1993).

1.6.1.7 Plasma Reforming

There is a growing interest in electricity-assisted generation of syngas and hydrogen. In these processes, electricity alone or a mixed source of energy (i.e., electrical and chemical) can be used to provide the syngas generation process with the required energy input. Use of electricity allows for better control and useful modularity of the syngas generation equipment. Hydro Quebec's technology research laboratory (LTE) developed an approach based on the joint use of electron transport and catalysis. A kinetic model was developed to analyze gaseous, homogeneous, and heterogeneous complexes in the reactive system. The model allows the computation of reaction rate in the search for a material with desired catalytic properties in electricity-assisted syngas production. Depending on the type of electric arc used and the chemical environment in the reformer reactor, electricity-assisted systems for hydrogen production can be categorized as follows: thermal versus nonthermal plasma and oxidative versus oxidant-free plasma system. In this section, we consider oxidative plasma (both thermal and nonthermal) systems only, leaving oxidant-free plasma systems for the discussion in Section 1.6.2.

1.6.1.7.1 Thermal Plasma Reforming

Thermal plasma consists of an electric arc (with temperatures exceeding 5000°C) through which a gaseous feedstock diffuses at a high velocity, generating ionized species. A thermal plasma reformer consists of two electrodes (an anode and a cathode) spatially arranged within a so-called plasmatron. Thermal plasmas operate at very high power densities and they can catalyze chemical reactions through the intermediate formation of active radicals and ionized species (Kovener 1983). Thus, thermal plasma reformers use less hydrocarbon fuel since reactant heating is provided by the electric plasma torch. Other advantages of thermal plasma reformers include high conversion efficiencies, a rapid response, compactness, fuel flexibility, and no need for the use of catalysts (thus, no catalyst deactivation problem).

The heat generation is independent of reaction chemistry and optimum operating condition can be maintained over a wide range of feed rates and gas composition. Disadvantages associated with the plasma reformers are as follows: the difficulty in sustaining a high-pressure operation, cooling electrodes (to reduce their thermal erosion) needed, high energy requirement, and reliance on electrical energy. Because of the high energy intensity of the plasma, the processes energetic may be less favorable than the purely thermal processes, especially for the endothermic reactions such as steam reforming.

Thermal plasma-assisted POx and steam reforming of methane (SRM) was studied by Bromberg and his coworkers. The process involved a combination of air and steam as an oxidizer. The authors demonstrated that hydrogen-rich gas (50%–75% H_2 and 25%–50% CO, for steam reforming mode) could be efficiently produced in compact plasma reformers (2–3 kW). In the POx regime, specific energy consumption of the plasma reformers is 40-MJ/kg H_2. Plasma catalytic reforming of methane was also conducted in the air–water mixture and the reformer catalyst. The composition of reformer gas is (vol%): H_2-40, N_2-38, CH_4-3.4, CO-3.4, CO_2-13.5. Plasma and plasma-catalytic modes of operation showed the significant advantage over the homogeneous plasma mode that increased two to three times hydrogen yields by consuming only one third power of the latter.

1.6.1.7.2 Nonthermal Plasma Reforming

Nonthermal plasma (known as cold or nonequilibrium plasma) reformers are operated under nonequilibrium thermal condition (where electrons are at much higher temperature than the ions, radicals, and neutral molecules, exist at room temperature). An electric discharge produces the chemically active species, such as, electrons, ions, atoms, free radicals, excited state molecules, and photons, which can catalyze chemical reactions involving hydrocarbons and oxidants (e.g., oxygen, steam, and CO_2). The benefits of the nonthermal plasma reformer includes: low electric current operation, relatively low reaction temperature than thermal plasma operations, low erosion of electrode, compactness, and so on.

Various Nonthermal plasma (known as cold or nonequilibrium plasma) reformers are operated under nonequilibrium thermal condition (where electrons are at much higher temperature than the ions, radicals, and neutral molecules, exist at room temperature) nonthermal plasma systems have been reported in the literature for reforming of hydrocarbons to hydrogen-rich gas:

1. Gliding arc technology
2. Corona discharge
3. Microwave plasma
4. Dielectric barrier discharge

1.6.1.8 Photoproduction of Hydrogen from Hydrocarbons

Photocatalytic production of hydrogen is a potentially attractive approach in converting solar photon energy to chemical energy of hydrogen. Owing to the high dissociation energy of CH3–H bond (4.48 eV) methane absorbs irradiation in vacuum ultraviolet (UV) region. The absorption spectrum of methane is continuous in the region from

1100 to 1600 Å (Okabe 1978). Unfortunately, wavelengths shorter than $\lambda = 160$ nm are present neither in the solar spectrum nor in the output of most UV lamps. Therefore, the production of hydrogen and other products by direct photolysis of methane does not seem to be practical. However, the use of special photocatalysts allows the activation and conversion of hydrocarbons to H_2 under the exposure to wavelengths extending well into the near UV area (300–360 nm) that is present in the solar spectrum. Hashimoto et al. (1984) reported the photocatalytic production of hydrogen from aliphatic and aromatic hydrocarbons using a powdered Pt/TiO_2 photocatalyst suspended in deaerated water. The hydrocarbon water photocatalytic system was exposed to UV irradiation and was converted to a mixture of H_2 and CO_2 as follows (Equation 1.14):

$$C_{16}H_{34} + 32\,H_2O + \text{Light} \rightarrow 49\,H_2 + 16\,CO_2 \quad \Delta G° = 1232 \text{ kJ/mol} \quad (1.14)$$

The rate of hydrogen production from organic aqueous solutions was found to be a function of several parameters, namely, light intensity, pH, tungstate, organic concentrations, temperature, catalytic additives, and so on. Along the main hydrogen product of the photoreaction, alkanes were also produced corresponding to the alcohols, ketones, and in some cases, dimerized products. The mechanism of isopoly tungstate (IPT)-catalyzed conversion of alkanes involves the photoinduced charge transfer in the photoactive $W^{VI} = O$ group in the octahedral moiety; this (1) leads to the formation of a reactive electron-deficient radical-like species, and (2) is capable of abstracting the H atom from organic substrates.

Recently, a German patent filed by Guemter et al. describes a photocatalytic composite material that can be used for the cleavage of hydrogen from H-containing compounds (hydrocarbon). This composite contains a porous support on which a photoactive semiconducting material (i.e. metal oxides of Cd, Cu, Ti, and Zn, etc). The composite can photolytically decomposed the methane into hydrogen and carbon, and might be used for the decomposition of organic contaminants (present in wastewater).

1.6.2 Nonoxidative Process

In most cases, the nonoxidative process is endothermic. Therefore, it requires energy before the decomposition process. Usually, high temperature ($\geq 500°C$) required to accomplish the process.

$$C_nH_m + [\text{Energy}] \rightarrow xH_2 + yC + zCpHq \quad (1.15)$$

where $C_n H_m$ is the original hydrocarbon ($n \geq 1$, $m \geq n$), $C_p H_q$ ($z \geq 0$, $p \geq 1$, $q \geq n$) is the stable hydrocarbon product of the hydrocarbon cracking, and [Energy] is input energy (thermal, electrical, plasma, or radiation).

1.6.2.1 Thermal Decomposition

When hydrocarbons are heated at high temperature, they thermally decompose into their constituent elements, that is, hydrogen and carbon.

$$C_nH_m \rightarrow nC + m/2H_2 \qquad (1.16)$$

The amount of energy required to carry out this process depends on the nature of the hydrocarbons, which is the highest for saturated hydrocarbons (alkanes, cycloalkanes) and stumpy for unsaturated and aromatic hydrocarbons because the decomposition of acetylene and benzene are exothermic reactions. Methane is one of the most thermally stable organic molecules due to its highest dissociation energy for the C–H bond among all organic compounds. It is an electronic structure that lacks polarity and any functional group makes it extremely difficult to thermally decompose into its constituent elements. The methane decomposition reaction is a moderately endothermic process with $\Delta H^0 = 75.6$ kJ/mol.

$$CH_4 \rightarrow C + 2H_2 \qquad \Delta H^\circ = 75.6 \text{ kJ/mol} \qquad (1.17)$$

At temperatures above 800°C, the molar ratio of hydrogen and carbon products approach their maximum values. The effect of pressure on the molar fraction of H_2 at different temperatures shows that the rate of H_2 production is favored by low pressure. The energy requirement per mole of hydrogen produced (37.8-kJ/mol H_2) along with clean carbon is significantly less than that required for the endothermic SMR reaction (68.7-kJ/mol H_2). Therefore, less than 10% of methane combustion is needed to drive the process. Since the 1950s, carbon black (CB) and byproduct hydrogen have been generated by the thermal decomposition of NG (Othmer 1992). The process (called the thermal black process) is operated in a semicontinuous mode using two tandem reactors running at high operational temperatures. One of the reactors is heated to the pyrolysis temperature (approximately 1400°C) of hydrocarbon, the air is cut off, and hydrocarbon is pyrolyzed over the heated contact to hydrogen and CB particles. Simultaneously, another reactor is heated to pyrolysis temperature, followed reversing the flow of hydrocarbon feedstock from the pyrolysis reactor to the heated reactor, and the process continuing in a cyclic mode.

Steinberg (1999) proposed a methane decomposition reactor consisting of a molten metal bath. The methane bubbles through molten tin or copper bath at high temperatures (900°C and higher). The advantages of this system are the efficient heat transfer to a methane gas stream and the ease of carbon separation from the liquid metal surface by density difference. The determined reaction activation energy of $E_a = 131.1$ kJ/mol is substantially lower than the E_a reported in the literature for the homogeneous methane decomposition (272.4 kJ/mol), which points to a significant contribution of the heterogeneous process caused by the submicron-sized carbon particles adhered to the reactor surface (Shpilrain et al. 1999). The reactor was combined with a steam turbine to increase the overall efficiency of the system.

The mechanism of thermal decomposition (pyrolysis) of methane has been extensively studied by Chen et al. (1976) and Dean (1990). Because C–H bonds in the methane molecule are significantly stronger than C–H and C–C bonds of the products, the secondary and tertiary reactions contribute at the very early stages of the reaction, which obscures the initial processes. According to Holmen et al. (1995) the overall methane thermal decomposition reaction at high temperatures can be described as a stepwise dehydrogenation by Equation 1.18 and the overall stepwise mechanism is illustrated by Equations 1.19 through 1.21:

$$2CH_4 \rightarrow C_2H_6 \rightarrow C_2H_4 \rightarrow C_2H_2 \rightarrow 2C \tag{1.18}$$

$$CH_4 \rightarrow CH_3^{\cdot} + H^{\cdot} \tag{1.19}$$

$$CH_4 + H^{\cdot} \rightarrow CH_3^{\cdot} + H_2 \tag{1.20}$$

$$2CH_3^{\cdot} \rightarrow C_2H_6 \tag{1.21}$$

The homogeneous dissociation of methane is the only primary source of free radicals and it controls the rate of the overall process. The thermal black process, advanced thermolysis, catalytic decomposition, metal catalyzed, carbon catalyzed reactions are used to produce free radicals.

1.6.2.2 Metal-Catalyzed Decomposition of Methane

In 1966, Pohlenz and Scott of Universal Oil Products Co. (UOP) developed the HYPRO™ process for the continuous production of hydrogen by catalytic decomposition of gaseous hydrocarbon streams. Methane decomposition was carried out in a fluidized-bed catalytic reactor in the temperature range from 815°C to 1093°C. Supported Ni, Fe, and Co catalysts were used in the process. The cocked catalyst was continuously removed from the reactor to the regeneration section where carbon was burned off by air, and the regenerated catalyst was recycled to the reactor. However, the system with two FBRs and the solids circulation system was too complex and expensive and could not compete with the SMR process.

The National Aeronautics and Space Administration (NASA) conducted studies on the development of catalysts for the methane decomposition process for space life support systems. A special catalytic reactor with a rotating magnetic field to support Co catalyst at 850°C was designed. In the 1970s, Callahan, a U.S. Army researcher, (1974) developed a fuel processor to catalytically convert different hydrocarbon fuels to hydrogen, which was used to feed a 1.5-KW FC. He screened a number of metals for the catalytic activity in the methane decomposition reaction including Ni, Co, Fe, Pt, and Cr. Alumina-supported Ni catalyst was selected as the most suitable for the process. The following rate equation for methane decomposition was reported:

$$-d[CH_4]/dt = K'S(1 - \phi)[CH_4] \tag{1.22}$$

where K' is an intrinsic rate constant, ϕ is the fraction of the active sites covered by carbon, and S is the surface area.

However, the energy efficiency of the processor was relatively low (<60%) and it produced CO_2 byproduct in quantities comparable to that of the SMR and POx processer. The kinetic curve of methane decomposition over the alumina-supported (10%) Fe_2O_3

at 850°C could be broken down to four zones (Muradov 1998). Zone A corresponds to the reduction of iron oxide to metallic Fe accompanied by the formation of CO_x. In zone B, a quasisteady process of methane decomposition occurred with the peak H_2 concentration in the effluent gas approaching 98 vol%. This was followed by the drastic reduction in the catalytic activity of iron catalyst in zone C and, finally, the steady-state methane decomposition and yield of about 900 mL of gas per milliliter of gasoline (Muradov 1996). The rate of methane activation in the presence of transition metals followed the order Co, Ru, Ni, Rh > Pt, Re, Ir > Pd, Cu, W, Fe, Mo.

1.6.2.3 Simultaneous Production of Hydrogen and Filamentous Carbon

Owing to the high value and practical importance of carbon filaments, catalytic decomposition of methane and other hydrocarbons as a means of production of different types of filamentous carbon (carbon nanotubes [CNTs], carbon nanofibers, carbon whiskers, and so on) has been a very active area of research for several decades. Carbon filaments with their unique mechanical and electrical properties have important practical applications in areas such as composite materials, electronics, catalysis, space, and military. NG is an abundant feedstock for the production of hydrogen and filamentous carbon. In 1889, a U.S. Patent was filled by Hughes and Chambers for vapor-grown carbon fibers (Hughes and Chambers 1889; Radushkevich and Lukyanovich 1952) for observing the growth of relatively thick carbon filaments from the CH_4–H_2 mixture in an iron crucible. Other researchers have made significant contributions to the understanding of the role of catalytic particles in producing carbon filaments. According to a widely accepted mechanism of carbon filament formation, the reaction product carbon dissolves into the metal particle, diffuses through it, and precipitates at the rear of the metal crystallite with the formation of a carbon filament. The growth of the carbon filament is inhibited when the catalyst particle is encapsulated in carbon layers to prevent further decomposition of the hydrocarbons. Depending on the carbon filament growth conditions, they may possess different structures, namely, "platelet," "herringbone," and "ribbon"; it should be noted that most of the early works were concerned with the production of carbon filaments only, and not much consideration was given to hydrogen as a reaction product.

A series of kinetic studies on carbon filament formation by methane decomposition over Ni catalysts was reported by Snoeck et al. (1997). The authors derived a rigorous kinetic model for the formation of the filamentous carbon and hydrogen by methane cracking. The model is represented by the following steps in Equations 1.23 through 1.31:

1. Surface reaction:

$$CH_4 + * \Leftrightarrow CH_4^{-*} \qquad K_{CH4} \qquad (1.23)$$

$$CH_4^{-*} + * \Leftrightarrow CH_3^{-*} \qquad (rds)K_2 \qquad (1.24)$$

$$CH_3^{-*} + * \Leftrightarrow CH_2^{-*} + H^{-*} \qquad K_3 \qquad (1.25)$$

$$CH_2^{-*} + * \Leftrightarrow CH^{-*} + H^{-*} \qquad K_4 \qquad (1.26)$$

$$CH^{-*} + * \Leftrightarrow C^{-*} + H^{-*} \qquad K_5 \qquad (1.27)$$

$$2H^{-*} \Leftrightarrow H_2 + 2^* \qquad 1/K_H \qquad (1.28)$$

2. Carbon dissolution/segregation:

$$C - 1 \Leftrightarrow C_{Ni,f} + 1 \qquad (1.29)$$

3. Diffusion of carbon through Ni:

$$C_{Ni,f} \Leftrightarrow C_{Ni,r} \qquad (1.30)$$

4. Precipitation of carbon:

$$C_{Ni,r} \Leftrightarrow C_w \qquad (1.31)$$

where "*" is an active site and rds is a rate-determining step.

1.6.2.4 Carbon-Catalyzed Decomposition of Methane

Although transition metal-based catalysts enjoy a high catalytic activity at moderate temperatures (500°C–750°C), there are still some technical challenges, particularly with regards to (1) catalyst deactivation, (2) sulfur poisoning, and (3) separation of the carbon products from the metal catalyst. It was suggested to remove the combusted carbon from the catalyst surface to regenerate its original activity. A series of surface stepwise dissociation reactions, Equations 1.32 through 1.34, lead to the formation of the elemental carbon and hydrogen.

$$(CH_{3-x})_a \rightarrow (CH_{2-x})_a + (H)_a \qquad (1.32)$$

$$(C)_a \rightarrow \frac{1}{n}(C_n)_c \quad \text{(carbon crystallite growth)} \qquad (1.33)$$

$$2(H)_a \rightarrow (H_2)_g \qquad (1.34)$$

where $0 < x < 2$, and subscripts a, c, and g denote adsorbed, crystalline, and gaseous species, respectively.

Moliner et al. (2005) investigated the effect of textural properties (pore size distribution) and surface chemistry of activated charcoal (AC) on the thermocatalytic decomposition of methane. Microporous carbons with a high content of oxygenated surface groups exhibit a high initial activity, but they rapidly deactivate. In contrast, mesoporous carbons with a high surface area showed better stability. The production of hydrogen by thermal decomposition of methane over a fluidized bed reactor (FBR) was reported by Dunker et al. (2006). They studied the effect of a catalyst, temperature, and residence time on the rate of methane decomposition over a fluidized bed of CB particles at the range of temperature of 810°C–980°C and space velocities of 95–210/h. Under optimum conditions, the process produced an effluent gas with hydrogen concentration ≥40% by volume. There was an initial rapid decrease in the catalyst activity during the first 50 minutes, followed by a slower decrease in the beginning at about 1000 minutes after the start of the experiment. The reaction in the hydrogen production yield is explained by a loss of micropores and filling of internal cavities in the catalyst by deposited carbon. Kushch et al. (1997) have found that fullerene black is a very active catalyst for the dehydrogenation of methane. Hydrogen and pyrocarbons were formed during the first hours of methane pyrolysis at 1000°C. Methane conversion decreased with time due to the formation of carbon. The reaction also produced hydrodimerization products (C_2+) such as ethylene, propane, and propylene. During pyrolysis of a CH_4–Ar (50–50) mixture, a stable methane conversion (4.2%) was observed with a selectivity to ethylene of 89%–94%. Changes in surface properties of carbon before and after the reaction were investigated. The pore size change in the course of methane decomposition over ACs indicated that the catalytic reaction occurs mainly in the micropores.

1.6.2.5 Catalytic Decomposition of Methane for FC Applications

Recently, catalytic decomposition of methane has attracted a great deal of interest as a simple and relatively efficient route to produce CO_x-free hydrogen suitable for FC applications. Some types of FCs do not tolerate CO_x impurities in hydrogen gas even at very low levels. For example, the CO concentration in the hydrogen fuel for a polymer electrolyte membrane fuel cell (PEMFC) is restricted to <10 ppmv; an alkaline FC does not tolerate CO_2 in hydrogen because it reacts with the electrolyte. Conventional hydrocarbon reformers can potentially deliver hydrogen gas with the required levels of the impurities, but they are very complex and bulky systems because they involve several gas conditioning and purification steps. In contrast, methane decomposition systems do not involve oxidants, and thus, do not produce CO and CO_2. In particle systems, however, it would be very difficult to completely avoid CO_x because it may originate from the traces of moisture or air in the feedstock or from water and air entrained in the catalyst pores. In this case, the addition of a methanation unit could reduce the concentration of CO_x in hydrogen gas down to parts per million by volume levels.

Pourier and Sapundzhiev (1997) described the concept of a fuel processor for PEMFC applications based on the catalytic decomposition of NG. In this concept, NG is decomposed over a ceramic-supported Pd catalyst; hydrogen-rich gas is produced and carbon is deposited on the catalyst surface. Once the catalytic bed is filled with carbon, it was regenerated by burning the carbon in air. A hydrogen gas stream of purity 95% with a trace amount of CO and CO_2 was produced. The residual CO was removed by the methanation reaction. Nakagawa et al. (2010) reported the catalytic decomposition of methane over oxidized diamond-supported Ni and Pd catalyst at 600°C for the formation of nanostructured carbon and hydrogen. Whisker-type carbon nanomaterials were produced using Ni- and Pd-loaded oxidized diamond. It was concluded that high catalytic activity could be attributed to the chemical interaction between the metal and the support surface.

Shah et al. (2001) investigated catalytic decomposition of undiluted methane into hydrogen and carbon over alumina-supported nanostructured binary Fe–M catalysts as a means of producing H_2 for FC applications. The authors observed that all the Fe–M (where M = Pd, Mo, or Ni) exhibited significantly higher catalytic activity than Fe or any secondary metals alone. At the reaction temperatures of 700°C–800°C and space velocities of 600 mL/g/h, the product stream is composed of over 80 vol% of H_2. Multiwalled CNTs were the dominant form of carbon produced. No CO_x or C_2 hydrocarbons were observed in the hydrogen gas, which makes it potentially suitable for PEMFC. Choudhary et al. also demonstrated hydrogen production by catalytic decomposition of methane over supported Ni catalysts.

1.6.2.6 Methane Decomposition Using Nuclear and Solar Energy Input

Decomposition of methane is a high-temperature endothermic process. The thermal energy for the process is usually provided by combusting some portion of methane. Alternatively, the heat input to the process could be provided by nonfossil energy sources such as high-temperature nuclear reactors and concentrated solar radiation. The above-mentioned energy sources have twofold advantages: first is in eradication of the CO_2 level in atmosphere and second is to utilize the waste heat that generated from high temperature nuclear reactor. Serban et al. (2004) proposed the concept of utilizing the "waste" heat from the generation IV high-temperature nuclear reactors to produce hydrogen and carbon from methane or NG by direct contact pyrolysis.

The generation IV nuclear reactors used heavy liquid metal coolant (Pb or Sn). According to the concept, high-temperature gas (He) is used for the heat transfer from the nuclear reactor to the thermal cracking unit. NG was bubbled through a bed of low melting point metals (Pb or Sn) at temperatures ranging from 600°C to 900°C. No other gaseous products beside hydrogen were observed in the effluent stream. The main advantage of the proposed system is the ease of separation of the effluent stream. The main advantage of the proposed system is the ease of separation of the generated carbon byproduct from the heat transfer media. At 750°C, methane conversion was about 9%. Two types of carbon were produced: soot and pyrocarbon. These experiments lay the groundwork for developing technical expertise in producing pure H_2 cost effectively by utilizing heat energy contained in the liquid metal coolant in the generation IV nuclear reactors.

1.6.2.7 Plasma-Assisted Decomposition of Hydrocarbons

The objective of plasma-assisted decomposition of hydrocarbons is to produce hydrogen and carbon in an oxidant-free environment (as opposed to plasma-assisted POx and steam reforming that produce hydrogen and CO_2), according to the following generic reaction (Equation 1.35):

$$C_nH_m + Energy \rightarrow nC + m2H_2 \tag{1.35}$$

The plasma decomposition process is applicable to any hydrocarbon fuel, from methane to heavy hydrocarbons. Similar to oxidative plasma reforming, plasma decomposition processes fall into two major categories: thermal and nonthermal plasma systems.

1.6.2.7.1 Thermal Plasma Systems

In thermal plasma reactors, the energy required to accomplish an endothermic hydrocarbon decomposition process is provided by high-temperature plasma. The scheme was illustrated by the simplified schematics of a thermal plasma reformer for the decomposition of methane to hydrogen and carbon. In the 1990s, Kvaerner Company of Norway developed the Kvaemer CB&H process for the production of CB and hydrogen by thermal plasma decomposition of NG. In Kvaemer's reactor, a plasma torch supplies the necessary energy to pyrolyzc methane. The plasma gas is hydrogen, which is recirculated from the process. The characteristics of the Kvaerner CB&H process are as follows: maximum output per generate, 8 MW; thermal efficiency of the process, in excess of 90%; hydrogen purity, 98%; energy consumption, 12.5 KW/Nm^3 H$_2$; hourly capacity, 270 kg of CB and 1000-Nm^3 H$_2$. The advantages of the CB&H process include (a) feedstock flexibility, (b) nearly 100% conversion of the feedstock, (c) no need for catalysts, and (d) easy scale-up due to modular concept. The CB product of the Kvaerner CB&H process was evaluated and found to be suitable for rubber, plastic, and metallurgical industries. The authors demonstrated that excitation and dissociation of molecular hydrogen and methane into atomic hydrogen and methyl radicals in plasma play an important role in the hydrocarbon cracking reactions. Thermodynamic analysis indicated that hydrogen atoms appear in significant quantities at about 3200°C–3700°C, and methane dissociation occurs at about 1230°C–2730°C according to the following equations:

$$H_2 \rightarrow 2H^{\cdot} \qquad \Delta H = 436 \text{ kJ} \tag{1.36}$$

$$CH_4 \rightarrow H^{\cdot} + CH_3^{\cdot} \qquad \Delta H = 419 \text{ kJ} \tag{1.37}$$

The experimental results obtained from the plasma fluidized bed cracking devices were in agreement with the theoretical predictions taking into account the role of reactive species production by the plasma.

1.6.2.7.2 Nonthermal Plasma Systems

Czernichowski et al. in 1994 studied nonthermal plasma decomposition of NG in a gliding arc reactor. The methane decomposition reaction was carried out in a steel tube reactor re-equipped with two or six symmetrically located steel electrodes. The gliding arc plasma reactor was powered by voltage in the range 0.5–10 kV and currents from 0.1 to 5 A. The gaseous feed input was in the range 0.13–2.3 Nm^3/h. The electric power dissipated in the gliding arc plasma reactor and reached 1.6 kW. No chemical corrosion or thermal erosion of the electrodes was observed during the plasma reactor operation.

At these conditions, up to 34% of the methane was converted into soot, hydrogen, and acetylene. The energy efficiency is calculated in terms of the amount of energy consumption to produce H_2 and O_2 (4.1 kWh to produce 1.0-Nm^3 H_2 and 0.22-Nm^3 C_2H_2).

The most probable mechanism of the methane decomposition reaction catalyzed by the microwave plasma impulses is similar to the ion molecular Winchester mechanism as follows (Equations 1.38 through 1.41):

Initiation:

$$CH_4 + e^- \rightarrow (CH_3)^- + H \qquad (1.38)$$

Cluster growth:

$$(CH_3) + CH_4 \rightarrow (C_2H_5)^- + H_2 \qquad (1.39)$$

$$(C_nH_{2n+1})^- + CH_4 \rightarrow (C_{n+1}H_{2(n-q)+3})^- + (1-q)H_2 \; where \; q = 0.1 \qquad (1.40)$$

Termination:

$$(CH_4)^+ + (C_nH_m)^- \rightarrow C_{n+1}H_{4+m-x} \qquad (1.41)$$

The authors demonstrated that this ion molecular mechanism explains the experimental results from the viewpoints of the energetic and reaction kinetics of the process.

1.7 HYDROGEN AND ITS APPLICATIONS

By virtue of the supreme qualities of hydrogen, it was strongly established as an energy carrier that has enough potential to reduce energy dependence on gasoline and also reduce pollution and greenhouse gas emissions. Therefore, advancement in hydrogen technologies is under way. The hydrogen economy can provide RE solutions to the energy crisis, such as emergency backup power, heating and electricity for commercial and residential purpose, and hybrid electric vehicles (onboard storage of hydrogen still remains a "critical path" barrier, and it is one of the primary areas of focus). PEC water splitting for hydrogen production is a promising approach for terrestrial (versatile applications in portable, stationary, and transportation) and space applications. To advance the commercialization of PEC-generated hydrogen technologies, market transformation activities aim to promote their adoption in stationary, portable, and specialty vehicle applications, such as forklifts, municipal vehicles, lawn mowers, and so on. Space photovoltaics is another field of possibilities for design and fabrication techniques of PEC device for hydrogen production. Following are the regions where hydrogen can be used as an energy carrier and fuel. The largest industrial consumption of hydrogen in the United States is ammonia production (40.3%), refinery (37.3%), methanol production (10%), with the rest of the hydrogen consumed in the food process, metal manufacturing, and chemical process industries, as well as other fields, as shown in Figure 1.5a and b.

1.7.1 PORTABLE

The clean fuel hydrogen is used for the portable supply of energy to run electrical and other utility appliances: TVs, refrigerators, heaters, air conditioners, fans, and so on. Among them, the hydrogen burner is one interesting example of a portable hydrogen appliance and can be used indoors safely without risk (H_2ydropole+ 2010). The reaction product is water, which is actually beneficial for the room climate. Fundamental technical experience will be attained through a thoughtful review of standalone power systems (SAPs), the use of hydrogen as a future energy carrier, durability, storage systems, as well as use of hydrogen for cooking and heating purposes. A commercial barbecue has been adapted by replacing the propane burner by a series of hydrogen burners connected to the hydrogen source. A panel of evaluators confirmed that hydrogen roasted meat was undistinguishable in taste from that of propane roasted meat. Likewise, an eco-friendly portable hydrogen house is another use of PEC hydrogen, in which hydrogen is used as an energy source for most of the appliances.

1.7.2 STATIONARY

Technological advancement drives meaningful reductions in cost such as in the installation of a UNI-SOLAR solar energy system and will reduce the cost of generating solar electricity by over 20%. Ultimately, the company expects to offer solar energy systems that are capable of providing electricity at a cost below the utility grid. In Comparison to conventional photovoltaic modules, which required an electrolyzer as an additional component of the system, affects cost factor for electricity plants. Direct feed of sunlight in photovoltaic system, generates electricity at day time and accumulator store the harvested surplus electricity for night time. As well as an innovative PEC cells with long-term hydrogen storage and gas engine-driven generator, was used for the seasonal electricity compensation. Direct solar hydrogen production using dynamic PEC cells is newly available, which eases and simplifies both plant design and operation. These PEC cells utilize their great potential in exemplary application of the autonomous electricity supply for industrial plants. Another most promising application of hydrogen produced by PEC cells is in supply of micro combined heat and power (CHP) to residential houses by exploiting gas-engine generated electric power and waste heat. It seems to be an interesting perspective for which the employment of PEC cells is currently being investigated. First, results prove that for a typical single family house in Germany with up-to-date thermal insulation, a share of approximately 9% (1 MWh/year) of yearly NG consumption could be covered by hydrogen produced by 14 m^2 of PEC cells installed on the roof. Under the present prototype conditions, the investment cost, which is supposed to be considerably reduced in the future, still is questionable under a pure economic aspect as the PEC cells are at the very beginning of their development phase.

1.7.3 TRANSPORTATION

Highly efficient clean fuel hydrogen is the most famous fuel for spacecraft. In 1990, the world's first solar-powered hydrogen production plant (a research and testing facility) became operational, at Solar-Wasserstoff-Bayern, in southern Germany.

In 1994, Daimler Benz demonstrated its first NECA I (New Electric CAR) FC vehicle at a press conference in Ulm, Germany. In 1999, Europe's first hydrogen fueling stations were opened in the German cities of Hamburg and Munich. United Solar's technology roadmap takes this groundbreaking technology into commercial production in 2012. Hydrogen-driven city buses, cars, lawn mowers (Yvon and Lorenzoni 2006), municipal vehicles, and so on are some commercial means of transportation which utilize the PEC hydrogen.

Challenges to expand the uses of hydrogen gas for stationary power production include better education and training of local codes and standard officials on the processes for hydrogen system permitting, continued efforts to bring down the costs of electrolyzers to enable renewable hydrogen production, improved efficiency and performance of steam methane reformers (particularly in smaller sizes), the current relatively high costs for hydrogen storage and piping systems, and improvements in other scientific processes and technologies for producing hydrogen with low to zero emissions of greenhouse gases and costs that can ultimately be competitive (Lipman 2011). It can also be utilized as a raw material in many chemical industries, and as fuel for house warming, FC, power production, high-temperature atomic welding, and running vehicles.

1.7.4 Uses as a Chemical

It is primarily used to create water. Hydrogen gas can be used for reduction of metallic ore, production of various chemicals such as water, ammonia, methanol, hydrochloric acid, radioactive isotope tritium, unhealthy unsaturated fats to saturated oils and fats, and so on. Hydrogen can be used as a coolant in the electrical rotor of generators that use the gas as a rotor coolant. The element is relied upon in many manufacturing plants to check for leaks. Hydrogen can be used on its own or with other elements. It is a radioactive isotope used to make H-bombs. It can also be used as a luminous paint radiation source. Tritium is used in biosciences for isotopic labeling and to trace the mechanism of the reaction.

1.8 ENVIRONMENTAL EFFECTS OF HYDROGEN

1.8.1 Health Hazards

Although hydrogen is not a toxic gas, when its high concentration absorbed into the body of the sufferer by inhalation, it may cause an oxygen-deficient environment and turn the victim's skin blue. A patient may experience symptoms including headaches, ringing in ears, dizziness, drowsiness, unconsciousness, nausea, vomiting, and depression of all the senses. Under extreme circumstances, death may occur. It is not expected to cause mutagenicity, embryotoxicity, teratogenicity, or reproductive toxicity. Preexisting respiratory conditions may be aggravated by overexposure to hydrogen. In case of hydrogen fire, first shut off the supply. If this is not possible and there is no risk to the surroundings, then let the fire burn itself out; in other cases, extinguish it with water spray or CO_2. In case of fire, one should keep the cylinder cool by spraying water. Affected persons should inhale fresh air and artificial respiration may be provided, under medical supervision, if required.

1.8.2 Physical Hazards

Hydrogen can easily form an explosive mixture with air, oxygen, halogens, and strong oxidants. Moreover, it has one of the widest explosive/ignition mix ranges with almost all of the gases with a few exceptions such as acetylene, silane, and ethylene oxide. This means that at almost every mix proportion between air and hydrogen, a hydrogen leak will most likely lead to an explosion, when a flame or spark ignites the mixture. This makes the use of hydrogen particularly dangerous in enclosed areas such as tunnels or underground parking. It is suggested to check the oxygen level before entering in hydrogen zone. The gas is lighter than air, which might cause leakage problems. On the loss of containment, a harmful concentration of this gas in the air will be reached very quickly. However, currently the effect of these leakage problems may not be prominent. The amount of hydrogen that leaks today is much lower (by a factor of 10–100) than the estimated 10%–20% figure conjectured by some researchers; for example, in Germany, the leakage rate is only 0.1% (less than the NG leak rate of 0.7%). At most, such leakage would likely be no more than 1%–2% even with widespread hydrogen use, using present technology.

1.8.3 Chemical Hazards

The high chemical reactivity of hydrogen toward air, oxygen, halogens, and strong oxidants causes fire and explosion hazards. Furthermore, heating or the presence of a metal catalyst (platinum and nickel, etc.) greatly enhance the rate of these reactions. Pure hydrogen–oxygen flames burn with the UV color range and are nearly invisible to the naked eye, so a flame detector is needed to detect the burning of hydrogen leakage. Molecular hydrogen leaks slowly from most containment vessels. As a transportation fuel, H_2 is mainly used in FCs that do not emit greenhouse gases. But the transport and storage of this elemental gas may cause a big problem if care is not taken in its handling.

1.8.3.1 Effect to Ozone Layer

A California Institute of Technology (Caltech) study also estimates the potential of hydrogen fuel effects in the form of inevitable hydrogen leaks on the planet's atmosphere as a 10% decrease in the ozone layer. Due to the moisture buildup from hydrogen on combining with oxygen, there is cooling of the upper atmosphere that leads to indirect destruction of the ozone layer. Increased hydrogen fuel emissions may add moisture to the stratosphere, which raises the rate and size of polar stratospheric clouds and aerosols. These both lead to ozone reduction.

1.8.3.2 Greenhouse Effect

A British science study names hydrogen an indirect greenhouse gas. According to the European Commission DG Environment News Alert Service, the research concludes that the emitted hydrogen in atmosphere, reacted with the atmospheric CO_2 to add the volumes of methane in atmosphere that significantly contributes in global warming.

Despite all the limitations, liquid hydrogen shows much promise for the future.

1.8.4 Environmental Hazards of Hydrogen

Since 2010, hydrogen is growing in popularity as a promising sustainable alternative fuel source. Unlike conventional fuels, which produce CO_2 on combustion, hydrogen produces only benign water and can cut nitrogen emissions up to 80%. The effects of hydrogen on the atmosphere may not be as good as anticipated. If significant amounts of hydrogen gas (H_2) are released in the atmosphere, it was scientifically assumed that in the stratosphere, under UV radiation exposure, radical hydrogen (H) would be formed. These free radicals would act as the catalysts for ozone layer depletion. An ICE running on hydrogen fuel may produce some nitrous oxides and other pollutants, which results in smog ("Photochemical Ozone Production or Summer Smog," http://www.scienceinthebox.com/en_UK/sustainability/summersmog_en. html, retrieved 2005). Nitrogen oxide (NO_x) emissions increase exponentially with combustion temperature. Therefore, these can be influenced through appropriate process control. Because hydrogen has sufficient energy to break an N–N bond (binding energy of ≈ 226 kcal/mol) of N_2 gas present in the combustion cylinder as an air component. This can produce unwanted toxic components such as nitric acid (HNO_3) and hydrogen cyanide gas (HCN) (Wofsey 2009) But the particulate and sulfur emissions in ICE are limited to small quantities of lubricant remnants. Still, research is going on toward determining whether hydrogen fuel can reverse the effects of decades of continued fossil fuel use. In addition, the global auto industry is designing and manufacturing hybrid vehicles since hydrogen is a highly efficient and burnable fuel for ICEs. Or it can be used as an energy carrier in hydrogen FC propulsion systems with low temperature, where H_2 combines with O_2 (air) to produce the energy to run the engine of the vehicle and emit water, completely eliminating all polluting emissions. But the long-term impact of the use of hydrogen as a global fuel on plants, animals, soil, water, and humans remains speculation. In 2006, Caltech studied the consequences of hydrogen affecting the planet's atmosphere; the results of anticipated leakage of hydrogen from vehicles, hydrogen manufacturing, and transport of the elemental gas are not all good news for the planet. They investigated the H_2 emissions into the atmosphere and found out that they will grow from 60 to 120 trillion grams a year by replacing fossil fuel with hydrogen globally. That will be four to eight times the current hydrogen generation from human/ecological actions released into the air. The study predicts the results of the continued use of the hydrogen as the replacement for fossil fuel will double or triple the normal impact on the atmosphere. These figures remain uncertain until scientists gain more understanding of the effects of using hydrogen as an energy source.

1.9 HYDROGEN SAFETY

The safety of any energy source is a prime concern for a good fuel. Even though the major physical and chemical characteristics of hydrogen are different from conventional petroleum-based fuels, its use is as safe as gasoline, diesel, or kerosene. The hazards associated with the use of hydrogen can be characterized as physiological (frostbite, respiratory ailment, and asphyxiation), physical (phase changes, component failures, and embrittlement), and chemical (ignition and burning). A combination of hazards occurs in most instances. The primary hazard associated with any form of hydrogen is inadvertently producing a flammable or detonable mixture, leading to a fire or detonation. Safety will be improved when the designers and operational personnel are aware of the specific hazards associated with the handling and use of hydrogen. Hydrogen is the odorless, nontoxic, nonpoisonous, smoke-free, and high buoyancy gas (diffusion coefficient 0.61 cm^3/s) that burns with an invisible flame of low radiant heat, all of which makes it undetectable by human senses. Hydrogen's wide explosive range (13%–79%), coupled with its very low ignition energy (0.02 mJ), gives it a potential disadvantage since an accumulation of hydrogen in a poorly ventilated vehicle interior may explode easily. Furthermore, hydrogen is 14 times lighter than air and rises at a speed of almost 20 m/s, six times faster than NG, which means that when released, it rises and disperses quickly. For these reasons, hydrogen systems are designed with ventilation and leak detection. Hydrogen is an odorless and colorless gas; therefore, a sulfur-containing odorant is added so that people can detect it. Unfortunately, there is no known odorant light enough to "travel with" hydrogen at an equal dispersion rate, so it has significantly less radiant heat than a hydrocarbon fire (a fire fueled by hydrocarbon products such as petroleum and NG). Although a hydrogen flame is just as hot as a hydrocarbon flame, the levels of heat emitted from the flame are lower. This decreases the risk of secondary fires. Combustion cannot occur in a hydrogen vessel or any contained location that contains only hydrogen—an oxidizer, like oxygen, is required. Hydrogen burns very quickly. Atmospheric moisture absorbs thermal energy radiated from a hydrogen fire and can reduce the value. The radiation intensity of a hydrogen flame at a specific distance depends profoundly on the amount of water vapor present in the atmosphere and is expressed as:

$$I = I_0 e^{0.0046\,wr} \tag{1.42}$$

where I_0 = initial intensity (energy/time·area), w = water vapor (percent by weight), and r = distance (meters).

Under the most favorable combustion conditions, the energy required to set off hydrogen combustion is notably lower than that required for combustion of other common fuels, such as NG or gasoline. The energy required to initiate combustion of low concentrations of hydrogen in the air is similar to that of other fuels. Hydrogen flames have very low-radiation heat. N2, CH3Br, and CF3Br are well-known fire-extinguishing materials for hydrogen diffusion flames in air. These inhibitors were more effective when added to the air stream; nitrogen was more effective when added to the fuel stream. Unlike oxygen of high concentration, H_2 never causes suffocation because it rises and disperses very rapidly in the atmosphere. Hydrogen is a nontoxic and nonpoisonous gas. Therefore, it will not contaminate groundwater. Discharge of hydrogen does not contribute to either atmospheric or water pollution, under normal atmospheric conditions. Consequently, hydrogen can be used as safely as other common fuels we use today with proper guidelines and understanding of its behavior. To ensure the safety of combustion engine operation, hydrogen vehicles and hydrogen FC material are handled just like NG vehicles, with gas refueled using a closed-loop system.

1.10 SUMMARY

Energy is the highway to our future development. This energy is riding on the saddle of fuel that is currently generated from conventional fossil fuels. The present energy scenario was discussed in this chapter by keeping our past, present, and future energy needs in mind. Due to public and governmental awareness and interest, satisfactory results are coming from the RE side and they are contributing smartly to green energy generation. Some light was shed on hydrogen as a RES. The chemical properties of hydrogen distinguished it from conventional fuels and established it as a chemical fuel. The day-by-day growing inclination toward a hydrogen economy make it popular among all stakeholders. Various techniques of hydrogen production were also elaborated in this chapter. Furthermore, application of hydrogen in portable and stationary devices, transportation, laboratories, and industries, are discussed. Although there are rumors about hydrogen gas leakage accidents/explosions, to date no such incident has been experienced. Some light was shed on the chemical and physical effects on the environment due to hydrogen diffusion/leakage in the atmosphere. Safety issues of storage and handling of hydrogen are also discussed in this chapter.

REFERENCES

Armor, J. N. 1999. The multiple roles for catalysis in the production of hydrogen. *Appl. Catal. A- Gen.* 176: 159–76.

Callahan, M. A., and W. R. Haas. 1976. Hydrocarbon fuel conditioner for a 1.5 KW fuel cell power plant. In: *Proceedings of the 26th Power Sources Symposium,* held on June 21–24. Red Bank, NJ, p. 180.

Chen, C., M. Back, and R. Back. 1976. Mechanism of the thermal decomposition of methane. In: *Proceedings of the ACS Symposium on Industrial Laboratory Pyrolysis. Series* 1, p. 91.

Chen, J., and F. Wu. 2004. Review of hydrogen storage in inorganic fullerene-like nanotubes. *Appl. Phys. A* 78: 989–94.

Coley, G. D., and J. E. Field. 1973. Role of cavities in the initiation and growth of explosions in liquids. *Proc. R. Soc. London, Ser. A.* 335 (1600): 67–86.

College of the Desert. 2001. *Hydrogen Fuel Cell Engines and Related Technologies, Revision 0.* Palm Desert, CA: College of the Desert.

Coward, H. F., and G. W. Jones. 1952. *Limits of Flammability of Gases and Vapors.* Bureau of Mines Bulletin 503. Washington, DC: United States Government Printing Office.

Czernikowski, A. 1994. Gliding arc applications to engineering and environment control. *Pure Appl. Chem.* 66: 1301–10.

Dean, A. 1990. Detailed kinetic modeling of autocatalysis in methane pyrolysis. *J. Phys. Chem.* 94: 1432–39.

Derwent, R., P. Simmonds, S. O'Doherty, A. Manning, W. Collins, and D. Stevenson. 2006. Global environmental impacts of the hydrogen economy. *Int. J. Nucl. Hydrogen Prod. Appl.* 1(1): 57–67.

Dillon, A. C., K. M. Jones, T. A. Bekkedhal, C. H. Kiang, D. S. Bethune, and M. J. Heben. 1997. Letters to nature. Storage of hydrogen in single wall carbon nanotubes. *Nature* 386: 377–79.

Dunker, A. M., S. Kumar, and P. A. Mulawa. 2006. Production of hydrogen by thermal decomposition of methane in a fluidized-bed reactor—Effects of catalyst, temperature, and residence time. *Int. J. Hydrogen Energy* 31: 473–84.

Global Strategy Institute, Center for Strategic and International Studies. 2005. *Hydrogen: The Fuel of the Future? Future Watch*. Washington, DC: Global Strategy Institute, Center for Strategic and International Studies.

GM. 2007. Fuel Cell Technology: Prospects, Promises and Challenges. Available at: http://www.gm.com/company/gmability/adv_tech/400_fcv/fc_challenges.htm.

Guemter, S. et al. 2003. Photocatalytic composite element for cleavage of hydrogen containing compounds, especially, for hydrocarbon decomposition and wastewater treatment. German patent No. DE 10210465.

H$_2$ydropole+. 2010. Hydrogen Report Switzerland 2010/2011. Major Achievements. May 12, 2010. Available at: http://issuu.com/hydropole/docs/hsr11. Accessed May 12, 2010.

Hacker, V. 1998. Usages of biomass gas for fuel cell by the SIR process. *J. Power Source* 71: 226–300.

Hairston, D. 1996. Hail hydrogen. *Chem. Eng.* 103: 59.

Hashimoto, K., T. Kawai, and T. Sakata. 1984. Photocatalytic reactions of hydrocarbons and fossil fuels with water: Hydrogen production and oxidation. *J. Phys. Chem.* 88(18): 4083–88.

Hinrich-Rahlwes, R., S. Teske, and J. Wijnhoven. 2013. Energy [R]evolution: A sustainable Netherlands energy outlook. Report 2013 Netherlands energy scenario. Utrecht, the Netherland: Utrecht University Repository.

Hirano, S., K. S. Young, A. Kwabena, and J. A. Schwarz. 1993. The high surface area activated carbon hydrogen storage system, In: *Proceedings of the First International Conference on New Energy Systems and Conversions*, Frontiers Science Series No. 7, pp. 67–72, June 1993, Yokohama, Japan.

Holmen, A., O. Olsvik, and O. Rockstad. 1995. Pyrolysis of natural gas: Chemistry and process concepts. *Fuel Process Technol.* 42: 249–67.

Hu, Y. H., and E. Ruckenstein. 2002. Binary MgO-based solid solution catalysts for methane conversion to syngas. *Catal. Rev.-Sci. Eng.* 44: 423–53.

Hubbert, M. K. 1956. Nuclear Energy and Fossil Fuels, Paper presented before the Spring Meeting of the Southern District Division of Production, American Petroleum Institute Plaza Hotel, March 7–9, San Antonio, TX, Publication No. 95, pp. 22–27.

Hubbert, M. K. 1971. The energy resources of the Earth. *Sci. Am.* 225: 60–70.

Hughes, T. V., and C. R. Chambers. 1889. Manufacture of carbon filaments, U.S. Patent No. 405, p. 480.

Jiang, L., and B. C. O'Neill. 2004. The energy transition in rural China. *Int. J. Global Energy* 21(1–2): 2–26.

Jones, L. W. 1970. Toward a liquid hydrogen fuel economy. Presented at the University of Michigan Environmental Action for Survival Teach In. Ann Arbor, MI: University of Michigan. Available at: http://hdl.handle.net/2027.42/5800.

Kovener, G. S. 1983. Use of a RF plasma for thermal pyrolysis of CH$_4$ and heavy oils. *ISPC-6 Montreal* 1: 258–63.

Kushch, S. D., A. P. Moravskii, V. Y. Muradyan, and P. V. Fursikov. 1997. Activation of methane on fullerene black. *Pet. Chem.* 37: 112–18.

Lehrhofer, J. 1996. Integrated analysis of the sponge iron reactor and fuel cell system. In: *Proceedings of the Fuel Cell Seminar*, held on 17–20 November. Orlando, FL, p. 710.

Lipman, T. 2011. An overview of hydrogen production and storage systems with renewable hydrogen case studies, Clean Energy States Alliance Report, pp. 1–32.

McCarty, R. D., J. Hord, and H. M. Roder. 1981. *Selected Properties of Hydrogen (Engineering Design Data)*. NBS Monograph 168. Boulder, CO: National Bureau of Standards.

Moliner, R., I. Suelves, M. J. Lázaro, and O. Moreno. 2005. Thermocatalytic decomposition of methane over activated carbons: Influence of textural properties and surface chemistry. *Int. J. Hydrogen Energy* 30(3): 293–300.

Muradov, N. 1996. Thermocatalytic decomposition of methane using fixed reactor. In: *Proceedings of the 1996 U.S. DOE Hydrogen Program Review.* May 1–3, 1996, Miami, FL.

Muradov, N. 1998. CO_2 free production of hydrogen by catalytic pyrolysis of hydrocarbon fuel. *Energy Fuel* 12: 41–48.

Muradov, N., and Z. Nazim. 2009. Production of hydrogen from hydrocarbon. In: *Hydrogen Fuel: Production, Transport and Storage* (ed. R. B. Gupta). Washington, DC: CRC Press, Taylor and Francis, p. 39.

Nakagawa, K., Y. Tanimoto, T. Okayama, K.-I. et al. 2010. Sintering resistance and catalytic activity of platinum nanoparticles covered with a microporous silica layer using methyltriethoxysilane. *Catal. Lett.* 136(1–2): 71–76.

Newton, E., T. H. Haletter, P. Hottinger, F. Von Roth, W. H. Scherrer, and T. H. Schuan. 1998. Seasonal storage of hydrogen in stationary systems with liquid organic hydrides. *Int. J. Hydrogen Energy* 23(10): 904–09.

Nezhad, H. 2007. *Software Tools for Managing Project Risk: Management Concepts.* Vienna, VA: Management Concepts, pp. 1037–55.

Nezhad, H. 2009. *World Energy Scenarios to 2050: Issues and Options.* Minnesota, MN: State University of Minneapolis.

Okabe, H. 1978. *Photochemistry of Small Molecules,* Chapter 7. New York, NY: Wiley, p. 126

Othmer, K. 1992. *Encyclopedia of Chemical Technology,* 3rd Edn, Vol. 4. New York, NY: Wiley, p. 631.

Pakhare, D., and J. Spivey. 2014. A review of dry (CO_2) reforming of methane over noble metal catalysts. *Chem. Soc. Rev.* 43: 7813–37.

Pohlenz, J., and N. Scott. 1966. Method for hydrogen production by catalytic decomposition of a gaseous hydrocarbon stream. U.S. Patent No. 3284161 (UOP).

Pourier, M., and C. Sapundzhiev. 1997. Catalytic decomposition of natural gas to hydrogen for fuel cell application. *Int. J. Hydrogen Energy* 22: 429–33.

Radushkevich, L. V., and V. M. Lukyanovich. 1952. The structure of carbon forming in thermal decomposition of carbon monoxide on an iron catalyst. *Zh. Fiz. Khim.* 26: 88–95.

Rostrup-Nielsen, J. R. 2000. New aspects of syngas production and use. *Catal. Today* 63:159–64.

Rostrup-Nielsen, J. R., and J.-H. Bak Hansen. 1993. CO_2 reforming of methane over transition metals. *J. Catal.* 144: 38–39.

Rühl, C., P. Appleby, J. Fennema, A. Naumov, and M. E. Schaffer. 2012. Economic development and the demand for energy: A historical perspective on the next 20 years. *Energy Policy* 50: 109–16.

Schlapbach, L., and A. Zuttel. 2001. Hydrogen-storage materials for mobile applications. *Nature* 414: 353–54.

Schmitt, M. L., J. E. Shelby, and M. M. Hall. 2006. Preparation of hollow glass microspheres from sol-gel derived glass for application in hydrogen gas storage. *J. Non Crystalline Solids.* 352:626–31.

Serban, M., M. A. Lewis, and J. K. Basco. 2004. Kinetic study of the hydrogen and oxygen production reactions in the copper-chloride thermochemical cycle in AIChE. In: *2004-Spring National Meeting,* held on 25–29 April. New Orleans, LA.

Shah, N., D. Panjala, and G. P. Huffman. 2001. Hydrogen production by catalytic decomposition of methane. *Energy & Fuels* 15: 1528–34.

Shell, V. 2004. Lighthouses for Hydrogen Shell. Available at: www.shell.com/static/hydrogen-en/downloads/speeches/interview_lighthouses.pdf.

Shpilrain, E., V. Shterenberg, and V. Zaichenko. 1999. Comparative analysis of different natural gas pyrolysis methods. *Int. J. Hydrogen Energy* 24: 613–24.

Snoeck, W., G. F. Froment, and M. Fowles. 1997. Kinetic study of the carbon filament formation by methane cracking on a nickel catalyst. *J. Catal.* 169: 250–62.

Solbken, A. 1991. *Synthesis Gas Production in Natural Gas Conversion* (eds A. Holmen et al.) Amsterdam, the Netherlands: Elsevier.

Sørensen, B. 1975. Energy and resources. *Science* 189: 255–60.

Steinberg, M. 1999. Fossil fuel decarbonization technology for mitigating global warming. *Int. J. Hydrogen Energy* 24: 771–77.

U.S. Environmental Protection Agency, Light-Duty Automotive Technology, Carbon Dioxide Emissions, and Fuel Economy Trends: 1975 Through 2012. March 2013 US UPA- 420-R-13-01 Copyright 2012@ United State Environmental Protection Agency.

van Ruijven, B., J. -F. Lamarque, D. P. van Vuuren, et al. 2011. Emission scenarios for a global hydrogen economy and the consequences for global air pollution. *Glob. Environ. Chang.* 21: 983–94.

Wang, H., Y. Cong, and W. Yang, 2002. Oxygen permeation study in a tubular $Ba_{0.5}Sr_{0.5}Co_{0.8}Fe_{0.2}O_{3-\delta}$ oxygen permeable membrane. *J. Memb. Sci.* 210: 259–71.

Wofsey, D. 2009. The Hidden Danger of Hydrogen Power. Available at: http://www.sparkplugengineering.com/, retrieved June 4, 2009.

Yang, W., H. Wang, X. Zhu, and L. Lin. 2005. Development and application of oxygen permeable membrane in selective oxidation of light alkanes. *Topics Catal.* 35(1): 155–67.

Yüksel, I. 2008. Global warming and renewable energy sources for sustainable development in Turkey. *Renewable Energy* 33(4): 802–12.

Yvon, K., and J.-L. Lorenzoni. 2006. Hydrogen-powered lawn mower: 14 years of operation. *Int. J. Hydrogen Energy* 31(12): 1763–67.

2 Concepts in Photochemical Water Splitting

2.1 INTRODUCTION

In looking to our everyday growing clean-energy needs and to subsidize the environmental damage done by the emissions of convention fuels, the statement of Prof. James Barber seems quite contemporary, that is, our sun is the champion of energy sources, delivering more energy to Earth in an hour than we currently use in a year from fossil, nuclear, and all renewable sources combined. Its energy supply is inexhaustible in human terms, and its use is harmless to our environment and climate (Barber 2009). Therefore, a large number of principal investigation projects are devoted worldwide to generate chemical fuels from sunlight. In this age of light, carbon-neutral water can be considered a fuel of the future. Because when the water is coupled with sunlight at the platform of a photocatalyst, they can produce clean fuels, hydrogen and oxygen. These products are carbon-neutral molecules, which never produce carbon-containing systems and therefore can be treated as clean fuels. Hydrogen can especially be considered as a good energy carrier or fuel for transport, industry, and electricity generation. In this chapter, we discuss the fundamental elements related to efficient hydrogen generation through water splitting. These include artificial photosynthesis, the electrochemistry of water splitting, criteria for the selection of photocatalytic material, overpotential, band gap and band edge position in photocatalytic materials, band edge bending of semiconductor efficiency (solar to hydrogen conversion), efficiencies (turnover number [TON], quantum yield, photoconversion efficiency [incident photon-to-current efficiency {IPCE} {%}, absorbed photon-to-current efficiency {APCE}]), excitonic binding energy, diffusion length and carrier mobility in photocatalyst, and so on. For each of these elements, we discuss the figures of merit, the critical length scales associated with each term, and the way in which these length scales must be balanced for efficient generation of hydrogen. We have to develop the right materials to make it work efficiently. For this, scientists of all backgrounds are coming together to beat nature at her own game by treating/setting all of the parameters well, such as light absorption, photogenerated carrier collection, photovoltage, electrochemical transport, and catalytic behavior.

2.2 ARTIFICIAL PHOTOSYNTHESIS

Photosynthesis is nature's own way to make fuel (i.e., carbohydrates) in a sustainable manner by utilizing solar energy. Although the total amount of solar energy (every hour almost 4.3×10^{20} J solar energy reaches the surface of Earth) reaching Earth's surface is large, the energy incident on any given square meter of the planet's surface over time is diffuse and low due to dilution. In the United States, the average insolation is approximately 200 W/m² (Cogdell et al. 2010). In this process, plants, algae, and some bacteria use solar energy to convert carbon dioxide and water into carbohydrates and oxygen. Artificial photosynthesis (Figure 2.1) is a photochemical process that mimics natural photosynthesis for producing sustainable fuels. In general, this term refers to the scheme that is used to arrest and accumulate the energy from sunlight in the chemical bonds of a fuel. This procedure is capable enough to use solar energy for producing clean fuels at attractively high efficiencies. Most common reactions fall under these categories: (1) light-induced carbon dioxide and water reduction (carbon fixation) into chemical fuels and (2) photocatalytic water splitting into hydrogen and oxygen.

2.2.1   CARBON DIOXIDE REDUCTION

After estimating the risks and reasons of global climatic change and depleting energy resources, carbon dioxide is found as one of the great culprits. The efforts for the prevention of further increase in CO_2 concentrations in atmosphere are undoubtedly an important task that includes carbon taxation, increase in plantation, sequential CO_2 deposition in the deep earth core, fruitful utilization of carbon dioxide in higher chemicals, and so on. Furthermore, the world population and their corresponding energy and electricity demands for a luxurious life had added volumes to the emission levels of CO_2. Figure 2.2a and b exhibit the worldwide trends in atmospheric and anthropogenic contribution in the emission of carbon dioxide up to 2025 (Energy Information Administration 2007). Excessive atmospheric carbon dioxide along with its greenhouse gas partners are responsible for the drastic changes in global climate. To reduce the risks that would follow a global climatic change, time-based systematic strategies/action plans are needed to suppress atmospheric CO_2 concentrations. If we tied up this CO_2 emission with a renewable energy technology, then it would be a win–win situation from the environmental and economic points of view.

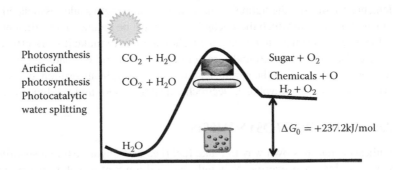

Photosynthesis
Artificial
photosynthesis
Photocatalytic
water splitting

$CO_2 + H_2O$ Sugar + O_2

$CO_2 + H_2O$ Chemicals + O

$H_2 + O_2$

H_2O

$\Delta G_0 = +237.2kJ/mol$

FIGURE 2.1   Schematic presentation of the photosynthesis, artificial photosynthesis, and photocatalytic water splitting process.

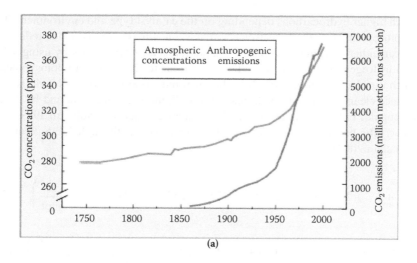

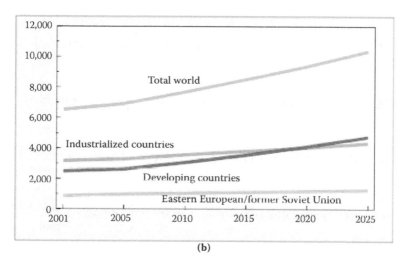

FIGURE 2.2 (a) Carbon emission contribution by atmosphere (yellow) and anthropogenic (blue) emission from 1725 to 2025 and (b) worldwide zonal contribution in total anthropogenic emission from 2001 to 2025. [(a) From Oak Ridge National Laboratory, Carbon Dioxide Information Analysis Center, http://cdiac.esd.oml.gov/. (b) From Energy Information Administration, International Energy Outlook, Washington, DC, EIA, http://www.eia.doe.gov/oiaf/1605/ggccebro/chapter1.html, 2003.]

To resolve this problem in an eco-friendly and sustainable manner, reduction of CO_2 is a good solution. In this issue, sunlight along with a photocatalyst can be treated as an attractive and renewable solution that can be used to produce fuels from carbon dioxide on combination with water. These fuel products include a variety of chemicals such as carbon monoxide, formic acid, methanol, methane, oxalate, and even higher hydrocarbons depending on the catalyst type and environment. The whole process involves multielectron chemistry, known as carbon dioxide reduction, which can either be achieved directly in a photoreactor or indirectly by using solar-derived hydrogen as a "reductant." This is similar to the artificial photosynthesis, in which the electrochemical or photochemical reduction of carbon dioxide is used to produce clean, green, and sustainable fuels (Figures 2.3 and 2.4a and b). A few of the popular products of carbon dioxide reduction reactions (Sullivan 1989) with their energetic parameters are mentioned in Equations 2.1 through 2.7 (Scibioh and Viswanathan 2004):

$$CO_2 + 2e^- + 2H^+ \rightarrow CO + H_2O \quad E = -0.76\text{ V} \quad \left(\Delta G_f^0 = -137\frac{kJ}{mol}\right) \quad (2.1)$$

$$CO_2 + 2e^- + 2H^+ \rightarrow HCOOH \quad E = -0.85\text{ V} \quad \left(\Delta G_f^0 = -361\frac{kJ}{mol}\right) \quad (2.2)$$

$$2CO_2 + 2e^- \rightarrow (COO)_2^{2-} \quad E = -1.14\text{ V} \quad \left(\Delta G_f^0 = -671\frac{kJ}{mol}\right) \quad (2.3)$$

$$2CO_2 + 2e^- \rightarrow CO + CO_3^{2-} \quad E = -0.79\text{ V} \quad \left(\Delta G_f^0 = -527\frac{kJ}{mol}\right) \quad (2.4)$$

$$CO_2 + 4e^- + 4H^+ \rightarrow HCHO + H_2O \quad E = -0.72\text{ V} \quad \left(\Delta G_f^0 = -102\frac{kJ}{mol}\right) \quad (2.5)$$

$$CO_2 + 6e^- + 8H^+ \rightarrow CH_3OH + H_2O \quad E = -0.62\text{ V} \quad \left(\Delta G_f^0 = -166\frac{kJ}{mol}\right) \quad (2.6)$$

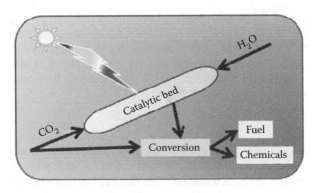

FIGURE 2.3 Schematic representation of the light-induced photocatalytic reduction of carbon dioxide for fuel and chemical production.

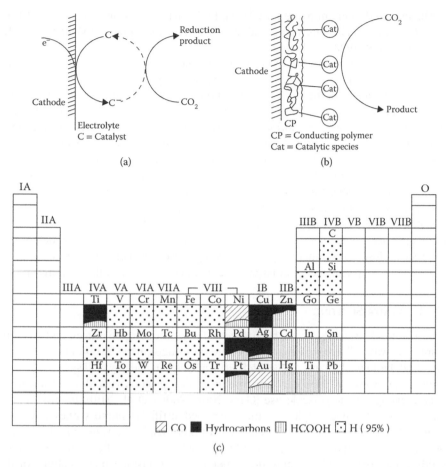

FIGURE 2.4 Reduction of CO_2 using electrocatalyst (a) platinum group of metallic cathode, (b) catalytic species loaded conducting polymer, and (c) predict the elements of periodic table suitable for different CO_2 reduction products. (From Hori, Y. et al., Electrochemical reduction of carbon dioxides to carbon monoxide at a gold electrode in aqueous potassium hydrogen carbonate. *J. Chem. Soc. Chem. Commun.*, 728–729, 1987.)

$$CO_2 + 8e^- + 8H^+ \rightarrow CH_4 + H_2O \qquad E = -0.48 \text{ V} \qquad \left(\Delta G_f^0 = -51\frac{kJ}{mol}\right) \qquad (2.7)$$

The field of homogeneous catalysis in compressed CO_2 will attract major interest in the future. But the full potential of these hydrogenation catalytic reactions has yet to be explored. Therefore, the finding of an efficient and highly selective catalyst for the reduction of carbon dioxide has been a gigantic scientific challenge. There are two catalytic ways to reduce CO_2: (1) by the use of photocatalyst and (2) electrocatalysis, which can be further divided into two parts, that is, electrocatalysis in solution and molecular electrocatalysis.

Prominent photocatalysts used for CO_2 reduction are Fe supported on MY-zeolite (M = Li, Na, K, Rb and Y = Lanthanum) for hydrocarbon production (Nam et al.

2000), anchored titanium oxide for CH_4 and CH_3OH production (Mori et al. 2012), and CdSe (Anpo et al. 1995; Wang et al. 2002) Ti oxide/Y-zeolite for MeOH and CH_4 (Yamashita et al. 1998) production.

Usually, the reduction of carbon dioxide to fuels and chemicals can be accomplished by the use of the electrocatalysts of the platinum group metals. Figure 2.4c suggests the elements from the periodic table and their corresponding suitable products for CO_2 reduction. Three classes of the electrocatalysts have been identified: catalytic metal surfaces, monomeric solution complexes, and chemically modified electrodes, the latter being a novel hybrid of solid state catalyst, surface, and homogeneous solution chemistry. Recently, a major breakthrough occurred when a research group of Prof. A. Salehi-Khojin of University of Illinois, Chicago, Illinois, developed a unique two-step process using MoS_2 in ionic liquids (96 mol% water and 4 mol% 1-ethyl-3-methylimidazorium tetrafluoroborate [EMIM-BF$_4$] solution) (Figure 2.5a–g) that reduces the waste carbon dioxide into syngas (Asadi 2014) (carbon monoxide and hydrogen), a precursor of gasoline and other energy-rich products, which bring the process closer to commercial viability. Syngas uses inexpensive catalysts to produce another source of energy at large scale, while making a healthier environment.

2.2.2 WATER SPLITTING

Since the 1950s, scientists have been successfully producing hydrogen and oxygen by splitting water. There were many major breakthroughs in the forms of the catalysts using both biological (plants, algae, and certain bacteria) and physical systems (semiconductors). However, these early systems were cost prohibitive and not efficient or durable enough for large-scale use (Gary 2008; Walter et al. 2010). Figure 2.6a–c shows some of the examples of man-designed artificial photosynthesis, which require photocatalyst material, water, and sunlight to split the water into hydrogen and oxygen in 2:1 ratio. An ample number of metal oxides, sulfides, nitrides, (oxy) sulfides, and (oxy)nitrides containing either typical transition metal cations of d^0 electronic configuration (e.g., Ti^{4+}, Ce^{4+}, Zr^{4+}, V^{5+}, Nb^{5+}, Ta^{5+}, Mo^{2+}, W^{2+}, Nd^{2+}, etc.) or representative cations of d^{10} electronic configuration (e.g., Cu^{2+}, Ag^+, Cd^{2+}, Zn^{2+}, Ga^{3+}, Ge^{4+}, In^{3+}, Sb^{3+}, Sn^{4+}, etc.) as principal cationic components or their combinations have been reported as active photocatalysts for overall water splitting. There are three ways to split water: photocatalytic (Figure 2.6a), photoelectrochemical (PEC) using a photoanode (Figure 2.6b), and PEC using a photoanode and photocathode (tandem) (Figure 2.6c).

All three types includes three basic steps: (1) photocatalyst absorbs more photon energy (sunlight) than the band gap energy of the photocatalyst and generates photoexcited electron–hole pairs (charge carrier); (2) these photoexcited charges separate out and migrate toward different sites of the photocatalyst's surface without recombination; and (3) at these sites, water is reduced and oxidized by the photogenerated electrons and holes to produce H_2 and O_2, respectively. The first two steps are strongly dependent on the structural and electronic properties of the photocatalyst, whereas the third step is promoted by the presence of a solid cocatalyst (Figure 2.6a). The cocatalyst is typically either noble metals or transition metal oxide or a combination of both (e.g., Pt, NiO_x, RuO_2, RhO_x, IrO_2,

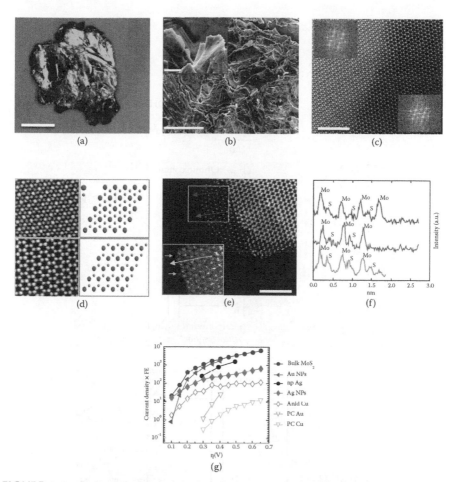

FIGURE 2.5 Structural and elemental analysis of MoS_2. (a) Optical image of bulk MoS_2 used as catalyst (scale bar, 2 mm). (b) Scanning electron microscopy images of MoS_2 displaying the stacked-layered structure and sharp edges of the MoS_2 flakes. Scale bars are 50 and 5 µm (for inset). (c) high-angle annular dark-field (HAADF) images (scale bar, 5 nm) showing both the 1T (blue) and 2H (red) phases of MoS_2, along with their respective Fast Fourier Transforms (FFTs) (inset). (d) Higher magnification HAADF images show clearly distinct atomic configuration corresponding to the 1T (top, blue) and 2H (bottom, red) type of MoS_2. The related schematic atomic models have also been shown on the right side. (e) Raw grey scale HAADF and false-color low-angle annular dark-field (LAADF) image (inset) of MoS_2 edges (scale bar, 5 nm). (f) the line scans (red and blue towards edges) identifying Mo atoms to be the terminating atoms in the general case. In limited instances, an additional light atom (grey line scan) occupying what should be a Mo-position, most probably a carbon atom, from the STEM substrate. (g) CO_2 reduction performance of the different catalysts' with respect to the bulk MoS_2 at different overpotentials, in 1-ethyl-3-methylimidazorium tetrafluoroborate (EMIM-BF$_4$) solution. Bulk MoS_2, oxidized Au nanoparticles (Au NPs), nanoporous Ag (np Ag), Ag nanoparticles (Ag NPs), Annealed copper (Anld Cu), polycrystalline Au (PC Au), PC Cu bulk Ag results were taken from the present study, where electrochemical experiments were performed in similar conditions. (From Asadi, M. et al., *Nat. Commun.* 5, Article no. 4470, 2014.)

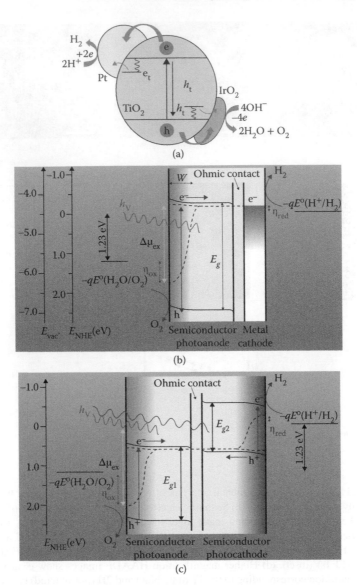

FIGURE 2.6 (a) One semiconductor-assisted photocatalytic water splitting reaction aided by Pt and IrO_2 cocatalysts. (From Kamat, P. V., *J. Phys. Chem. Lett.*, 3(5), 663–672, 2012.) (b) Two photon-assisted photoelectrochemical water splitting using a photoanode. (c) Four photon-assisted photoelectrochemical water splitting using a photoanode and photocathode in tandem. (From Sivula, K., *J. Phys. Chem. Lett.*, 4, 1624–1633, 2013.)

and $RhCr_2O_3$), loaded onto the surface of a photocatalyst to produce active sites and reduce the activation energy for gas evolution in photocatalytic overall water splitting. Details of the water-splitting process and relevant examples are discussed in Chapters 3 through 7.

2.3 ELECTROCHEMISTRY OF WATER SPLITTING

2.3.1 THERMODYNAMIC AND ELECTROCHEMICAL ASPECTS OF WATER SPLITTING

Solar water splitting is an energetically uphill chemical reaction that requires 1.23 V for the complete decomposition of pure water into hydrogen and oxygen. Usually, due to very low ionization ($K_w = 1.0 \times 10^{-14}$) power, water splitting becomes thermodynamically (Gibbs free energy $\Delta G_0 = 237$ kJ/mol, 2.46 eV per molecule) unfavorable at standard temperature and pressure. Where ΔG calculated at 25°C using the thermodynamic parameters (delta H, T, and delta S) that required for the water-splitting process:

$$\Delta G = \Delta H - T\Delta S = 285.83 \text{ kJ} - 48.7 \text{ kJ} = 237.13 \text{ kJ} \qquad (2.8)$$

In the case of water electrolysis, Gibbs free energy represents the minimum *work* necessary for the reaction to proceed and the reaction enthalpy is the amount of energy (both work and heat) that has to be provided so the reaction can proceed at the same temperature as the reactant. It also requires energy to overcome the change in entropy of the reaction (Figure 2.7a and b). Here, ΔG, the electrical energy demand of water splitting, decreases with temperature. Thus, the useful electrical work requirement reduces on increasing the reaction temperature T. Heat demand ($T\Delta S$) gradually increases with temperature. Therefore, the total energy demand ΔH does not increase significantly with temperature.

Therefore, the process cannot proceed below 286 kJ/mol if no external heat/energy is added. At this condition, an electrolyzer operating at 1.48 V would be 100% efficient. The positive value for ΔG (a measure of the thermodynamic driving force that makes a reaction occur) indicates that a reaction cannot proceed spontaneously without any large external inputs. The standard cell potential $E°$ of any reaction, related to the Gibbs free energy ($\Delta G° = -nF \cdot E°$), is represented by the following equation:

$$E° = -(\Delta G°/nF) \qquad (2.9)$$

where n is the number of electrons transferred in the reaction (in this case, 2) and F is a proportionality constant, F is the Faraday units (96,485 C/mol). Using Equation 2.9, the standard potential of the water electrolysis can be calculated as -1.229 V at 25°C. This cell potential belongs to the difference in potentials of the two half-cell reactions occuring at the cathode (reduction; hydrogen evolution reaction [HER]) and anode (oxidation; oxygen evolution reaction [OER]). The Nernst equations (Equations 2.10 through 2.12) for the half-cell reactions of water splitting are mentioned below at pH 0 and 7:

Anode (oxidation): $4OH^- + 4h^+ \rightarrow O_2 + 2H_2O \quad E = -1.23$ V versus NHE (2.10)

Cathode (reduction): $2H_2O + 2e^- \rightarrow H_2 + 2OH^- \quad E = 0.00$ V versus NHE (2.11)

(a)

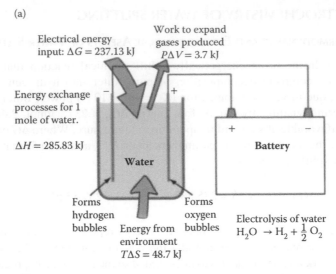

(b)

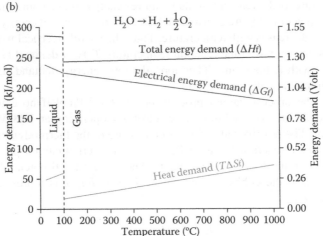

FIGURE 2.7 (a) Thermodynamics of water splitting and (b) effect of temperature versus energy demand for water splitting (https://esc.fsu.edu/documents/lectures/ECS2006/Hydro genProduction.pdf).

Overall reaction :

$$2H_2O(l) + 4e^- + 4h^+ \rightarrow O_2(g) + 2H_2(g) \quad E = -1.23 \text{ V} \quad \text{versus NHE} \quad (2.12)$$

where NHE is the normal hydrogen electrode.

2.3.2 Oxygen Evolution Reaction

Of the two half-cell reactions, oxidation is the most energetically demanding (needs higher overpotential) step because it requires the coupling of four electrons and four holes that tend to the formation of an oxygen–oxygen bond. Systems always suffer from an overpotential due to activation barriers, concentration effects, or voltage drops due to resistance. The activation barriers or activation energies are associated with high-energy transition states that are reached during the electrochemical process of OER. The lowering of these barriers would allow OER to occur at lower overpotentials with a fast rate of reactions. Heterogeneous OERs are sensitive to the pH of the solution and surface, where the reaction takes place. Under acidic conditions, water binds to the surface with the irreversible removal of one electron and one proton to form a metallic hydroxide (Conway and Liu 1990). In an alkaline solution, a reversible binding of hydroxide ion coupled to a one-electron oxidation is thought to precede a turnover-limiting electrochemical step involving the removal of one proton and one electron to form a surface oxide species (Birss et al. 1986). The electron transfer mechanisms for acidic and alkaline solutions are shown in Figures 2.8 and 2.9.

OER has been studied on a variety of materials including platinum surfaces, transition metal oxides (Matsumoto and Sato 1986), and first-row transition metal spinels (Parmon et al. 1983; Bockris and Otagawa 1983). Recently, metal–organic framework (MOF)–based materials have shown to be a highly promising candidature for water oxidation with first-row transition metals (Nepal and Das 2013; Hansen and Das 2014). Some of them include amorphous molecular sieves of MnO (Iyer et al. 2012), Ru(110) single crystal oxide surfaces (Castelli et al. 1986), compact films (Lodi et al. 1978), titanium-supported films (Trasatti 2000), RuO_2 films (Castelli et al. 1986), and so on.

2.3.3 HYDROGEN EVOLUTION REACTION

For reduced overpotential and increased efficiency of the electrochemical process (Walter et al. 2010), an advanced catalyst is required for electrochemical HERs. The most effective HER electrocatalysts are Pt group metals and these are facing challenges in developing highly active HER catalysts based on abundantly available materials of low cost (Trasatti and Petri 1991). In comparison to oxidation of water, reduction of water is energetically more viable through the multielectron pathway (1.23 V vs. NHE) than the single-electron pathway (5 V vs. NHE), which makes the designing of a complete water-splitting system complicated. Therefore, in half-cell reactions, either an oxidation or reduction process that will be in the presence of sacrificial electron acceptor (EA) or electron donor (ED), respectively, must be better understood. Cathodic HER in alkaline water electrolysis follows the subsequent reaction steps in Equations 2.13 through 2.15 (Rosalbino et al. 2003):

$$\text{Discharge step: } M + H_2O + e^- \leftrightarrow M_{Hads} + OH^- \quad \text{(Volmer reaction step)} \quad (2.13)$$

$$\text{Desorption: } M_{Hads} + H_2O + e^- \leftrightarrow H_2 + M + OH^- \quad \text{(Heyrovsky reaction step)} \quad (2.14)$$

$$\text{Recombination: } M_{Hads} + M_{Hads} \leftrightarrow H_2 + 2M \quad \text{(Tafel reaction step)} \quad (2.15)$$

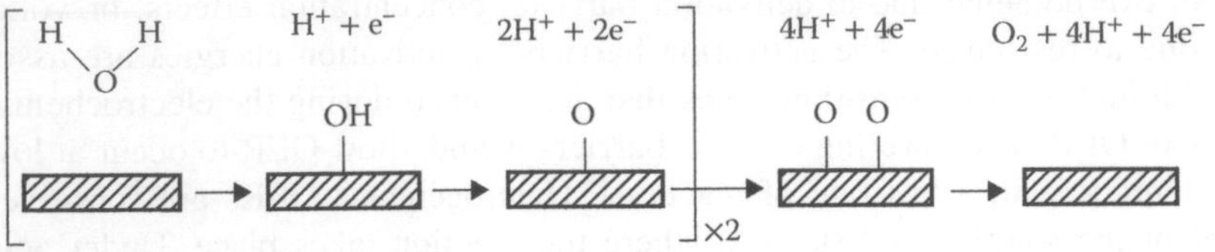

FIGURE 2.8 Oxygen evolution reaction (OER) at catalytic surface under acidic medium. (From Birss, V. I. et al., *J. Electrochem. Soc.,* 133, 1621–1625, 1986.)

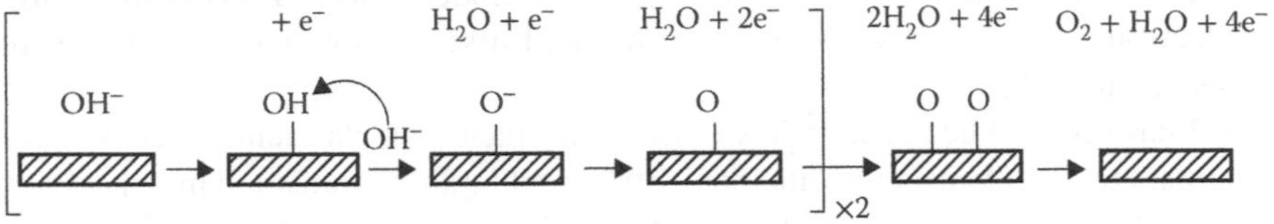

FIGURE 2.9 OER at catalytic surface under basic conditions. (From Birss, V. I. et al., *J. Electrochem. Soc.,* 133, 1621–1625, 1986.)

Similarly, three possible reaction steps (Equations 2.16 through 2.18) have been suggested for HER in acidic media by Conway and Tilak (2002), including a primary discharge step, followed either by an electrochemical desorption step or the recombination step:

$$\text{Discharge step: } M + H_3O^+ + e^- \leftrightarrow M_{Hads} + H_2O \text{ (Volmer reaction step)} \tag{2.16}$$

$$\text{Desorption: } M_{Hads} + H_3O^+ + e^- \leftrightarrow H_2 + M + H_2O \text{ (Heyrovsky reaction step)} \tag{2.17}$$

$$\text{Recombination: } M_{Hads} + M_{Hads} \leftrightarrow H_2 + 2M \text{ (Tafel reaction step)} \tag{2.18}$$

The designing of a perfect catalytic system that can effectively break water to generate hydrogen is a mega challenge of this era. Recently, mammoth efforts have been made to elucidate the charge transfer processes and to improve the efficiency of water splitting in terms of hydrogen production or photocurrent density. A few notable systems that include the use of sunlight and the light harvesting assemblies to make this challenge easy are Cr- or Fe-doped TiO_2 (Dholam et al. 2009), with an H_2 production rate of 15.5 µmol/h for Fe-doped TiO_2 and 5.3 µmol/h for Cr-doped TiO_2; quantum dot-sensitized hybrid-TiO_2 (QD/H-TiO_2) with a photocurrent density of ~16.2 mA/cm^2 (Kim et al. 2013); ZnO-based photoelectrodes (Chen et al. 2011); CdS/Si–Au nanoparticles (Torimoto et al. 2011); GaN (Kamat and Bisquert 2013); PbS-based materials (Trevisan et al. 2013); graphene-based photocatalyst (Xiang and Yu 2013); N-doped $Ba_5Ta_4O_{15}$ (Mukherji et al. 2011); MoS_2 catalysts supported on Au (Jaramillo et al. 2007); activated carbon (Hinnemann et al. 2008); carbon paper (Bonde et al. 2008);

or graphite (Jaramillo et al. 2008). Such systems were prepared by physical vapor deposition or annealing of molybdate in H_2S (Sasikala et al. 2010).

2.4 CRITERIA FOR THE SELECTION OF PHOTOCATALYTIC MATERIAL

The development of photocatalytic materials that can utilize maximum solar energy inputs to convert into either electric or chemical energy is one of the holy grails of future advancement in material research and technology that can direct the way in obtaining clean energy. At the molecular level, several physicochemical functions need to be integrated into one stable chemical system that can set the criteria, which must be satisfied simultaneously:

1. **Band gap** of the semiconducting material should lie between 1.6 eV (1.23 eV + overpotential) and 2.5 eV (larger than 2.43 eV). Then, the material can harvest the visible part of the sunlight and enhance the efficiency of the water splitting.
2. **Band edge positions** means that band edges must straddle between the redox potentials of H_2O (0.00 eV and 1.23 eV), as illustrated in Figure 2.10. Semiconductor materials must satisfy the minimum band gap requirement (~1.4 eV). The materials of low band gap are visible (Vis) light active but susceptible to photocorrosion, and the stable materials with a wider band gap absorb light only in the ultraviolet (UV) region (i.e., 5% of the whole sunlight spectrum). It has been found that the valence band (VB) holes are powerful oxidants (+1.0 to +3.5 V vs. NHE depending on the semiconductor and pH), while the conduction band (CB) electrons are good reductants (+0.5 to −1.5 V vs. NHE).
3. **Charge transfer** is necessary at the photocatalytic surface and it must be fast enough to prevent photocorrosion and shifting of the band edges, which result in the loss of photon energy that can provide efficient oxidation and reduction sites on the surface of the material.

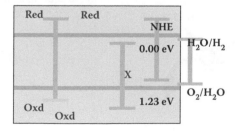

FIGURE 2.10 Band gap and band edge positions of the semiconductors, which can decide the nature of semiconductor, that is, redox (Red/Oxd), oxidative (Oxd), reductive (Red), and none (X) during water splitting.

4. **Stability** of material in aqueous medium is essentially required (at least for 20 years). The hydrophilic or hydrophobic surface of the photocatalysts under light radiation is one of the important issues to maximize the photocatalytic efficiency.

5. **Aid of a cocatalyst** (Li 2013) for hydrogen generation is necessary, but some of the cocatalysts are highly active and induce a reverse reaction, that is, the generation of water from molecular oxygen and hydrogen, which must be reduced. For example, Pt@ TiO$_2$–anatase (Figure 2.11) produce both reactions at surface, but addition of an iodine layer on the Pt surface prevents this backward reaction (Abe et al. 2003).

6. **Abundant availability** can reduce the cost of the material and assure sustainability.

7. **A complementary metal–oxide semiconductor (CMOS) and biocompatibility** of the photocatalytic material should be considered as a parameter for their selection.

8. **Nontoxic** and easy to handle materials with good compatibility to sensitizers are important.

The photocatalytic process is quite complicated because the reactions occur at the photocatalytic surface and on the interface of the photocatalysts and reactant solution. A simple arbitrary description of the photocatalytic process involves the following steps:

1. **Photon absorption:** Photocatalysts absorb photons and generate electrons and holes at the surface. Semiconductors (e.g., TiO$_2$, ZnO, Fe$_2$O$_3$, WO$_3$, AgPO$_4$, CdS, and ZnS) can act as photocatalysts for light-induced redox processes due to their electronic structure, which is characterized by a filled VB and an empty CB with a suitable gap between them. When a photon with energy hv matches or exceeds the band gap energy, E_g, of the semiconductor, an electron jumps from the VB to the CB, leaving a hole in the VB.

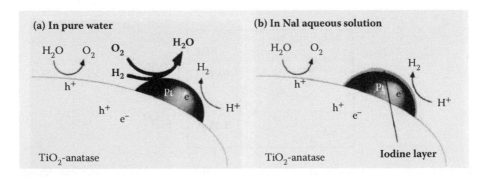

FIGURE 2.11   Pt-loaded TiO$_2$–anatase photocatlyst surface. (a) A platinum cocatalyst can also catalyze the reverse reaction. (b) This can be prevented by adding an iodine layer on the platinum surface. (From Abe, R. et al., *Chem. Phys. Lett.*, 371, 360–364, 2003.)

2. **Relaxation:** Electrons and holes release energy (heat) and move the CB to minimum position and VB to maximum position, respectively.
3. **Recombination:** Excited-state CB electrons and VB holes can recombine and dissipate the input energy as heat, get trapped in metastable surface states, or react with EDs and EAs adsorbed on the semiconductor surface or within the surrounding electrical double layer of the charged particles. In the absence of suitable electron and hole separation forces, the stored energy is dissipated within a few nanoseconds in the recombination process. If a suitable scavenger or surface defect state is available to trap the electron or hole, recombination is prevented/reduced and subsequent redox reactions may occur.
4. **Charge transport:** The excited electrons and holes will move to the cathode (electrons) and anode (holes), respectively. For bulk semiconductors, either a hole or an electron can be accumulated at the cathode or anode for further photocatalytic reaction. In a very small semiconductor particle suspension, both electrons and holes are present at the same surface. Electron transport in nanostructured semiconductors shows key differences with respect to their bulk counterparts due to their low dimensionality and high surface to volume ratio and lower number of defects. The main significance of the disorder in the electronic structure of the material is the appearance of localized states or trap centers at the material surface.
5. **Water absorption on the surface**: The hydrophobic surface of the photocatalysts that merged in water and exposed under light radiation is one of the important issues to maximize the photocatalytic efficiency.

Various strategies are adopted for acquiring the aforementioned qualities simultaneously. Materials attain the required band gap along with appropriate band positions after some modifications at the molecular level, such as morphological advancement at the surface level that improves the shape and size of the particle; way of synthesis; multilayer systems (coupled semiconductors); doping of cations/anions into the main catalyst; and solid solutions.

A cationic dopant like indium can contribute at additional levels with the CB of the semiconductor, which decreases its band gap (E_g). As the additional levels mix with the CB, they do not act as electron traps and contribute in enhancing photocatalytic activity of the material. Doping of oxide semiconductors with anionic dopants like N or S (Figure 2.12a and b) leads to the mixing of the N 2p or S 3p orbitals with the O 2p orbitals and due to the repulsion among electrons of N 2p or S 3p orbitals and O 2p orbitals, an expansion in the VB is observed. This process results in decrease in E_g without affecting the CB level and makes a material capable of entrapping Vis. As a result, band edge potentials become suitable for overall water splitting. Similarly, when a large band gap semiconductor afire with short band gap semiconducting material at high temperatures, then a short band gap solid solution substance (in comparison to the large band gap material) is produced, as shown in Figure 2.12a. Beside the band gap engineering (Figure 2.12a), high surface to volume ratio can be controlled by the shape and size of the functional nanomaterial and contribute in creating more active sites on surface, facilitating in the electron/charge transfer.

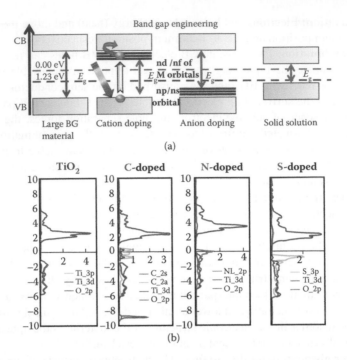

FIGURE 2.12 (a) Band gap engineering of large band gap material using cation doping, anion doping, and solid solution formation by solid state reaction of a large band gap semiconductor with a short band gap semiconductor. (b) Band gap and band positioning effect of C, N, and S doping on TiO_2 nanostructures. (From Yang, K. et al., 2012. Review of the structural stability, electronic and magnetic properties of nonmetal-doped TiO_2 from first-principles calculations, arXiv:1202.5651v1 [cond-mat.mtrl-sci] February 25, 2012.)

Furthermore, the perspective focuses on the photocatalytic performance of the semiconductor–metal and semiconductor–semiconductor nanostructure composite assemblies that influence the rate of photoinduced electron transfer at the interface. The storage and discharge of charged species in metal nanoparticles plays an important role in confirming the photocatalytic performance of the semiconductor–metal composite assemblies. In the same way, the coupled semiconductors of well-matched band energies are useful to improve charge separation. Both electron and hole transfer across the interface with comparable rates are important in maintaining high photocatalytic efficiency and stability of the semiconductor assemblies. Moreover, the semiconductor and metal nanoparticles accumulated on the reduced graphene oxide sheets offer new ways to design a multifunctional catalyst matrix, which is extremely important in getting a fundamental understanding of charge transfer processes in light harvesting assemblies. In light harvesting molecular assemblies, the specific property effect accomplishes several changes in composite materials. Some of them are mentioned below:

1. Small particle size: High surface area and high adsorption of light
2. High crystalline material: Single site structure and homogeneity

3. Engineering band gap: More efficient light absorption
4. Presence of adequate cocatalysts: Long lifetime of the charge separation, preferential migration along a certain direction, efficient charge separation for low recombination, and efficient chemical reactions with a single product
5. High crystallinity: High mobility of charge carriers and more efficient charge separation
6. Quantum dots (QD) sensitizers: Promising "sensitizers" due to the tunability in band gap across the UV/Vis/infrared (IR) region; broad absorption with high molar absorptivity, robustness against photobleaching, and multiple exciton generation and energy transfer–assisted charge collection (Chang and Lee 2007; Diguna et al. 2007; Ruland et al. 2011; Jin et al. 2013)

QDs based on the above promising characteristics along with wide band gap n-type metal oxides (photoanode) have been used as sensitizers for PEC water-splitting applications. Efficient electron injection from a Vis light–active QD to a metal oxide requires that the CB edge of the QD is located above that of the metal oxide. The electron transfer between the QD and the metal oxide can facilitate the charge separation and inhibit recombination through a potential gradient at the interface.

2.5 OVERPOTENTIAL

Principally, the electrolysis of fresh water requires 1.23 V for the generation of hydrogen and oxygen, but on practical grounds it needs higher voltages for the reaction to proceed because of the different kinds of loss and nonideality present in the electrochemical process. The part that exceeds the limit of 1.23 V is called overpotential or overvoltage. In the case of water splitting, reduction of water evolved two-electron process and needs very low overpotential (Equation 2.10), but a considerable overpotential is needed to cross the kinetic barriers associated with the four-electron process of water oxidation (Equation 2.11) to compete with the electron–hole recombination. To satisfy the overpotential needs of the system, extra energy is required that can be obtained from either solar energy or electrical energy (applied potential) and the catalytic material with or without cocatalyst. Water oxidation to some extent can be achieved/controlled by coupling the photon absorption (in the form of sunlight) with catalytic functions of the electrode loaded with cocatalyst. For practical energy-conversion applications, applied potential should be reduced to lower the cost. In the water-splitting process, a large overpotential is required in support of the four-electron oxidation of water to oxygen at the anode, for a well-designed cell; on the other hand, reduction of water to hydrogen requires a two-electron process at the cathode (Equation 2.10) that can be electrocatalyzed with almost no overpotential using platinum or platinum alloys, and in theory hydrogenase enzyme. Thus, costly platinum alloys are the state-of-the-art electrocatalysts for oxidation, but in the case of other less-effective materials used for the cathode (e.g., graphite), large overpotentials are required. Therefore, the efforts to develop a cheap, effective electrocatalyst for this reaction would be a great advance. There are many approaches to facilitate this reaction, and a few of them include molybdenum sulfide (Kibsgaard et al. 2014), graphene QDs (Fei et al. 2014), carbon nanotubes (CNTs) (Gong et al. 2014), and perovskite (Luo et al. 2014).

Charge transfer at electrodes can be characterized by an equilibrium potential. At equilibrium, the rate of the forward reaction is equal to the rate of the reverse reaction so that there is no net reaction or an absence of faradaic current. But for a noticeable cathodic reaction, the rate of the forward reaction must be higher than that of the reverse reaction. In this condition, the applied potential at the cathode must be of a higher magnitude (more negative) than that implied by the equilibrium potential. The difference in the electrode potential from the equilibrium value is termed as polarization. The extent of polarization is measured by the overpotential, η, which is a distinguished characteristic of a particular electrode–electrolyte system, and is expressed by Equation 2.19:

$$\eta = E_{appl} - E_{eq} \tag{2.19}$$

where E_{appl} = applied potential and E_{eq} = equilibrium potential. Usually, the electrode reactions are identified by the current density as a function of potential, which will be driven by a certain overpotential:

$$\eta = a + b \log j = \frac{2.3RT}{\alpha nF} \log j_0 - \frac{2.3RT}{\alpha nF} \log j \tag{2.20}$$

where j = current density, b = Tafel constant, η = overpotential, j_0 = exchange current density, R = universal gas constant (8.314 kJ/mol/K), α = charge transfer coefficient, n = number of electrons exchanged during electrode reaction, and F = Faraday constant (96,485 C/mol). From the theory of square wave voltammetry (Nuwer 1991; Fatouros and Krulic 1998; Miles and Compton 2000), one-electron oxidative charge transfer (α) is given as follows:

$$\alpha = 1.66 \, RT/WF \tag{2.21}$$

where W is the half width on the differential current voltammogram at mid-height with respect to the peak potential and T is the absolute temperature of the reaction.

The absorption of a photon ($h\nu$) by the semiconductor with a band gap of E_g creates an electron–hole pair that can be separated by the space charge layer, W, to generate a free energy of $\Delta\mu_{ex}$. This free energy must be greater than the energy needed for water splitting (1.23 eV) plus the overpotential losses at both the anode and cathode, η_{ox} and η_{red}, for the water-splitting reaction to occur. Two photons must be absorbed to produce one H_2, while four photons are needed for the half O_2 production.

We can compare the electrode process of different reactions by the overpotential versus $\log j$ plots (Equation 2.20 and Figure 2.13), from which it is evident that the electrode processes are similar in alkaline water electrolysis and black liquor, that is, eucalyptus, wheat straw, and bagasse electrolysis. It is also observed that in the high overpotential range (η between −150 and −400 mV), the polarization plots are linear with positive slope, as expected from Equation 2.20.

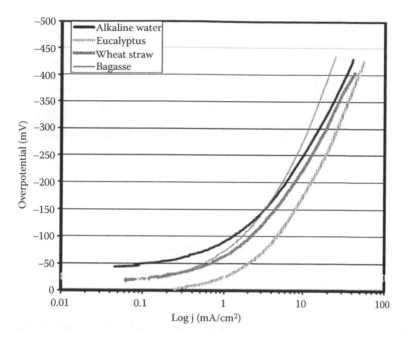

FIGURE 2.13 Steady-state polarization curves for hydrogen evolution reaction in alkaline water electrolysis and black liquor (eucalyptus, wheat straw, and bagasse) electrolysis. (From http://shodhganga.inflibnet.ac.in/bitstream/10603/3339/13/13_chapter%204.pdf.)

Hence, the actual interelectrode potential to be applied for a given overall rate of electrolysis, including other potential terms, is given below:

$$F_{appl} = F_{eq}(\text{cathode}) + \eta(\text{cathode}) + F_{eq}(\text{anode}) + \eta(\text{anode}) + F_{ohmic} \quad (2.22)$$

where E_{appl} = actual interelectrode potential applied, E_{eq}(cathode) = equilibrium potential at cathode, E_{eq}(anode) = equilibrium potential at anode, η(cathode) = overpotential losses at cathode, η(anode) = overpotential losses at anode, and E_{ohmic} = potential drop due to resistance offered by various components of the cell.

2.6 BAND GAP AND BAND EDGE POSITION IN PHOTOCATALYTIC MATERIALS

In material science, the electronic band structure of a solid describes those ranges of energy that an electron within the solid may have (called *CB/VB*, *allowed bands*, or simply *energy bands*) and ranges of energy that it may not have (called *band gaps* or *forbidden bands*) free carriers. On the basis of the forbidden bands, solids can be classified into four main parts, metals, semimetals, semiconductors (p-type, intrinsic, and n-type), and insulators, which are exhibited in Figure 2.14. Metals have a partially filled CB (conduction through electrons). Therefore, metal has an appreciable density of states at the Fermi level (Kittel 1996) and like semimetals, have a Fermi level inside one of the allowed bands. Semimetals, however, have a very small overlap between the bottom

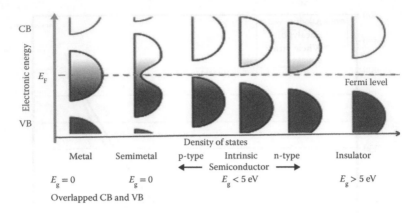

FIGURE 2.14 Fermi level, positions of conduction band (CB) and valence band (VB), and electronic density of states in metal, semimetal, semiconductors (p-type, intrinsic, n-type), and insulators at equilibrium along with their energy levels. (From https://en.wikipedia.org/ wiki/Fermi_level, last accessed 16 July, 2016; Sze, S. M., *Physics of Semiconductor Devices,* Wiley, New York, NY, 1964.)

of the CB and the top of the VB and they possess both kinds of charge carriers, that is, holes and electrons, but in fewer number than metals. Furthermore, a semimetal has no band gap just like metals but a negligible density of states at the Fermi level. Semiconductors or insulators having the Fermi level inside the band gap (forbidden zone) if it is close to the CB are known as n-type semiconductors, whereas those having Fermi level near the VB are known as p-type semiconductors. Semiconductors in which the Fermi level lies at same distance from the CB and VB are known as intrinsic semiconductors. A semimetal also differs from an insulator or semiconductor in that a semimetal's conductivity is always nonzero, whereas a semiconductor has zero conductivity at zero temperature and insulators have zero conductivity even at ambient temperatures (due to a wider band gap). The bands (CB and VB) have different widths, which depend on the degree of overlap in the atomic orbitals from which they arise. In semiconductors and insulators, the electron-filled VB is separated from an empty CB by a band gap. For insulators, the magnitude of the band gap is quite a bit larger (e.g., >5 eV) than that of a semiconductor (e.g., <5 eV) (Perry et al. 1969). A few representative examples of all four categories are mentioned as follows: metals (Na, Al, Au, Ag, Cu, etc.), semimetals (arsenic, antimony, bismuth, α-tin or graphite, and the alkaline earth metals), mercury telluride (HgTe), conductive polymers (Bubnova et al. 2014), semiconductors (Ge, Si, ZnO, GaN, InN, WO_3, etc), and insulators (glass, paper, Teflon, etc.).

Semiconducting materials are further divided into two types: direct band gap and indirect band gap. Direct band gap material has the same space momentum (k) for the lowest-energy state (VB) above the band gap and the highest-energy state (CB) beneath the band gap. For indirect band materials, the values of k are different for both states, as depicted in Figure 2.15. Direct or indirect band gaps are one of the major parameters in the application of photoconduction and electroluminescence processes of inorganic semiconductors (such as Si, Ge, and GaAs), which determines the efficiency of these processes. Typical examples of them are as follows:

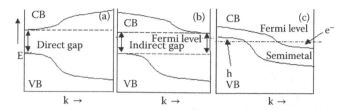

FIGURE 2.15 Schematic illustration of the lowest-energy conduction band and the highest-energy valence band in one dimension of momentum space (or k-space) for (a) a direct semiconductor, (b) an indirect semiconductor, and (c) a semimetal. (From https://en.wikipedia.org/wiki/Semimetal, last accessed 24 August, 2016; Burns, G., *Solid State Physics*, Academic Press, New York, NY, pp. 339–340, 1985.)

1. Direct band gap semiconductor: GaAs, ZnO, InAs, GaAs, CdTe, GaN, CuInSe$_2$, and so on
2. Indirect band gap semiconductor: Si, Ge, and so on

Metal or metal alloys are usually used for the reduction, whereas semiconducting materials are very important in exploiting the oxidation and reduction process at different sites of the surface.

Band gap E_g governs the optical absorption edge, playing a vital role in deciding the fate of the optoelectronic materials. There are several known methods for the estimation of band gaps using various physical properties: transmittance/reflectance of UV–Vis spectroscopy (Cheng and Wang 2009), voltammetry study (Shaukatali et al. 2008), soft x-ray spectroscopic techniques (Dong et al. 2004), and the tight-binding linear muffin-tin orbital (TB-LMTO) method to determine band gaps (E_g) (Singh et al. 2013).

1. Transmittance/reflectance of UV–Vis spectroscopy: In crystalline semiconductors, the following equation has been obtained to relate the absorption coefficient (α) to incident photon energy ($h\nu$) (Figures 2.16a and b and 2.17) (Tauc and Menth 1972):

$$\alpha(\nu)h\nu = B(h\nu - E_{gap})^m \tag{2.23}$$

where E_{gap}, B, and $h\nu$ are the optical band gap, constant, and incident photon energy, respectively. Generally, m is 2 for an indirect band gap and 1/2 for direct band gap materials. $\alpha(\nu)$ is the absorption coefficient defined by Beer–Lambert's law as

$$\alpha(\nu) = \frac{(2:303 \times Abs(\lambda))}{d} \tag{2.24}$$

where d and Abs(λ) are the film thickness and film absorbance, respectively.

2. Voltammetry study: Voltammetry experiments like cyclic voltammetry and differential pulse voltammetry are used to determine the anodic peak oxidation potential, V_{ox}, and the cathodic peak reduction potential, V_{red}, of the

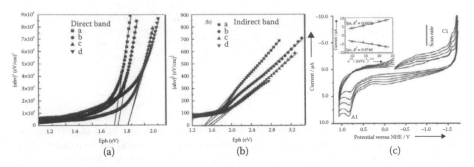

FIGURE 2.16 (a) $(\alpha h\nu)^2$ versus $h\nu$ and (b) $(\alpha h\nu)^{1/2}$ versus $h\nu$ plot of samples of chemically synthesized $AgIn_5S_8$ ternary system semiconductor films on indium tin oxide (ITO)-coated glass substrates at different thicknesses of $a = 205$ nm, $b = 423$ nm, $c = 632$ nm, and $d = 1070$ nm. (c) Scan-rate-dependent response of the cyclic voltagrams recorded on trioctylphosphine oxide (TOPO)-capped CdSe quantum dots (Q-dots). The arrow indicates an increasing scan rate from $n = 100$–500 mV·s^{-1}. The inset shows a linear fitting of peak currents for A1 and C1 versus $n^{1/2}$. (From (a) and (b) Cheng, K.-W., and S.-C. Wang, *Mater. Chem. Phys.*, 115, 14–20, 2009 and (c) Shaukatali, N. I. et al., *Chem. Phys. Chem.*, 9, 2574–2579, 2008.)

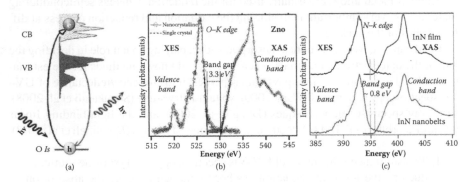

FIGURE 2.17 (a) A schematic representation of the soft x-ray absorption spectroscopy (XAS) and soft x-ray emission spectroscopy process. Soft x-rays cause an electron to be excited to CB (inorganic)/lowest occupied molecular orbital (organic) by absorption of photons; when another electron fills the resulting hole, X-rays of a different wavelength are emitted ($h\nu$: absorbed or emitted photons), (b) ZnO at O–K edge, and (c) InN film at N–K edge. (From (b) Guo, J.-H et al., *J. Phys. Condens. Matter*, 14, 28, 6969–74, 2002; Dong, C. L. et al., *Phys. Rev. B*, 70, 195325, 2004 and (c) Eisebitt, S. et al., *Phys. Stat. Sol. B*, 215, 803–808, 1999.)

studied material. These terms are utilized in determining the highest occupied molecular orbital and lowest unoccupied molecular orbital (HOMO–LUMO) or band gap. When we oxidize a molecule or material, we take the first electron out of the HOMO (or VB). When we reduce the molecule or material, we put an electron into the LUMO (or CB). Therefore, V_{ox} informs us about the HOMO and V_{red} about the LUMO. The energy gap (in electron volts) is simply the difference between these two potentials (in volts) multiplied by the charge on the electron $e = -1$ eV/V:

$$E(\text{eV}) = e(V_{red} - V_{ox})$$
(2.25) (Shaukatali et al. 2008)

If the geometric mean χ of the Mulliken electronegativities of the semiconductor constituents is known, the flat-band potential V_{fb} (V vs. NHE) for a *metal oxide* can be calculated as

$$V_{fb} = E_0 - \chi + \frac{1}{2}E_G \qquad (2.26)$$

Here, E_G is the semiconductor band gap (eV) and E_0 (+4.44 eV) is the energy of a free electron on the H_2 redox scale (Pleskov and Gurevich 1986; Kim et. al. 1993).

3. Soft x-ray spectroscopic techniques: Among the soft x-ray spectroscopic techniques, x-ray absorption spectroscopy (XAS) and x-ray emission spectroscopy (XES) probe the energy distribution of electronic states in atoms, molecules, and solid state materials. The basic concepts, as shown in Figure 2.17a, involve the interaction of x-rays and matter as explained by molecular orbital theory. In XAS, the absorption of photons excites electrons from deep core levels, such as 1s, of a selected atom in a molecule to unoccupied states, leaving behind a core hole. In XES, the core hole is filled by a valence electron causing the emission of an x-ray photon. XES gives information about chemical bonding in the molecule. By combining XAS and XES, one can obtain information about unoccupied states (CB) and occupied states (VB). The difference in energy between the CB and the VB is called the band gap. Figure 2.17b (Guo et al. 2002) and 2.17c (Eisebitt et al. 1999) exhibit the use of XAS–XES in determining the band gap of the ZnO and InN film.

2.7 BAND EDGE BENDING: SEMICONDUCTOR/ELECTROLYTE INTERFACE REACTIONS

The semiconductor–electrolyte or metal–semiconductor interfaces are attractive zones for electrochemists, where the steady state and dynamic aspects of the carrier transfer are considered. In an intrinsic semiconductor at ambient state, the Fermi level lies at the middle of the band gap and the carrier concentration is very low. It happens when a semiconductor electrode is not contacted with electrolyte and not under light illumination than no bending in bands was observed (Figure 2.18a). This provides the theoretical basis for several efficient semiconductor-redox couple. The region where the charge distribution differs from the bulk material is known as the space-charge layer/depletion region. When n-type-extrinsic semiconductors (doping of group 15 or VA elements to Si metal) immerse in electrolyte (Figure 2.18b), a depletion region arises due to the positive charge on the dopant, the Fermi level (close to the CB) moves down, and the process stops when the Fermi level achieves same position on either side of the interface. Electric current initially flows across the interface until electronic equilibrium is achieved, where the Fermi energy level of the electrons in the solid (E_F) is equal to the redox potential of the electrolyte (E_{redox}), as shown in Figure 2.18b. As a result, there will be an

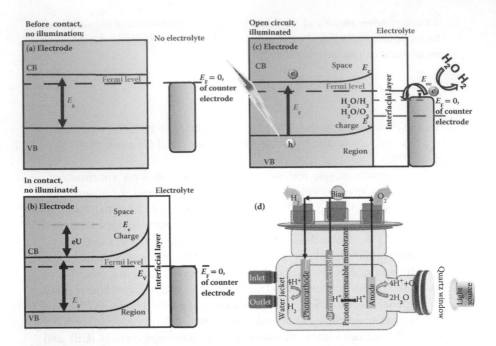

FIGURE 2.18 Band gap and band positions, under (a) no illumination-no electrolyte, (b) electrolyte, and (c) illumination-electrolyte conditions. (d) Schematic presentation of the PEC cell for water splitting.

appreciable concentration of the electron. The transfer of electric charge produces a region at each side of the junction. In the vicinity of this heterojunction, a band-bending (upward or downward) phenomenon is observed, depending upon the position of the Fermi level in the semiconductor (n-type or p-type). If the Fermi level of the electrode is equal to the flat-band potential, there is no excess charge on either side of the junction and hence the bands are flat.

On the electrolyte side, this corresponds to the familiar electrolytic double layer, that is, the compact (Helmholtz) layer followed by the diffuse (Gouy–Chapman) layer.

If electrons accumulate at the semiconductor side, one obtains an accumulation layer. However, when they deplete from the solid into the solution, a depletion layer is formed, leaving behind a positive excess charge formed by immobile ionized donor states. Finally, electron depletion can go so far that the concentration at the interface falls below the intrinsic level. Hence, the semiconductor is *p*-type at the surface and *n*-type in the bulk, corresponding to an inversion layer. The illustration in the figure refers to *n*-type materials, where electrons are the mobile charge carriers. In this case, even after bending the difference between CB and VB edges remains the same. However, if the electric field of this charge and, consequently, band bending are strong enough to provide quantum confinement of electrons or hole (forming two-dimensional [2D] electron gas), then the gap between electron and hole energies can be changed due to quantization of the energy band, which leads to splitting into sub-bands. The Fermi level should be constant for the structure in thermodynamic equilibrium. For p-type semiconductors, analogous considerations apply. Positive holes are the mobile charge carriers and the immobile negatively charged states of the acceptor dopant form the excess space charge within the depletion layer. Under illumination (Figure 2.18c), the electrode surface, whose photon energy is greater than the band gap, promotes electrons into the CB, leaving holes in the VB. In the case of a photoanode, small band bending is observed in the depletion region driven by any electron that is promoted into the CB into the interior of the semiconductor and holes existed in the VB region.

Figure 2.17d shows the schematic diagram of the three-electrode (working electrode/anode, reference electrode, and cathode/Pt wire) PEC cell, where the whole bending procedure occurs.

2.8 EFFICIENCY (SOLAR TO HYDROGEN CONVERSION, TURNOVER NUMBER, QUANTUM YIELD, PHOTOCONVERSION EFFICIENCY, INCIDENT PHOTON-TO-CURRENT EFFICIENCY [%], ABSORBED PHOTON-TO-CURRENT EFFICIENCY)

Water-splitting efficiency of the photocatalytic material can be defined in terms of the power to generate an electron–hole pair at the semiconductor surface by the consumption of sunlight and utilization of an electron–hole pair in reduction and oxidation of water without recombination. In other words, we can say the efficiency of the photocatalyst produces hydrogen and oxygen in 2:1 ratio under light exposure. The amounts of H_2 and O_2 released should be greater than the amount of the photocatalyst used and the ratio of these products must be 2:1, respectively. Otherwise it will be hard to judge whether the reaction proceeds photocatalytically (for truly catalytic reaction, the TON >1) or not because the reaction might occur due to some stoichiometric reactions. Several terms have been employed to describe the efficiency of water splitting for converting solar energy to a useful form of energy (electrical or chemical), namely TON, IPCE, APCE, solar-to-hydrogen conversion efficiency (STH), and quantum efficiency (QE).

2.8.1 TURNOVER NUMBER

This is usually defined by the ratio of the number of reacted molecules to the number of active sites (Equation 2.27):

$$TON = \frac{\text{Number of reacted molecules}}{\text{Number of active sites}} \tag{2.27}$$

However, it is often difficult to determine the number of active sites for photocatalysts. Therefore, the number of reacted electrons to the number of atoms in a photocatalyst (Equation 2.28) or on the surface of a photocatalyst (Equation 2.29) is employed as the TON:

$$TON = \frac{\text{Number of reacted electrons}}{\text{Number of atoms in a photocatalyst}} \qquad (2.28)$$

$$TON = \frac{\text{Number of reacted electrons}}{\text{Number of atoms at the surface of a photocatalyst}} \qquad (2.29)$$

The number of reacted electrons is calculated from the amount of evolved H_2. The TONs calculated using Equations 2.28 and 2.29 are smaller than the real TON measured using Equation 2.27 because the number of atoms is more than that of active sites. The TON approaches 10^6 per second for the enzymatic catalytic reversible reduction of N^5,N^{10}-methenyltetrahydromethanopterin with H_2 in the presence of hydrogenase enzyme to N^5,N^{10}-methylene-tetrahydromethanopterin (Vignais et al. 2001). On the other hand, nitrogenase-catalyzed hydrogen production is highly energy intensive due to the consumption of 16 ATP molecules. Hence, these are less efficient compared to the hydrogenase-based reaction, where the TON is less than 10 per second. Normalization of photocatalytic activity by weight of used photocatalyst (e.g., millimole per hour per gram [mmol/h/g]) is not acceptable because the photocatalytic activity is not usually proportional to the weight of photocatalyst and the amount of the photocatalyst should be optimized for each experimental setup. In this case, photocatalytic activity usually depends on the number of photons absorbed by a photocatalyst unless the light intensity is too strong.

2.8.2 INCIDENT PHOTON-TO-CURRENT EFFICIENCIES

This is a measure of the ratio of the photocurrent (converted to the electron transfer rate) versus the rate of incident photon (converted from calibrated power of light source) as a function of the wavelength. It is considered as the efficiency of the photon absorption/charge excitation and separation ($\eta_{e-/h+}$), charge transport within the solid to the solid–liquid interface ($\eta_{transport}$), and interfacial charge transport across the solid–liquid interface ($\eta_{Interface}$). IPCE(%) is calculated using Equation 2.31, which includes photocurrent density, wavelength of light, and irradiance:

$$IPCE = \eta_{\frac{e-}{h}} + \eta_{transport} \eta_{interface} \qquad (2.30)$$

$$IPCE\,(\%) = \frac{1240 \times \text{photocurrent density}}{\lambda \times \text{irradiance}} \qquad (2.31)$$

IPCE results are quite reproducible in laboratories, but discrepancies arise when spectral illumination power density varies from one test setup to another. So care should be taken for similar light intensity at each wavelength. For these differences in illumination, intensities and band pass at each wavelength must be taken into account since nonlinear effects start dominating when incidental light intensities change significantly. Therefore, it may be advantageous to use a tungsten lamp as intense emission spectral lines in the output of an arc lamp can produce spectral artifacts due to nonlinear photocurrent response. One way to eliminate this effect is to use white light bias in addition to the monochromated light. If bias is not used, then the measurement made for IPCE is assumed to be independent of the monochromated light level.

2.8.3 ABSORBED PHOTON-TO-CURRENT EFFICIENCY

This is usually used to characterize the photoresponse efficiency of a photoelectrode material under an applied voltage (Khan et al. 2002). Various factors including nanotube length, direction of growth, and constitute-composition are taken into account when APCE is calculated through the absorbance α. The APCE is obtained by dividing the IPCE by the fraction of incident photons absorbed at each wavelength by light harvesting efficiency (LHE) (Equations 2.32 and 2.33). APCE is usually used to characterize the photoresponse efficiency of a photoelectrode material under an applied voltage:

$$APCE = \frac{IPCE}{LHE\ (\lambda)} \quad (2.32)$$

where

$$LHE(\lambda) = TTCO(\lambda)\left(\frac{\alpha_{dye}}{\alpha_{film}}\right) \times (1 - e^{\alpha}) \quad (2.33)$$

where TTCO is the transmittance of the transparent conductive oxide, α_{film} is the absorption coefficient of the entire film, and α_{dye} is the absorption coefficient due to the dye/sensitizer molecules. This is a simplest approach for calculating APCE, where second-order reflectance terms are not considered. LHE is estimated by using the information about injection and collection efficiencies (Equation 2.33).

2.8.4 SOLAR-TO-HYDROGEN CONVERSION EFFICIENCY

To describe the true hydrogen production efficiency of a water-splitting reaction under sunlight, a term known as STH (Chen et al. 2010) is often used. The definition of STH conversion efficiency is shown in Equation 2.34. The STH of the water-splitting reaction can be determined as follows:

$$STH = \left[\frac{\left|J_{sc}\left(\frac{mA}{cm^2}\right)\right| \times (1.23\ V) \times \eta_F}{P_{Total}\left(\frac{mW}{cm^2}\right)}\right]_{AM1.5G} \quad (2.34)$$

where P_{total} represents the power density of incident simulative sunlight (air mass [AM] 1.5G) and the numerator is the product of photocurrent density (j_{sc}) at zero bias (short-circuit photocurrent), the thermodynamic voltage required for water splitting (1.23 V), and the faradic efficiency (η_F). Different forms of STH that can also be used are shown in Equations 2.35 and 2.36:

$$STH = \left[\frac{\left(\frac{mmol\ H_2}{s}\right) \times \left(\frac{237\ kJ}{mol}\right)}{P_{Total}\left(\frac{mW}{cm^2}\right) \times area\ (cm^2)}\right]_{AM1.5G} \quad (2.35)$$

$$\text{STH}(\%) = \frac{j_{p}(V_{\text{WS}} - V_{\text{Bias}}) \times 100}{eE_{s}} \qquad (2.36)$$

where j_p is the photocurrent density (mA/cm^2) produced per unit of irradiated area, V_W (1.23 eV) is the water-splitting potential per electron, V_{Bias} is the bias voltage applied between the working and the counter electrodes, E_s is the photon flux, and e is the electronic charge. According to Equations 2.35 and 2.36, every electron contributing to the current produces half an H$_2$ molecule. The photoconversion efficiency (η) is a percentage conversion of light energy into chemical energy in the presence of applied external voltage and it can be calculated using Equations 2.37 and 2.38. Such terms cannot be used to represent the true photoconversion efficiency for photocatalytic water splitting because of the involvement of the varied applied voltage during measurement:

$$\eta(\%) = \frac{\text{total power output} - \text{electrical power output}}{\text{light power input}} \times 100 \prod_{i=1}^{n} X_i \qquad (2.37)$$

$$= j_{p} \left[\frac{(E_{0,\text{rev}} - | E_{\text{app}} |)}{I_0} \right] \times 100 \qquad (2.38)$$

where $E_{\text{app}} = E_{\text{meas}} - E_{\text{aoc}}$, j_p is the photocurrent density in mA/cm^2, $E_{0,\text{rev}}$ is the standard reversible redox potential of the water and equal to the value 1.23 versus NHE, E_{meas} is the electrode potential (vs. Ag/AgCl or saturated calomel electrode [SCE]) of the working electrode at which photocurrent was measured under light illumination, E_{aoc} is the electrode potential (vs. *standard* Ag/AgCl or calomel electrode) of the same working electrode at open circuit conditions under the same illumination and in the same electrolyte solution, and I_0 is the intensity of incident light in mW/cm^2.

QE is one of the most important parameters used to evaluate the quality of a detector and is often called the spectral response, which reflects its wavelength dependence. QE is the fraction of photon flux that contributes to the photocurrent in a photodetector. It is defined as the number or percentage of signal electrons created per incident photon but does not include subsequent photons produced by amplification processes. If, over a period of time, an average of 10,000 photoelectrons are emitted as the result of the absorption of 100,000 photons of light energy, then the QE will be 10%. In some cases, it can exceed 100% (i.e., when more than one electron is created per incident photon). It is noted that QE neglects the energy loss of solar irradiance and the chemical conversion efficiency. Therefore, it is used to qualify the photoactive films/material but not to represent the water-splitting reaction conversion efficiency. The apparent QE, measured under an xenon arc lamp or four 420-nm light-emitting diodes (LEDs) (3 W), is used to trigger the photocatalytic reaction, using the following formula:

$$\text{QE}(\%) = \left(\frac{\text{number of evolved H}_2 \text{ molecules} \times 2}{\text{number of incident photons}} \right) \times 100 \qquad (2.39)$$

Figure 2.19 represents the H_2 evolution rate and STH energy conversion efficiency as a function of the band gap of the photocatalysts involves a single step photocatalysis with a quantum yield of 100% (Kubota and Domen 2013). It is assumed that photons with wavelengths shorter than the value of the band gap of the studied material are contributed to half H_2. If the photocatalysts utilize Vis light up to 600–700 nm with a quantum yield of 50%, the STH efficiency reaches ~10%, which is similar to that of a commercial photovoltaic system (solar cell). Vacuum processing required for photovoltaic device preparation make the system more demanding than photocatalytic water-splitting devices. If solar energy of 27 TJ/km²/d is available at the terrestrial equatorial area for a 7.6 h/d irradiation and an AM of 1.5G, and a 25-km²-scaled solar plant is available, then it will work with 10% STH efficiency and can produce hydrogen of 570 tons/d, which is equivalent to the output of typical natural gas plants.

2.8.5 QUANTUM EFFICIENCY

QE is often related to the term spectral responsivity, which shows how much current comes out of the device per incoming photon of a given energy and wavelength. One can measure the spectral responsivity in a similar measurement, but it has different units: amperes per watt (A/W). Responsivity (R_λ, in A/W) holds the following relationship with QE_λ (on a scale of 0–1):

$$QE_\lambda = \frac{R_\lambda}{\lambda} \times \frac{hc}{e} \approx \frac{R_\lambda}{\lambda} \times (1240 \text{ W} \cdot \text{nm/A}) \tag{2.40}$$

where λ is the wavelength in nanometers, h is Planck's constant, c is the speed of light in a vacuum, and e is the elementary charge.

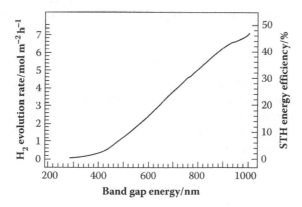

FIGURE 2.19 Theoretical relation of H_2 evolution rate and solar-to-hydrogen conversion efficiency to the band gap energy of photocatalysts in wavelength units. The solar light is assumed to be air mass (AM) 1.5 G14, and 100% quantum yield is assumed. (From Electricity, Resources, and Building Systems Integration Center, National Renewable Energy Laboratory, U.S. Department of Energy (DOE), Solar Resource Data, ASTM G-173 Air Mass 1.5.)

2.9 EXCITONIC BINDING ENERGY

Knowledge of the excitonic movement in different energy levels is very important for better understanding of the designing, functioning, and performance of opto-electronic devices. Optical excitation gives rise to excitons (photogenerated charge carriers i.e. electrons and holes) rather than free carriers. These excitons localize to the molecule, hence an extra amount of energy, termed as excitonic binding energy (E_{ex}), is needed to produce free charge carriers. Therefore, excitonic binding energy is the system's synergy. Sir Nevil Francis Mott developed a concept of excitons in semiconductor crystals (Wannier 1937; Mott 1938), where the rate of electron and hole hopping between different crystal cells much exceeds the strength of their Coulomb coupling with each other. This Coulomb-correlated electron–hole pair is known as an exciton and has a finite lifetime. The binding energy of an exciton (i.e., the energy of its ionization to a noncorrelated electron hole couple) can be of the order of 100–300 meV. Frenkel treated the crystal potential as a perturbation to the Coulomb interaction between an electron and a hole which belong to the same crystal cell system. Potential energy related to this columbic interaction between the hole and the electron at r distance is given by the following equation:

$$E_{coul} = \sum \frac{q_e q_h}{\epsilon r} \tag{2.41}$$

Then the excitonic binding energy is

$$E_{ex} = E_{ION} - \frac{E_{coul}}{n^2} \tag{2.42}$$

where q_e or q_h = elementary charge, ε = the electrical permittivity in the space, E_{coul} = coulombic potential energy (13.6 eV M_{red}/M_e), n (excitonic level) = 1, 2, 3, 4 ..., E_{ION} = energy required to ionize the molecule, M_e = effective mass of the electron, M_h = effective mass of the hole, and M_{red} = reduced mass of the exciton.

$$M_{red} = \frac{M_e \cdot M_h}{M_e + M_h} \tag{2.43}$$

Similar to a hydrogen atom, an exciton also has one positive and negative charge. The only difference between an exciton and a hydrogen atom is that the mass of the hole is much smaller than the mass of the proton (an exciton can only exist in a solid). Therefore, we can apply the solution of the relevant Schrodinger treatment of the hydrogen atom to an excitonic system, where the electron mass is not neglected with respect to the proton mass, which is as follows:

$$E_{ex}(k) = E_{gap} - \frac{M_{red} \cdot q^4}{8(n \cdot \epsilon_0 \epsilon_r \cdot h)^2} + \frac{h^2 \cdot k^2}{8\pi^2 (M_e + M_h)} \tag{2.44}$$

where the first term E_{gap} simply accounts for the crystal energy, the second one is taken straight from the hydrogen atom, and the third term is a correction if the two particles are not at the same place in k-space (it is zero for $ke = -kh$ or

$ke = kh = 0$ as for most direct semiconductors). Since ΔE_{ex} for a free exciton that moves through the crystal and transport energy, but not charge, is measured in a few milli-electron volts compared to the typical donor levels, it will not live very long at room temperature. In total, we have a system of energy levels right below the CB with the "deepest" level defined by the energy difference, as shown in Figure 2.20a.

Excitonic binding energy for a nonspherical object is calculated using the following equation:

$$\Delta E_{ex} = \frac{M_{red} \cdot q^4}{2 \, \epsilon_0 \epsilon_r \cdot h} \tag{2.45}$$

The key difference between organic and inorganic semiconductors is the dielectric constant. The dielectric constant shields the repulsion between electron–electron and coulombic interaction between holes and electrons. The dielectric constant is diminished by the increase in the size of QDs. Similarly, Zhou and Majumdar (2004) reported the scaling in the excitonic binding energy increases with the shortening of the tube diameter of the CNT. Figure 2.20b illustrates the role of the excitonic binding energy in organic molecule where, at initial state, the bounded hole and electron

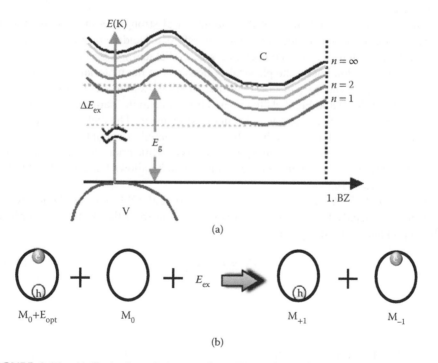

(a)

(b)

FIGURE 2.20 (a) Excitation of electron from VB to CB and their respective excitonic energy and band gap E_g and different energy levels beneath to CB, in an indirect band semiconductor, where BZ = Birillouin zone. (b) Schematic breaking of excitions. (From http://www.tf.uni-kiel.de/matwis/amat/semi_en/kap_5/backbone/r5_1_3.pdf; Knupfer, M., *Appl. Phys. A Mater. Sci. Process.*, 77(5), 623–626, 2003.)

pair is confined to the molecule. Thereafter, the excitonic binding energy utilized to take out the electron forms in a bound state (exciton) and is put onto another molecular unit far away (Knupfer 2003).

There is a huge difference in the excitonic binding energy of organic and inorganic materials, which is demonstrated in Figure 2.21. In general, inorganic semiconductor crystals or liquids with a small band gap and large dielectric constants possess *Wannier–Mott excitons* (Figure 2.21a). Consequently, electric field screening tends to reduce the Coulomb interaction between electrons and holes. The resulting Wannier–Mott excitons (Wannier 1937) have a radius larger than the lattice spacing. As a result, the effect of the lattice potential can be incorporated into the effective masses of the electrons and holes. Likewise, because of the lower masses and the screened Coulomb interaction, the binding energy is usually much less than that of a hydrogen atom, typically on the order of 0.01 eV. Frenkel excitons are typically found in alkali halide crystals and in organic molecular crystals composed of aromatic molecules, such as anthracene and tetracene, which have a very small dielectric constant (Figure 2.21b). The Coulomb interaction between an electron and a hole may be strong and the excitons thus tend to be small, of the same order as the size of the unit cell. Molecular excitons may even be entirely located on the same molecule, as in fullerenes, and have a typical binding energy on the order of 0.1–1 eV.

The physical size and shape of the nanomaterial strongly influences the nature and dynamics of the electronic excitation. Therefore, a deciding property of excitons in nanoscale systems is that the excitonic size is dictated not by the electron–hole Coulomb interaction but by the physical dimensions of the material or the arrangement of distinct building blocks. Therefore, in single-walled CNTs, excitons have both Wannier–Mott and Frenkel character. This is due to the nature of the Coulomb interaction between electrons and holes in one dimension. The dielectric function of the nanotube itself is large enough to allow for the spatial extent of the wave function to extend over a few nanometers along the tube axis, while poor screening in the vacuum or dielectric environment outside of the nanotube allows for large (0.4–1.0 eV) binding energies (Table 2.1).

Experimentally, E_{ex} can be determined by Hill et al. (2000), using the following equation (especially for organic semiconductors) and from the study of the optical absorption and emission spectra:

$$E_{ex} = E_t - E_{opt} \tag{2.46}$$

or

$$E_{ex} = E_g - E_x \tag{2.47}$$

where the difference between the two transport levels is referred to as the transport gap E_t and E_{opt} is the optical gap in organic semiconductors. For simple organic molecules, E_g is the band gap and E_x is the ground state to photoexcitonic energy, as predicted in Figure 2.22 (Scholes and Rumbles 2006).

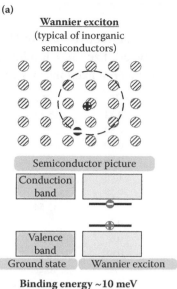

(a)

Wannier exciton
(typical of inorganic
semiconductors)

Semiconductor picture

Conduction band

Valence band

Ground state   Wannier exciton

Binding energy ~10 meV
Radius ~100 Å

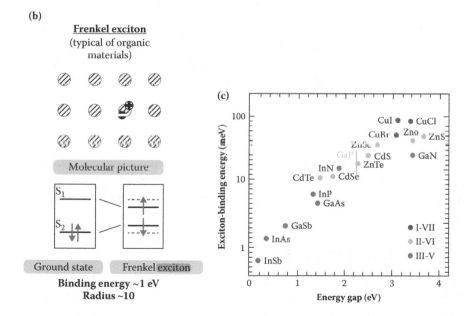

(b)

Frenkel exciton
(typical of organic
materials)

Molecular picture

S_1

S_2

Ground state   Frenkel exciton

Binding energy ~1 eV
Radius ~10

(c)

FIGURE 2.21   Schematic representation of the excitonic binding energy for (a) Wannier–Mott excitons in inorganic semiconductors and (b) Frenkel excitons in organic materials. (c) Exciton binding energy and the respective band gap of some typical inorganic materials. (From Pope M., and Swenberg, C.E., *Electronic Processes in Organic Crystals and Polymers*, Oxford University Press, New York, NY, 1999; http://www.tf.uni-kiel.de/matwis/amat/semi_en/kap_5/advanced/t5_1_3.html.)

TABLE 2.1

Effective Mass of Electron (m*e), Effective Mass of Hole (m*h) in Units of the Free Electron Mass, Primitivity (εr), Excitonic Binding Energy (E_{ex}), and Nominal Bohr Radius (r_{ex}) of the Free Excitons in Various Semiconductors

Material	m*e	m*h	εr	E_{ex} (meV)	r_{ex} (nm)
BN	0.752	0.38	5.1	131	1.1
GaN	0.20	0.80	9.3	25.2	3.1
InN	0.12	0.50	9.3	15.2	5.1
GaAs	0.063	0.50	13.2	4.4	12.5
ZnS	0.34	1.76	8.9	49.0	1.7
ZnO	0.28	0.59	7.8	42.5	2.2
Cu_2O	0.96		13.54	97.2	3.8
ZnSe	0.16	0.78	7.1	35.9	2.8
ZnTe	0.12	0.6	8.7	18.0	4.6
CdS	0.21	0.68	9.4	24.7	3.1
CdSe	0.11	0.45	10.2	11.6	6.1
CdTe	0.096	0.63	10.2	10.9	6.5
HgTe	0.031	0.32	21.0	0.87	39.3

$E_{ex} \approx e^2/(4\pi\varepsilon_0\varepsilon R)$ (Halliday et al. 2010), where ε is the dielectric constant and R is the equivalent radius of the molecule. However, if the molecule deviates from a spherical shape, a minor correction factor 0.9 should be added. In earlier quantum Monte Carlo (QMC) calculations, the excitonic binding energy was estimated using the Mott–Wannier formula:

$$E_{ex} \approx \frac{1}{2\,\varepsilon\, r} \quad (2.48)$$

The excitonic binding energy (E_{ex}) and the band gap energy (E_g) of polymers can be determined by the high-resolution measurements of the photoconductivity excitation profile as a function of light polarization, applied electric field, and temperature (Moses et al. 2001).

2.10 DIFFUSION LENGTH

The motion of a carrier can be caused by an electric field generated due to an externally applied voltage since the carriers are charged particles. We will refer to this transport mechanism as carrier mobility (drift). In addition, carriers also move from regions where the carrier density is high to regions where the carrier density is low. This carrier transport mechanism is due to the thermal energy and the associated random motion of the carriers (carriers are not sitting still in the lattice [they are only "sitting still" in k-space] but move around with some average velocity). We will refer to this transport mechanism as carrier diffusion. When excess of carriers is generated in a semiconductor, the minority carriers diffuse a distance, a characteristic length,

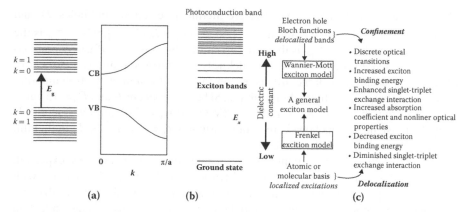

FIGURE 2.22 (a) Bands and molecular orbitals for extended systems. (b) Excitons in extended systems: bound electron–hole pairs. (c) General excitonic model in nanoscale materials. (From Scholes, G. D., and Rumbles, G., *Nat. Mater.*, 5, 683–696, 2006.)

over which minority carriers can diffuse before recombining with majority carriers. This is called as a diffusion length, L. Usually, the diffusion length of an electron is much greater than that of the holes. It is an average distance a carrier can move from the point of generation until it recombines. The minority carrier lifetime and the diffusion length depend strongly on the type and magnitude of recombination processes in the semiconductor. The method used to fabricate the semiconductor wafer and the processing also has a major impact on the diffusion length. Semiconductor materials that are heavily doped have greater recombination rates and consequently have shorter diffusion lengths. Higher diffusion lengths are indicative of materials with longer lifetimes, which is therefore an important quality to consider with semiconductor materials. For many types of silicon solar cells, Shockley-Read-Hall (SRH) recombination is the dominant recombination mechanism. The recombination rate will depend on the number of defects present in the material, so that while doping the semiconductor increases the defects in the solar cell; this will also increase the rate of SRH recombination. In addition, since Auger recombination is more likely in heavily doped and excited material, the recombination process is itself enhanced as the doping increases. The excess of minority carrier decays exponentially with diffusion distance. Diffusion length for electrons and holes is given by Equation 2.49. The parameter L depends on the carrier diffusion constant D, the carrier lifetime τ (Equation 2.2), and a dimensionality factor ($q = 2, 4, 6$ for one-dimensional [1D], 2D, or three-dimensional [3D] diffusion):

$$L_n = \sqrt[q]{qD_n\tau_n}; \quad L_h = \sqrt[q]{qD_h\tau_h} \tag{2.49}$$

$$\frac{1}{\tau_{\text{eff}}} = \frac{1}{\tau_{\text{bulk}}} + \frac{2s}{d} \tag{2.50}$$

where τ_n and τ_p are the respective effective lifetime of electrons and holes. D_n and D_p are diffusion coefficients at the n- and p-side $\left(D_{n/p} = D_0 e^{-Q/RT}\right)$, where τ_{eff} is the effective carrier lifetime, τ_{bulk} is the bulk carrier lifetime, s is the surface recombination velocity, and d is the film or wafer thickness. For intrinsic semiconductors, usually $L_n > L_h$ because of the larger diffusion constant D of the unit of electron mobility is cm²/(V.S) compared to holes. For example, Si has $D_n = 49$ cm²/s and $D_h = 13$ cm²/s (calculated from mobilities $m_n = 1900$ cm²/(V.S) and $m_h = 500$ cm²/(V.S) at 298 K using the Einstein–Smoluchowski relation), assuming $t_n = t_h = 10^{-6}$ s, $L_n = 98$ mm, and $L_h = 51$ mm for 1D diffusion ($q = 2$).

All of the above three quantities are temperature dependent. As expected, the magnitudes of the estimated lateral diffusion lengths generally increase with decreasing temperature, and they also decrease with increasing densities of dislocations in the absorption layer. The magnitude of the diffusion length is dependent on the crystalline quality of the epitaxy layer, which is a very important parameter for fundamental material chemistry. The impact of diffusion length in the calculation of the thermal activation energy of dark current in nPn photodiode is appreciable. In a single crystalline silicon solar cell, the lifetime can be as high as 1 msec and the diffusion length is typically 100–300 μm. These two parameters give a clear-cut indication of material quality and suitability for the solar cell.

The continuity equation describes a basic concept, namely that a change in carrier density over time is due to the difference between the incoming and outgoing flux of carriers plus the generation and minus the recombination. The flow of carriers and recombination and generation rates are illustrated in Figure 2.23 (Van Zeghbroeck 2011).

The rate of change of the carriers between x and $x + dx$ distance equals the difference between the incoming flux and the outgoing flux plus the generation and minus the recombination:

$$\frac{\partial_n(x,t)}{dt} A dx = \left(\frac{J_n(x)}{-q} - \frac{J_n(x + dx)}{-q}\right) A + (G_n(x,t) - R_n(x,t)) A dx \quad (2.51)$$

where $n(x,t)$ is the carrier density, A is the area, $G_n(x,t)$ is the generation rate, and $R_n(x,t)$ is the recombination rate. Using a Taylor series expansion

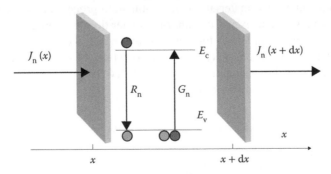

FIGURE 2.23 Electron currents and possible recombination and generation processes. (Van Zeghbroeck, B., Principles of Semiconductor Devices, http://ece-www.colorado.edu/~bart/book/, 2011.)

$$J_n(x + dx) = J_n(x) + \frac{dJ_n(x)}{dx} dx \qquad (2.52)$$

where this equation can be formulated as a function of the derivative of the current,

$$\frac{\partial_n(x,t)}{\partial t} = \frac{1}{q} \frac{\partial J_n(x,t)}{\partial x} + G_n(x,t) - R_n(x,t) \qquad (2.53)$$

and similarly for holes, one finds

$$\frac{\partial_p(x,t)}{\partial t} = -\frac{1}{q} \frac{\partial J_p(x,t)}{\partial x} + G_p(x,t) - R_p(x,t) \qquad (2.54)$$

A solution to these equations can be obtained by substituting the expression for the electron and hole current. This then yields two partial differential equations as a function of the electron density, the hole density, and the electric field. The electric field itself is obtained from Gauss's law.

Several methods are in practice to determine the diffusion length. It can be calculated by using current–voltage (I–V) data by assuming that the values of the two unknowns, LP (diffusion length of holes) and J (current density), are common to all devices fabricated from a single growth and measured under the same conditions; the diffusion length may be extracted from the current taken as a function of voltage (I–V) data of only two differently sized devices. The determination of LP relies on the measured current being a function of the effective area and on the measured current approaching zero as the effective area goes to zero. Spectrally resolved photoluminescence quenching (Tam et al. 2009), quenching of the triplet–triplet absorption in the presence of acceptor molecules (Samiullah et al. 2010), a wedge technique that does not require knowledge of the spectral absorption coefficient, doping, or surface recombination velocity of the sample (Pala et al. 2014) are some prominent techniques used to determine the diffusion length.

2.11 CARRIER MOBILITY AND PENETRATION IN PHOTOCATALYSTS

Any motion of free carriers in a semiconductor leads to current. In solids, at ambient condition, the carriers move in a random fashion in the absence of the applied electric field that results in zero overall motion of charge carriers in any particular direction over time. But under an applied electric field, the charged particles/carriers (i.e., electron or holes) can move quickly and smoothly through a metal or semiconductor. The transport of free charge carriers in a fluid under an applied electric field can be measured in terms of the electrical mobility, which is a function of material doping or impurity level and temperature. Mobility is higher for electrons than holes due to lower effective mass. Higher mobility leads to better device performance. For example, at room temperature (300 K) the carrier mobility for Si is $\mu_e = 1{,}400$ cm^2/(V·s) and $\mu_h = 450$ cm^2/(V·s) (Electrical properties of silicon, Ioffe Institute Database). Other values include GaAs, 7,500 cm^2/V·s; 6H-SiC, 400 cm^2/V·s;

low-dimensional systems including 2D electron gases, 35,000,000 cm²/(V·s) at low temperature (Umansky et al. 2009); CNTs, 100,000 cm²/(V·s) (Dürkop et al. 2004); and graphene, 200,000 cm²/(V·s) at low temperature (Bolotin et al. 2008). On the other hand, organic semiconductors have low carrier mobilities, that is, 10 cm²/(V·s).

Electrons in a vacuum would be accelerated with faster velocities (ballistic transport). A perfect periodicity is found in a perfect crystal and has no resistance to current flow; thus it behaves as a superconductor. The perfect periodic potential does not impede the movement of the charge carriers. However, in a real solid, the carriers repeatedly scatter off due to crystal defects, surface, the polar nature of semiconductors (piezoelectric effect), positioning of substituting atom species in a relevant sublattice (alloy scattering), acoustic (lattice) or optical phonons, ionized or neutral impurities, and so on, thus creating a resistance to current flow by changing its direction and/or energy. The presence of all these scatterers upset the periodicity of the potential seen by a charge carrier. If these scatterers are near the interface, the complexity of the problem increases due to the existence of crystal defects and disorders. In this case, scattering happens because after trapping a charge, the defect becomes charged and therefore starts interacting with free carriers. If scattered carriers are in the inversion layer at the interface, the reduced dimensionality of the carriers makes the case different from the case of bulk impurity scattering as carriers move only in two dimensions. Interfacial roughness also causes short-range scattering limiting the mobility of quasi-2D electrons at the interface (Ferry 2001). Therefore, scattering disrupts carrier acceleration in an applied electric field E across a piece of studied material; instead it moves with a finite average velocity, which is known as the drift velocity (v_d) and is represented by the following equation:

$$v_d = \mu E \tag{2.55}$$

Drift velocity is affected by a scattering process that leads to major changes in the mobility of the carriers. Scattering time (average time between scattering events) and mobility follows the relationship in Equation 2.58, which was derived by assuming that after each scattering event, the carrier's motion get randomized so that it has zero average velocity. After that, it accelerates uniformly in the electric field, until it scatters again. The value of the resulting average drift mobility (Yu and Cardona 2010) is shown by Equations 2.56 through 2.58:

$$\text{Drift velocity} = \text{acceleration} \times \text{mean free time} \tag{2.56}$$

$$v_d = (F/m^*) \times \tau = (qE/m^*) \times \tau \tag{2.57}$$

From Equations 2.56 and 2.57, we can get

$$\mu = \left(\frac{q}{m^*}\right) \times \tau \tag{2.58}$$

where q is the elementary charge in the electric field E, m^* is the carrier effective mass in the direction of the electric field, and τ is the average scattering time.

2.11.1 ELECTRICAL CONDUCTIVITY AND MOBILITY

Mobility is also an important parameter in defining the term electrical conductivity, especially in the case of semiconductors. Let N_D and N_A be respective shallow donor and acceptor concentrations, and n and p be the number density of the corresponding carrier electrons and holes, respectively; if q is the elementary charge, then the current density due to electron flow $J_e = nq\mu_e E$ and current density due to hole flow $J_h = pq\mu_h E$. Therefore, the total current density is

$$J = J_e + J_h = q \, (nE\mu_e + pE\mu_h) \tag{2.59}$$

$$J = q(N_D E\mu_e + N_A E\mu_h) \tag{2.60}$$

Electrical conductivity $(\sigma) = $ mobility $\times$ carrier concentration $= \mu \times nq \tag{2.61}$

$$\sigma = q(n\mu_e + p\mu_h) \tag{2.62}$$

2.11.2 TEMPERATURE DEPENDENCE OF MOBILITY

The increase in temperature directs the phonon concentration to increase and causes increased scattering phenomenon that lowers the carrier mobility at higher temperature. The mobility in nonpolar semiconductors, such as silicon and germanium, is dominated by acoustic phonon interaction and theoretically expected to be proportional to $T^{-3/2}$, while the mobility due to optical phonon scattering and the charged defects is expected to be proportional to $T^{-1/2}$ and $T^{3/2}$, respectively. The effect of ionized impurity scattering, however, decreases with increasing temperature because the average thermal speeds of the carriers also increase. Thus, the carriers spend less time near an ionized impurity as they pass and the scattering effect of the ions is thus reduced. The two effects (impurities and lattice scattering) operate simultaneously on the carriers through Matthiessen's rule (Chain et al. 1997). At lower temperatures, ionized impurity scattering dominates, while at higher temperatures, phonon scattering dominates, and the actual mobility reaches a maximum at an intermediate temperature (Figure 2.24a and b).

2.11.3 MOBILITY VERSUS DIFFUSION

The carrier transport is caused by two driving forces: an electric field generated due to an externally applied voltage and movement of the carriers from a high concentration region toward to a low concentration region due to the carrier density gradient. The first process is drift current and later is known as diffusion current. The carrier density gradient can be obtained by varying the doping density in a semiconductor or by applying a thermal gradient. The total current in a semiconductor equals the sum of the drift and the diffusion current. The Einstein relation relates the above two independent current transport mechanisms using mobility along with diffusivity (diffusion coefficient), temperature, and charge (Jeon and Burk 1989):

$$\frac{D_n}{\mu_n} = \frac{kT}{q} \text{ and } \frac{D_p}{\mu_p} = \frac{kT}{q} = 25 \text{ mV at RT, for electrons and holes} \tag{2.63}$$

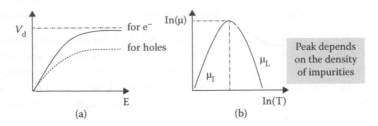

FIGURE 2.24 (a) Variation of drift velocity with applied potential for electrons and holes. (b) Temperature dependency of mobility in the presence of impurities μ_I and lattice scattering μ_L. (From Size, S. M., *Physics of Semiconductor Devices*, John Wiley and Sons, New York, NY, 1981.)

2.11.4 Doping Dependence of Electron Mobility and Hole Mobility

At low doping concentrations, the mobility of carriers is almost constant and primarily dominated by phonon scattering. At higher doping concentrations, the mobility decreases due to ionized impurity scattering with ionized doping atoms. The actual mobility also depends on the type of dopant and is linked to the total number of ionized impurities or the sum of the donor and acceptor densities. Mobility related to the free carrier density in the form of donor and acceptor concentration is described in Equation 2.63.

2.12 SUMMARY

Before going into detailed discussion on actual advancements in photochemical water-splitting processes, one should be aware of basic information and terminology related to the field. In this way, this chapter laid a foundation for the rest of the chapters as it includes the illustration of the basic concepts concerning the photochemical water-splitting phenomenon. These are artificial photosynthesis and its prominent categories, the electrochemistry behind water splitting, selection criteria of photocatalytic material, overpotential, band gap and band edge position, band edge bending, efficiency, excitonic binding energy, diffusion length, carrier mobility, penetration in photocatalysts, and so on. This chapter also dealt with the various methods used to determine the aforementioned terms.

REFERENCES

Abe, R., K. Sayama, and H. Arakawa. 2003. Significant effect of iodide addition on water splitting into H_2 and O_2 over Pt-loaded TiO_2 photocatalyst: Suppression of backward reaction. *Chem. Phy. Let.* 371: 360–64.

Anpo M., H. Yamashita, Y. Ichihashi, and S. Ehara. 1995. Photocatalytic reduction of CO_2 with H_2O on various titanium oxide catalysts. *J. Electroanal. Chem.* 396: 21–6.

Barber, I. J. 2009. Photosynthetic energy conversion: Natural and artificial. *Chem. Soc. Rev.* 38: 185–96.

Birss, V. I., A. Damjanovic, and P. G. Hudson. 1986. Oxygen evolution at platinum electrodes in alkaline solutions II. Mechanism of the reaction. *J. Electrochem. Soc.* 133(8): 1621–25.

Bockris, J. O'M. and T. Otagawa. 1983. Mechanism of oxygen evolution on perovskites. *J. Phys. Chem.* 87: 2960–71.

Bolotin, K., K. Sikes, Z. Jiang, et al. 2008. Ultrahigh electron mobility in suspended graphene. *Solid State Comm.* 146: 351–55.

Bonde, J., P. G. Moses, T. F. Jaramillo, J. K. Nørskov, and I. Chorkendorff. 2008. Hydrogen evolution on nano-particulate transition metal sulfides. *Faraday Discuss* 140: 219–31.

Bubnova, O., Z. U. Khan, H. Wang, et al. 2014. Semi-metallic polymers. *Nat. Mat.*, 13: 190–94.

Burns, G. 1985. *Solid State Physics.* New York, NY: Academic Press, Inc., pp. 339–40.

Castelli, P., S. Trasatti, F. H. Pollak, and W. E. O'Grady. 1986. Single crystals as model electrocatalysts: Oxygen evolution on RuO_2 (110). *J. Electroanal. Chem.* 210: 189–94.

Chain, K., J. H. Huang, J. Duster, P. K. Ko, and C. Hu. 1997. A MOSFET electron mobility model of wide temperature range (77–400 K) for IC simulation. *Semiconduct. Sci. Technol.* 12: 355–58.

Chang C. H. and Y. L. Lee.2007. Chemical bath deposition of CdS quantum dots onto mesoscopic TiO_2 films for application in quantum-dot-sensitized solar cells. *Appl. Phys. Lett.* 91(5): 053503 (3 p.).

Chen, H. M., C. K. Chen, C. C. Lin., et al. 2011. Multi-bandgap-sensitized ZnO nanorod photoelectrode arrays for water splitting: An x-ray absorption spectroscopy approach for the electronic evolution under solar illumination. *J. Phys. Chem. C* 115: 21971–80.

Chen, Z. B., T. F. Jaramillo, T. G. Deutsch, et al. 2010. Accelerating materials development for photoelectrochemical hydrogen production: Standards for methods, definitions, and reporting protocols. *J. Mater. Res.* 25: 3–16.

Cheng, K.-W. and S.-C. Wang. 2009. Effects of complex agents on the physical properties of Ag–In–S ternary semiconductor films using chemical bath deposition. *Mat. Chem. Phys.* 115: 14–20.

Cogdell, R. J., T. H. P. Brotosudarmo, A. T. Gardiner, P. M. Sanchez, and L. Cronin. 2010. Artificial photosynthesis—Solar fuels: Current status and future prospects. *Biofuels* 1(6): 861–76.

Conway, B. E. and B. V. Tilak. 2002. Interfacial processes involving electrocatalytic evolution and oxidation of H_2, and the role of chemisorbed H. *Electrochim. Acta* 47: 3571–94.

Conway, B. E. and T. C. Liu. 1990. Characterization of electrocatalysis in the oxygen evolution reaction at platinum by evaluation of behavior of surface intermediate states at the oxide film. *Langmuir* 6: 268–76.

Dholam, R., N. Patel, M. Adami, and A. Miotello. 2009. Hydrogen production by photocatalytic water-splitting using Cr- or Fe-doped TiO2 composite thin films photocatalyst. *Int. J. Hydrogen Energy* 34: 5337–46.

Diguna, L. J., Q. Shen, J. Kobayashi, and T. Toyoda. 2007. High efficiency of CdSe quantum-dot-sensitized TiO_2 inverse opal solar cells. *Appl. Phys. Lett.* 91(2): 023116.

Dong, C. L., C. Persson, L. Vayssieres, et al. 2004. Electronic structure of nanostructured ZnO from x-ray absorption and emission spectroscopy and the local density approximation. *Phys. Rev. B* 70: 195325.

Dürkop, T., S. A. Getty, E. Cobas, and M. S. Fuhrer. 2004. Extraordinary mobility in semiconducting carbon nanotubes. *Nano Lett.* 4: 35–39.

Eisebitt, S., J. Lüning, J.-E. Rubensson, and W. Eberhardt. 1999. Resonant inelastic soft x-ray scattering as a bandstructure probe: A primer. *Phys. Stat. Sol.* B 215: 803–808.

Energy Information Administration. 2003. International Energy Outlook. Washington, DC, EIA. Available at http://www.eia.doe.gov/oiaf/1605/ggccebro/chapter1.html.

Ferry, D. K. 2001. *Semiconductor Transport.* London, UK: Taylor & Francis.

Fei, H., R. Ye, G. Gonglan, et al. 2014. Boron- and nitrogen-doped graphene quantum dots/graphene hybrid nanoplatelets as efficient electrocatalysts for oxygen reduction. *ACS Nano.* 8: 10837–43.

Gary, H. 2008. Solar fuel II: The quest for catalysts. *Eng. Sci.* 46: 26–31.

Gong, M., W. Zhou, M.-C. Tsai, et al. 2014. Nanoscale nickel oxide/nickel heterostructures for active hydrogen evolution electrocatalysis. *Nat. Commun.* 5: Article No. 4695.

Guo, J.-H., L. Vayssieres, C. Persson, R. Ahuja, B. Johansson, J. Nordgren. 2002. Polarization-dependent soft-x-ray absorption of a highly oriented ZnO microrod-array. *J. Phys. Condens. Matter* 14(28): 6969–74.

Halliday, D., R. Resnick, and J. Walker. 2010. *Fundamentals of Physics*. New York, NY: John Wiley and Sons.

Hansen, R. E. and S. Das. 2014. Biomimetic di-manganese catalyst cage-isolated in a MOF: Robust catalyst for water oxidation with Ce(IV), a non-O-donating oxidant. *Energy Environ. Sci.* 7: 317–22.

Hill, I. G., A. Kahn, Z. G. Soos, and R. A. Pascal. 2000. Charge-separation energy in films of [pi]-conjugated organic molecules. *J. Chem. Phys. Lett.* 327: 181–88.

Hinnemann, B., P. G. Moses, and J. Bonde. 2005. Biomimetic hydrogen evolution: MoS_2 nanoparticles as catalyst for hydrogen evolution. *J. Am. Chem. Soc.* 127: 5308–9.

Hori, Y., A. Murata, K. Kikuchi, and S. Suzuki. 1987. Electrochemical reduction of carbon dioxides to carbon monoxide at a gold electrode in aqueous potassium hydrogen carbonate. *J. Chem. Soc. Chem. Commun.* 728–9.

Inamdar, S. N., P. P. Ingole, and S. K. Haram. 2008. Determination of band structure parameters and the quasi-particle gap of CdSe quantum dots by cyclic voltammetry. *Chem. Phys. Chem.* 9: 2574–79.

Iyer, A., J. Del-Pilar, and C. K. King'ondu. 2012. Water oxidation catalysis using amorphous manganese oxides, octahedral molecular sieves (OMS-2), and octahedral layered (OL-1) manganese oxide structures. *J. Phys. Chem. C* 116: 647483.

Jaramillo, T. F., J. Bonde, J. Zhang, et al. 2008. Hydrogen evolution on supported incomplete cubane-type [Mo3S4]4+ electrocatalysts. *J. Phys. Chem. C* 112: 17492–98.

Jaramillo, T. F., K. P. Jorgensen, J. Bonde, J. H. Nielsen, S. Horch, and I. B. Chorkendorff. 2007. Identification of active edge sites for electrochemical H_2 evolution from MoS_2 nanocatalysts. *Science* 317: 100–102.

Jeon, D.S. and D. E. Burk. 1989. MOSFET electron inversion layer mobilities—A physically based semi-empirical model for a wide temperature range. *IEEE Trans. Electron Devices* 36: 1456–63.

Jin, H., S. Choi, H. J. Lee, and S. Kim. 2013. Layer-by-layer assemblies of semiconductor quantum dots for nanostructured photovoltaic devices. *J. Phys. Chem. Lett.* 4(15): 2461–70.

Kamat, P. V. 2012. Manipulation of charge transfer across semiconductor interface. A criterion that cannot be ignored in photocatalyst design. *J. Phys. Chem. Lett.* 3(5): 663–72.

Kamat, P. V. and J. Bisquert. 2013. Solar fuels. Photocatalytic hydrogen generation. *J. Phys. Chem. C* 117: 14873–75.

Khan, S. U. M., M. Al-Shahry, and W. B. Ingler. 2002. Efficient photochemical water splitting by a chemically modified n-TiO2. *Science* 297: 2243–45.

Kibsgaard, J., T. F. Jaramillo, and B. Flemming. 2014. Building an appropriate active-site motif into a hydrogen-evolution catalyst with thiomolybdate [Mo3S13]2– clusters. *Nat. Chem.* 6(3): 248–53.

Kim, K., M.-J. Kim, S.-I Kim, and J.-H. Jang. 2013. Towards visible light hydrogen generation: Quantum dot-sensitization via efficient light harvesting of hybrid-TiO_2. *Sci. Rep.* 3: 3330.

Kim, Y. I., S. J. Atherton, E. S. Brigham, and T. E. Mallouk. 1993. Sensitized layered metal oxide semiconductor particles for photochemical hydrogen evolution from nonsacrificial electron donors. *J. Phys. Chem.* 97(45): 11802–10.

Kittel, C. 1996. *Introduction to Solid State Physics*. New York, NY: Wiley.

Knupfer, M. 2003. Exciton binding energies in organic semiconductors. *Appl. Phys. A Mat. Sci. Process.* 77(5): 623–26.

Krulic, D., N. Fatouros and D. E. Khoshtariya. 1998. Kinetic data for the hexacyanoferrate (II)/(III) couple on platinum electrode in various chlorides of monovalent cations. *J. Chim. Phys.* 95(3): 497–512.

Kubota J. and K. Domen. 2013. Photocatalytic water splitting using oxynitride and nitride semiconductor powders for production of solar hydrogen. In: *Proceedings of the Electrochemical Society Interface, Summer Meeting 2013* held between Oct. 27-Nov. 1 at San Francisco, CA, pp. 57–62.

Asadi, M., M. Kumar, B. Kumar, A. Behranginia, et al. 2014. Robust carbon dioxide reduction on molybdenum disulphide edges. *Nat. Commun.* 5: Article number 4470.

Li W. X. 2013. Photocatalysis of oxide semiconductors. *J. Aust. Ceram Soc.* 49: 41–6.

Lodi, G., E. Sivieri, A. D. Battisti, and S. Trasatti. 1978. Ruthenium dioxide-based film electrodes III. Effect of chemical composition and surface morphology on oxygen evolution in acid solutions. *J. Appl. Electrochem.* 8: 135–43.

Lunt, R. R., N. C. Giebink, A. A. Belak, J. B. Benziger, and S. R. Forrest. 2009. Exciton diffusion lengths of organic semiconductor thin films measured by spectrally resolved photoluminescence quenching. *J. Appl. Phys.* 105: 053711.

Luo, J., J.-H. Im, M. T. Mayer, et al. 2014. Water photolysis at 12.3% efficiency via perovskite photovoltaics and Earth-abundant catalysts. *Science* 345(6204): 1593–96.

Matsumoto, Y. and E. Sato. 1986. Electrocatalytic properties of transition metal oxides for oxygen evolution reaction. *Mater. Chem. Phys.* 14: 397–426.

Miles, A. B. and R. G. Compton. 2000. The theory of square wave voltammetry at uniformly accessible hydrodynamic electrodes, *J. Electroanaly. Chem.* 487(2): 75–89.

Mori, K., H. Yamashitaand M. Anpo. 2012. Photocatalytic reduction of CO2 with H2O on various titanium oxide photocatalysts. *RSC Adv.* 2: 3165–72.

Moses, D., J. Wang, A. J. Heeger, N. Kirovas, and S. Brazovski. 2001. Singlet exciton binding energy in poly(phenylene vinylene). *Proc. Natl. Acad. Sci.* 98: 13496–500.

Mott, N.F. 1938. Conduction in polar crystals. II. The conduction band and ultra-violet absorption of alkali-halide crystals. *Trans. Farad. Soc.* 34: 500-506.

Mukherji, A., C. H. Sun, S. C. Smith, G. Q. Lu, and L. Wang, 2011. Photocatalytic hydrogen production from water using N-doped $Ba_5Ta_4O_{15}$ under solar irradiation. *J. Phys. Chem. C* 115: 1567478.

Nam, S. S., G. Kishan, M. W. Lee, M. J. Choi, and K. W. Lee. 2000. Effect of lanthanum loading in Fe-K/La-Al2O3 catalyst for CO_2 hydrogenation to hydrocarbons. *Appl. Organometallic Chem.* 14: 794–98.

Nepal, B. and S. Das. 2013. Sustained water oxidation by a catalyst cage-isolated in a metal–organic framework. *Angew. Chem. Int. Ed.* 52: 7224–27.

Nuwer, M. J., J. J. O' Dea, and J. Osteryoung. 1991. Analytical and kinetic investigations of totally irreversible electron transfer reactions by square-wave voltammetry. *Analytica Chimica Acta* 251(1–2): 13–25.

Oak Ridge National Laboratory. Carbon Dioxide Information Analysis Center. Available at http://cdiac.esd.oml.gov/.

Pala, R. A., A. J. Leenheer, M. Lichterman, H. A. Atwater, and N. S. Lewis. 2014. Measurement of minority-carrier diffusion lengths using wedge-shaped semiconductor photoelectrodes. *Energy Environ. Sci.* 7: 3424–30.

Parmon, V. N., G. L. Elizarova, and T. V. Kim. 1983. Spinels as heterogeneous catalysts for oxidation of water to dioxygen by tris-bipyridyl complexes of iron(III) and ruthenium(III). *React. Kinet. Catal Lett.* 21(3): 195–97.

Perry, T. R. (1969). U.S. Patent No. 3,436,611. Washington, DC: U.S. Patent and Trademark Office.

Pleskov, Y. Y. and Gurevich, Y. V. 1986. *Semiconductor Photoelectrochemistry*. New York, NY: Consultants Bureau, p. 422.

Rosalbino, F., G. Borzone, E. Angelini, and R. Raggio. 2003. Hydrogen evolution reaction on Ni-RE (RE = rare earth) crystalline alloys. *Electrochim. Acta.* 48: 3939–44.

Ruland, A., C. Schulz-Drost, V. Sgobba, et al. 2011. Enhancing photocurrent efficiencies by resonance energy transfer in CdTe quantum dot multilayers: Towards rainbow solar cells. *Adv. Mater.* 23: 4573–77.

Samiullah, M., D. Moghe, U. Scherf, and S. Guha. 2010. Diffusion length of triplet excitons in organic semiconductors. *Phys. Rev. B* 82: 205211.

Sasikala, R., A. R. Shirole, V. Sudarsan, et al. 2010. Role of support on the photocatalytic activity of titanium oxide. *Appl. Catal. A Gen.* 390: 245–52.

Scholes, G. D. and G. Rumbles. 2006. Excitons in nanoscale systems. *Nat. Mater.* 5: 683–96.

Scibioh M. A. and B. Viswanathan. 2004. Electrochemical reduction of carbon dioxide: A status report. *Proc. Indn. Natl. Acad. Sci.* 70 A (3):407–62.

Shaukatali N. I., P. P. Ingole, and S.K. Haram. 2008. Determination of band structure parameters and the quasi-particle gap of CdSe quantum dots by cyclic voltammetry. *Chem. Phys. Chem.* 9: 2574–79.

Singh, P., M. K. Harbola, B. Sanyal, and A. Mookerjee. 2013. Accurate determination of band gaps within density functional formalism. *Phys. Rev. B* 87: 235110.

Size, S. M. 1981. *Physics of Semiconductor Devices,* 2nd Edn. New York, NY: John Wiley and Sons.

Stern, R. M., J. J. Perry, and D. S. Boudreaux. 1969. Low-energy electron-diffraction dispersion surfaces and band structure in three-dimensional mixed Laue and Bragg reflections. *Rev. Mod. Phys.* 41(2): 275.

Sullivan, B. P. 1989. Reduction of carbon dioxide with platinum metals electrocatalysts: A potentially important route for the future production of fuels and chemicals. *Platin. Met. Rev.* 33(1): 2–9.

Tauc, J. and A. Menth. 1972. States in the gap. *J Non-Cryst Solids* 569: 8–10.

Torimoto, T., H. Horibe, H. T. Kameyama, et al. 2011. Plasmon-enhanced photocatalytic activity of cadmium sulfide nanoparticle immobilized on silica-coated gold particles. *J. Phys. Chem. Lett.* 2: 2057–62.

Trasatti, S. 2000. Electrocatalysis: Understanding the success of DSA. *Electrochim. Acta* 45: 2377–85.

Trasatti, S. and O. A. Petri. 1991. Real surface area measurements in electrochemistry. *Pure & Appl. Chern.* 63(5): 711–734.

Trevisan, R., P. Rodenas, V. Gonzalez-Pedro, et al. 2013. Infrared photons for photoelectrochemical hydrogen generation. A PbS quantum dot based "quasi-artificial leaf." *J. Phys. Chem. Lett.* 4: 141–46.

Umansky, V., M. Heiblum, Y. Levinson, et al. 2009. MBE growth of ultra-low disorder 2DEG with mobility exceeding 35×106 cm₂/(V·s). *J. Cryst. Growth* 311: 1658–61.

Van Zeghbroeck, B. 2011. *Principles of Semiconductor Devices.* Denver, CO: University of Colorado. Available at http://ece-www.colorado.edu/~bart/book/.

Vignais, P. M., B. Billoud, and J. Meyer. 2001. Classification and phylogeny of hydrogenases. *FEMS Microbiol Rev.* 25: 455–501.

Walter M. G., E. L. Warren, J. R. McKone, et al. 2010. Solar water splitting cells. *Chem. Rev.* 110: 6446–73.

Wang, L. G., S. J. Pennycook, and S. T. Pantelides. 2002. The role of the nanoscale in surface reactions: CO_2 on CdSe. *Phys. Rev. Lett.* 89(7): 075506.

Wannier, G. H. 1937. The structure of electronic excitation levels in insulating crystals. *Phys. Rev.* 52(3): 191. Wigner, E. V. 1938. Effects of the electron interaction on the energy levels of electrons in metals *Trans. Faraday Soc.* 34: 678–85.

Xiang, Q. J. and J. G. Yu, 2013. Graphene-based photocatalysts for hydrogen generation. *J. Phys. Chem. Lett.* 4: 753–59.

Yamashita H., Y. Fujii, Y. Ichihashi, et al. 1998. Selective formation of CH_3OH in the photocatalytic reduction of CO_2 with H_2O on titanium oxides highly dispersed within zeolites and mesoporous molecular sieves. *Catal. Today* 45: 221–27.

Yang, K., Y. Dai, and B. Huang. 2012. Review of the structural stability, electronic and magnetic properties of nonmetal-doped TiO_2 from first-principles calculations. arXiv:1202.5651v1 [cond-mat.mtrl-sci] February 25, 2012.

Yu, Y. P. and M. Cardona. 2010. *Fundamentals of Semiconductors: Physics and Materials Properties.* Berlin, Germany: Springer Science and Business Media, pp. 205–40.

Zhou, H. and S. Mazumdar. 2004. Electron–electron interaction effects on the optical excitations of semiconducting single-walled carbon nanotubes. *Phys. Rev. Lett.* 93: 157402.

3 Water-Splitting Technologies for Hydrogen Generation

3.1 INTRODUCTION

Hydrogen plays a crucial role of an energy carrier as well as a primary fuel with the highest fuel efficiency on planet Earth. Moreover, it is the most abundant element (95%) of Earth but it cannot be mined and rarely exists in elemental form. It is always found in a bonded state as metal hydride, water, or simple/complex organic compounds. We need to extract the hydrogen from various sources. Currently, it is produced mostly from natural gas and coal, which contributes highly in releasing carbon dioxide into the atmosphere. Therefore, there is a need for generating hydrogen from a much more benign source on a large scale and carbon neutral water is a good choice. Here, the question arises as to how we break water to generate hydrogen. This chapter deals with the popular techniques used to cleave water.

3.2 ELECTROLYTIC WATER SPLITTING

To hit the target of delivering hydrogen at the cost of $2.00–$3.00 kg^{-1} gasoline (H$_2$) for passenger vehicles, electrolysis of water is the standard commercial technology (Turner et al. 2008). Electrolytic splitting of water is a promising option for hydrogen and oxygen production using electricity. Only 0.25% of hydrogen worldwide is produced by electrolysis. Hydrogen produced via electrolysis can result in zero greenhouse gas emissions, depending on the source of the electricity used. If renewable resources such as biomass, wind, solar, and so on are used for producing the electricity for this process, then the process becomes ecofriendly. Electrolysis of water has attracted ample attention of energy producers/researchers due to the so-called hydrogen economy. Electrolysis can occur only if the water is acidic or alkaline and not as pure water because pure water is highly resistive to electricity (Kreuter and Hofmann 1998). Therefore, the presence of ions in the water is required to conduct electricity for the water electrolysis process. The benefits and economic viability of hydrogen production via electrolysis have been evaluated on the basis of their cost and the efficiency of the electricity production, as well as emissions resulting from electricity generation. Therefore, the power grid is not counted as an ideal electricity provider because of the low efficiency and high greenhouse gas releases. In this way, renewable (wind) and nuclear energy are the better options for hydrogen generation through water electrolysis due to the virtually zero greenhouse gas emissions. Therefore, we can say electrolysis is the most promising method of hydrogen production from water due to the high efficiency of conversion and relatively low consumption of energy compared with thermochemical and photocatalytic methods.

The electrolyzer is the unit where electrolysis of water takes place. An electrolyzer consists of an anode and a cathode separated by an electrolyte. Water oxidizes at the anode to form oxygen and four positively charged hydrogen ions (protons) and an equal number of electrons. The electrons flow through an external circuit to a cathode, and the hydrogen ions selectively move across the polymer electrolyte membrane (PEM) to the cathode. At the cathode, the hydrogen ions combine with the electrons that come from the external circuit to generate hydrogen gas, as shown in Equations 3.1 and 3.2:

$$\text{Anode reaction:} \quad 2H_2O \rightarrow O_2 + 4H^+ + 4e^- \tag{3.1}$$

$$\text{Cathode reaction:} \quad 4H^+ + 4e^- \rightarrow 2H_q \tag{3.2}$$

Electrolysis of water that proceeds at 298 K (25°C) requires $\Delta H = 285.83$ kJ/mol (electric energy $[\Delta G] = 237.2$ kJ/mol and heat energy $[T\Delta S] = 48.6$ kJ/mol at $\Delta S = 163.14$ J/mol/K). Here, ΔS_0 (H_2) = 130.6, ΔS_0 (O_2) = 205.1, ΔS_0 (H_2O) (l) = 70 J/mol/K, ΔS_0 (total) = 130.6 + ½ (205.1 × 70) = 163.14 J/mol/K, and $\Delta S_0 = 233.1$ (kJ/mol). So, the minimum necessary cell voltage is $E_{cell} = 1.21$ V. In the case of an open cell, $E_{ocell} = -\Delta G_0/nF = 1.23$ V, with $\Delta G_0 = \Delta H_0$. $T\Delta S_0 = 237.2$ kJ/mol (at standard conditions: 1 bar pressure, 25°C temperature) (Neagu et al. 2000). The theoretical voltage of electrolysis or reversible potential is written as follows:

$$E_r = E_0 - RT/2F \log P/P_0 \tag{3.3}$$

where E_0 is the standard electrolysis potential, P_0 and P are the respective vapor pressure of pure water and electrolyte, respectively, R is the gas constant, and T is the absolute temperature. The theoretical potential required for electrolysis of water is 2.94 kWh for 1 m³ of hydrogen production, which is the minimum voltage required to continue the reaction and may be coupled with ohmic losses associated with electrolytes/diaphragms and overvoltage (overpotential) due to electrode reactions. But actually, the current efficiency is very high (almost more than 95%). Therefore, energy efficiency can be treated as the voltage efficiency.

In order for a reaction to get started, it is necessary to overcome an (extra) energy barrier, the activation energy E_{act}. The number of molecules able to overcome this barrier is the controlling factor for the reaction rate, r, and this is expressed by the statistical Maxwell–Boltzmann relation, which has exponential behavior: $r = r_0 \exp(-E_{act}/RT)$. So, the activation energy can be expressed as the speed with which a reaction takes place. The maximum possible efficiency of an ideal closed electrochemical cell is defined by the following equation:

$$\epsilon_{max} = \frac{\Delta H}{\Delta G} = -\frac{\Delta H}{nFE_{cell}} \text{(theoretical)} = -\frac{\Delta H}{nFE_{elec}} \text{(real)} \tag{3.4}$$

where ΔE_{elec} is the voltage to drive the electrochemical cell reactions at current I:

$$\Delta E_{elec} = \Delta A + IR + \Sigma\eta \tag{3.5}$$

R = the total ohmic series resistance in the cell, which includes the external circuit resistance of electrolytes, electrodes, and membrane material, and $\Sigma\eta$ = the sum of the overpotentials (activation overpotential at the two electrodes and the concentration overpotential due to the mass transport of the gaseous products away from the anode and cathode surfaces). The energy balance per mole during water electrolysis is exhibited in Figure 3.1a (Neagu et al. 2000). The activation overpotential increases by increasing the current density and can be lowered using electrodes with a catalytic action, such as platinum.

Under ideal reversible conditions, the maximum theoretical efficiency of the water electrolysis with respect to the electrical energy source would be ε_{max} = 120%. Therefore, heat would have to flow into the cell from the surroundings. When the value of ΔG in the denominator of Equation 3.4 becomes 1.48 nF (with overpotential ~0.25 V), the electrochemical cell would perform with 100% efficiency. Under these

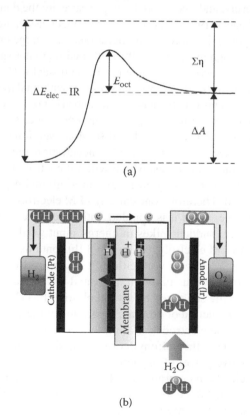

(a)

(b)

FIGURE 3.1 (a) Energies involved in an energy balance during water electrolysis reactions. (b) Schematic representation of the polymer electrolyte membrane (PEM) electrolyzer. (From Neagu, C. et al., *Mechatronics*, 10(4–5), 571–81, 2000.)

conditions ($\Delta S = 0$, $\Sigma\eta = 0$, so $\Delta G = \Delta H$), the cell does not heat or cool and the value of $E_{max} = \Delta H/nF = 1.48$ V is denoted as the thermoneutral potential (Neagu et al. 2000). The electrochemical cell produces heat at potentials above 1.48 V and takes heat in at potentials below this value. Under every condition, the cell temperature is maintained as a constant. In real practice, the IR drop may be equal to 0.25 V. The overpotential η should be kept low to maximize the efficiency and to minimize the production of heat. On the other hand, the lower the overpotential, the slower the reaction, so we have to maintain the balance in the overpotential, IR, and thermoneutral potential. One of the best ways to increase the current without increasing the overpotential is by enhancing the contact areas (with active sites) between the electrodes and the liquid (Neagu et al. 2000). Electrolyzers are commonly classified on the basis of the electrolyte material used in the cell.

3.2.1 PEM ELECTROLYZER

The PEM electrolyzer was first introduced by the General Electric in the 1960s to overcome the drawback of alkaline electrolyzers. The issues related to the partial load, low current density, and operational low pressure are the demerits of the alkaline electrolyzer (Gregory 1975). The major advantages associated with the PEM are fast dynamic response times, reduced operational cost, large operational ranges, high efficiencies, very high gas purities (99.999%) with greater safety, and reliability (since no caustic electrolyte is circulated in the cell stack). The PEM membrane should be able to sustain the high differential pressure without damaging membrane. Additionally, it efficiently prevent the gas mixing and also possess quite high current-bearing capacity, that is, several amps per square centimeter for the membrane of a few millimeters thickness (Carmo et al. 2013). Moreover, it has the noncorrosive, invariant (not mobile) electrolytes with fixed concentration and with no possibility of acid carryover into the effluent gas, which will not control or leak in the system. Furthermore, the PEM electrolysis unit results in a minimum power requirement per unit of gas generated. Therefore, one can say PEM electrolysis is a leading alternative for energy storage, when it is coupled with renewable energy sources such as wind and solar, where sudden spikes in energy input would otherwise result in uncaptured energy. A PEM electrolyzer (Figure 3.1b) contains a cell with a cathode (porous graphite current collector with either Pt or a mixed oxide as the electrocatalyst) and anode (Ir or porous titanium, activated by a mixed noble metal oxide catalyst) electrodes that are separated by a solid plastic/PEM (Nafion™ 117), which is responsible for the conduction of protons, separation of the gaseous products, and electrical insulation of the electrodes. Nafion is a copolymer of tetrafluorethylene and a perfluorinated vinyl-ethersulfonyfluoride that shows an extremely resistant character to the oxidative power of oxygen and even ozone evolving anodes. Prior to applying in the cell, the membranes are hydrated in an autoclave at 130°C for 2 hours in ultrapure water. After cooling, the membranes are treated with 1 M HCl for 30 minutes and subsequently rinsed with demineralized water. These cells operate at quite moderate temperatures, that is, 70°C–90°C, and use a solid plastic/polymer (perfluorinated sulfonic acid) material as the electrolyte. The presence of a polymer electrolyte allows the PEM electrolyzer to operate with a very thin membrane

(~100–200 μm), which can tolerate high pressures and result in low ohmic losses that are caused by the conduction of protons across the membrane (0.1 S/cm) and a compressed hydrogen release. An excess of water is usually supplied to the system and recirculated to remove any waste heat (Slade et al. 2002). Because of the solid structure of the PEM, a low gas crossover rate was observed that resulted in very high purity of product gases, which is very important for storage safety and for the direct usage of produced H_2 in a fuel cell (Carmo et al. 2013). The safety limit for H_2 in O_2 mixture is 4 mol% H_2 in O_2, at standard conditions (Schröder et al. 2004). The initial performances yielded 1.88 V at 1.0 A/cm², which was very efficient compared with the alkaline electrolysis technology of that time. In the late 1970s, the alkaline electrolyzers were reporting performance around 2.06 V at 0.215 A/cm² (LeRoy et al. 1979). When the gaseous products of the electrolyzer, that is, hydrogen and oxygen, permeate across the membrane, the phenomena is referred to as crossover. When O_2 reaches the platinum surface of the cathode through crossover, the oxygen can be catalytically reacted with hydrogen to form water. But at the anode, hydrogen and oxygen do not react in the presence of the iridium oxide catalyst (Schalenbach et al. 2013). Thus, safety hazards expected due to explosive anodic mixtures (hydrogen in oxygen) can occur. Moreover, the amount of the produced hydrogen is reduced due to its reaction with oxygen at the cathode and permeation from the cathode across the membrane to the anode compartment, which may lead to Faradaic losses. Faradaic losses describe the efficiency losses that are correlated to the ratio of the amount of lost and produced hydrogen. At pressurized operation of the electrolyzer the crossover and the correlated Faradaic efficiency losses increase.

3.2.2 ALKALINE ELECTROLYZERS

Alkaline electrolyzers have been used in chemical industries for a long time. In these electrolyzers, the hydroxide ions (OH⁻) are transported through an electrolyte solution of 20%–40% potassium hydroxide or sodium hydroxide from the cathode to the anode and hydrogen is generated on the cathode side, as is represented by Equations 3.6 and 3.7:

$$\text{At cathode:} \quad 2H_2O + 2e^- \rightarrow 2OH^- + H_2 \qquad (3.6)$$

$$\text{At anode:} \quad 2OH \quad H_2O + 2e + 1/2O_2 \qquad (3.7)$$

The electrodes are separated by a diaphragm (Murray 1982), which can be used to separate the gaseous products and transport the hydroxide ions. NiO, Ni/Co/Fe, and Ni/C–Pt are the respective catalytic materials used in construction of diaphragm, anode, and cathode, for which stainless steel, Ti/Ni/zirconium, and stainless steel mesh are bipolar/separator plate materials, respectively. Actual cell potential is calculated using Equation 3.8 and as shown in Figure 3.2a:

$$E = E_r + E_{ir/an} + E_{ir/conc} + E_{ohm/electrolyte} + E_{ohm/electrode} \qquad (3.8)$$

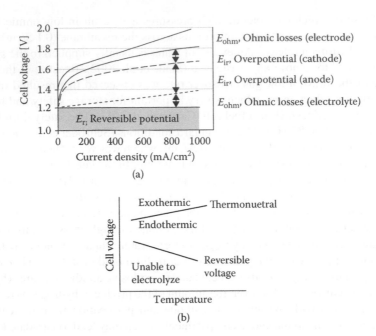

FIGURE 3.2 (a) Current density–cell voltage and (b) cell voltage–temperature plots of the alkaline electrolyzer. (From Schalenbach, M. et al., *Int. J. Hydrogen Energy*, 38(35), 14921–14933, 2013.)

The voltage required for the reversible water-splitting process can be determined by the Nernst equation (Equation 3.9):

$$E_r = E_0 + \frac{nF}{RT} \ln \left(\frac{pH_2 \sqrt{(pO_2)}}{pH_2O} \right) \quad (3.9)$$

where the ideal standard potential $E_0 = \Delta G/nF$; $\Delta G_0 = 242,000 - 45.8T$ [J/mol]; n = number of electrons transferred per hydrogen molecule; R is the universal gas constant; F is the Faraday constant; p_i ($i = H_2$, O_2, and H_2O) is the partial pressure of species i; $E_{ir/an}$ is the overvoltage due to the resistance of the chemical reaction rate of hydrogen evolution; $E_{ir/cat}$ is the overvoltage due to the resistance of the chemical reaction rate of oxygen evolution; concentration overpotential (with mass transport limitation) $E_{ir/conc} = m \exp(nj)$ where m and n are two constants that are independent of the current density; $E_{ohm/electrode} = (RT/\alpha nF)\ln(j/j_0)$ is the activation overpotential losses related to the electrochemical kinetics of circuitry/electrodes (Kreuter and Hofman 1998), where α is the transfer coefficient, j is the operating current density, and j_0 is the exchange current density (it can be reduced by shortening the distance between the cathode and anode); and $E_{ohm/electrolyte} = jt_{ele}/k$ is the ohmic overpotential losses mainly caused by the resistance of electrolytes since the resistance of electrodes is negligible compared with that of electrolyte, where t_{ele} is the thickness of the acid electrolyte, and k is the specific conductivity of the acid solutions.

A minimum of 2.948 kW/h energy is required for one cubic hydrogen generation. It is not practical to assume an alkaline water electrolyzer will work below thermoneutral voltage. It is the point beyond which all of the electricity given to the cell is converted into heat that must be removed to maintain the cell temperature. Reversible voltage decreases with increase in temperature but thermoneutral voltage does not show much change since it corresponds to the energy of hydrogen generation, as is depicted in Figure 3.2b. The commercially available alkaline electrolyzers are operated at 100°C–150°C. Recent approaches involve new promising solid alkaline exchange membranes as the electrolytes become more practical.

3.2.3 ACID ELECTROLYZERS

The electrolysis of aqueous acids is referred to as a potential corridor for hydrogen production (Chettiar 2004). Deniele et al. (1996) studied the hydrogen evolution process from aqueous strong and weak acids under pseudo-first- and pseudo-second-order kinetic conditions. Weaver et al. (2006) discussed the industrial production possibilities and technological aspect for large scale-up, economical hydrogen production by using sulfuric acid as the electrolyte. Marinovic and Despic (1999) studied the pyrophosphate solution on silver as a function of pH to investigate hydrogen evolution. The acid electrolysis process is just a reverse reaction of a PEM fuel cell (PEM FC) process, as shown in Figure 3.3. However, the catalytic materials used in acid electrolyzers are different from the PEM FC. Acid electrolyzers use H^+ charge carriers in water that comes from the acids that are used as electrolytes such as citric acid, butyric acid, sulfuric acid, or phosphoric acid and arranged with Pt wire as a cathode electrode at 150°C. Equations 3.10 and 3.11 represent the electrode reactions at the cathode and anode:

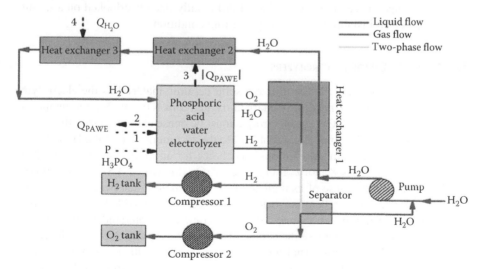

FIGURE 3.3 Schematic diagram of a photochemical acidic water electrolysis (PAWE) system for hydrogen production. (From Zhang, H. et al., *Int. J. Hydrogen Energy*, 35, 10851–10858, 2010.)

$$\text{At cathode:} \quad 2\text{H}^+ + 2\text{e}^- \rightarrow \text{H}_2 \tag{3.10}$$

$$\text{At anode:} \quad \text{H}_2\text{O} \rightarrow 1/2\,\text{O}_2 + 2\text{e}^- + \text{H}_2 \tag{3.11}$$

It was observed that an optimal value for the circulating electrolyte concentration at which the input voltage of the electrolyzer attains its minimum in given conditions. Joule heat resulting from the irreversibilities inside the electrolyzer is larger than the thermal energy needed in the water-splitting process. The max efficiency of the system can be simplified to Equation 3.12:

$$\eta = \frac{H_{\text{H}_2,\text{out}}\text{HHV}}{P + W_{\text{com}} + Q_{\text{AWE}}\left(1 - \dfrac{T_0}{T}\right) + Q_{\text{H}_2\text{O}}\left(1 - \dfrac{T_0}{T}\right)} \tag{3.12}$$

where HHV is the higher heating value of H_2 and P is the electric power required by the PAWE at different temperatures. The electrical power consumption of the water pump is not considered in Equation 3.12 because it is very minute compared with the input electric energy of the electrolyzer, and W_{com} is the power consumption of the two compressors (Zhang et al. 2012). Many acid electrolytes such as citric acids, sulfuric acid, phosphates, biogenic butyric acid, and pyrophosphoric acid were used as the electrolyte with photocatalytic material as the anode and Pt or steel or silver as the cathode, to split water (Mohilner et al. 1962; Marinovic and Despic 1997; Liu 2005; De Silva Munoz et al. 2010). Recently, Bloor et al. (2014) investigated the first-row transition metals to mediate heterogeneous electrolytic water splitting in acidic media by exploiting these metastable electrocatalysts based on abundant elements for a wide range of reactions, which have traditionally been overlooked on account of their perceived instability under the prevailing conditions.

3.2.4 SOLID OXIDE ELECTROLYZERS

Solid oxide electrolyzers (SOEs) utilize a solid ceramic material as the electrolyte that selectively conducts negatively charged oxygen ions (O^{2-}) at elevated temperatures (about 700°C–800°C). The electrolyte must be dense enough to stop the diffusion of the steam and hydrogen gas that lead to the recombination of the H_2 and O^{2-}. The dense ionic conductors are an obvious choice for an SOE. One of the notable examples of an SOE is 8 mol% Y_2O_3-doped ZrO_2 (known as YSZ). Here, zirconia dioxide is used because of its high strength, high melting temperature (approximately 2700 C), and excellent corrosion resistance. Y_2O_3 is added to mitigate the phase transition from the tetragonal to the monoclinic phase on rapid cooling, which can lead to cracks and decrease the conductive properties of the electrolyte by causing scattering (Bocanegra-Bernal and De la Torre 2002). Other common choices for SOEs are scandia-stabilized (Bi 2015) and ceria-based electrolytes, lanthanum gallate materials, scandium-doped chromate, and so on (Ishihara et al. 2010; Chen et al. 2015).

Ni-doped YSZ is most commonly used as a cathode material but it degrades under high steam partial pressures and low hydrogen partial pressures, which leads to irreversible degradation (Laguna-Bercero 2012). Another interesting cathode material is perovskite-type lanthanum strontium manganese (LSM) or scandium-doped LSM; it shows higher performance at low temperatures than traditional LSM (Yue et al. 2008). New cathode materials such as lanthanum strontium manganese chromate (LSCM) with high redox stability, and high ionic conductivity (scandium-doped LSCM), are being researched, and have been proven to be stable under electrolysis conditions and to demonstrate high efficiency at temperatures as low as 700°C (Yang and Irvine 2008). LSM is a noticeable high-performance anode electrode material. LSM offers high performance under electrolysis conditions due to generation of oxygen vacancies under anodic polarization that aids oxygen diffusion (Wan and Jiang 2010). In addition, impregnating an LSM electrode with Gd/Ce nanoparticles was found to increase the cell life by preventing delamination at the electrode/electrolyte interface (Chen et al. 2010). But the exact mechanism by how this happens needs to be explored further. In 2010, it was found that neodymium nickelate as an anode material provided 1.7 times the current density of typical LSM anodes when integrated with a commercial solid oxide electrolyzer cell (SOEC) and operated at 700°C, and approximately four times the current density when operated at 800°C. The increased performance resulted due to higher "overstoichiometry" of oxygen in the neodymium nickelate that made it a successful conductor of both ions and electrons (Chauveau et al. 2010). The beauty of SOEs is that they can effectively use the heat that is available at these elevated temperatures (from various sources, including nuclear energy) to decrease the consumption of electrical energy needed to produce hydrogen from water. At the cathode, water molecules react with electrons of the cathode and generate oxygen ions and hydrogen. On the other hand, the oxygen ions pass through the solid ceramic membrane and react at the anode to form oxygen gas and generate electrons for the external circuit, as represented by Equations 3.13 and 3.14 (Ni et al. 2008):

$$\text{Cathode reaction:} \quad 2H_2O + 4e^- \rightarrow 2H_2 + 2O^{2-} \tag{3.13}$$

$$\text{Anode reaction:} \quad 2O^{2-} \rightarrow O_2 + 4e^- \tag{3.14}$$

SOE cells are superior to batteries in terms of hydrogen generation due to their high storage capacity and minimum waste material disposal (Ni et al. 2007). Other advantages associated with SOEs are their high efficiencies (not limited by Carnot efficiency), long-term stability, fuel flexibility, low emissions, and low operating costs. However, they suffer from the greatest disadvantage, that is, the high operating temperature, which results in long start-up times, short lifetime, and mechanical compatibility issues that arise when there is a thermal expansion mismatch and diffusion between layers of material in the cell, which causes chemical instability. However, the heat produced due to the Joule heating of an electrolysis cell may be utilized to reduce the energy demand in the water-splitting process at high temperatures. Concentrating solar thermal collectors and geothermal sources are also considered good external heat sources to add heat in an SOE.

3.3   BIOPHOTOCATALYTIC WATER SPLITTING

A phenomenon of biophotocatalytic splitting of water is evident in a diverse group of living objects, such as bacteria, plants, enzymatic systems, and so on. These organisms process several anaerobic conversion reactions through photosynthesis, fermentation, and microbial electrolysis, for production of oxygen and hydrogen on water cleavage. It was first observed in green algae (deprived of sulfur) in the 1990s and utilized for water splitting. Photobiology seems to be the long-term ecofriendly, cost-effective, and pollution-free solution for hydrogen production from water by using renewable solar energy. There are several pathways to produce hydrogen from water and biomass, using solar energy. Fortunately or unfortunately, all of these technologies are in the development stage and they take a long time to achieve efficient and cheaper hydrogen energy (Turner et al. 2008). The photobiological hydrogen production process uses microorganisms (such as certain algae and cyanobacteria) and sunlight to split water (sometimes organic matter) into oxygen and hydrogen ions with low carbon release in atmosphere. The hydrogen ions can be converted into hydrogen gas by direct or indirect routes. Microbes like algae and bacteria could be grown in water (wastewater) that cannot be used either for drinking or for agriculture purposes. Not only may this provide economical hydrogen production from sunlight but also low to net-zero carbon emissions. For all of the aforementioned reasons, this technique can fall in the category of ecofriendly. Photobiological water splitting basically can be divided into two groups based on the microorganisms selected (green algal or cyanobacteria, etc.) and reaction mechanisms (photosynthesis or biophotolysis or photofermentative) involved. Currently, there are a number of challenges to be faced in both photolytic and photofermentative biological hydrogen production. This can be overcome by working along the following lines:

- Improving the activity of the enzymes that produce the hydrogen, as well as the depth of knowledge about the mechanism of the metabolic reactions, which is needed to increase the hydrogen production rates.
- Developing strategies that can efficiently use sunlight and other inputs to increase hydrogen yields.
- Developing strains and reactor configurations that can ultimately be used at large scales for commercial hydrogen production.

The National Renewable Energy Laboratory (NREL) researchers have created a bacterial system in cyanobacteria or green algae for synthesis of the key enzyme, hydrogenase or nitrogenase, that is responsible for photosynthetic hydrogen evolution, as illustrated in Figure 3.4 and Equations 3.15 through 3.18. Furthermore, their enzyme engineering is focused on blocking the access of oxygen at the catalytic site of the hydrogenase. Hydrogen production by using oxygenic cyanobacteria or green algae under light irradiation and anaerobic conditions is referred to as water biophotolysis. On the other side, under organic biophotolysis hydrogen gas is produced photosynthetically using anoxygenic bacteria under light irradiation

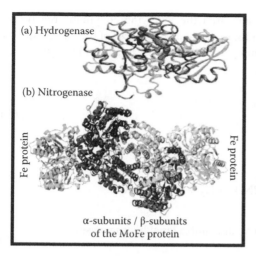

(a) Hydrogenase

(b) Nitrogenase

Fe protein

Fe protein

α-subunits / β-subunits
of the MoFe protein

(c)

FIGURE 3.4 (a) Hydrogenase enzyme structure with oxygen-diffusion channels (red and yellow ribbons) and backbone of the protein (gray). The oxygen-sensitive FeS catalytic centers are indicated in green, yellow, and red. (b) Structure of nitrogenase enzyme that contains two α-subunits of the MoFe protein in blue and red color, and the two β-subunits of the MoFe protein in blue and red color (middle) and two Fe proteins (top and bottom). (c) Green algae used in bioreactors for photobiological hydrogen production. (From Williams, T. et al., *Hydrogen Technologies: Photobiocatalytic Production of Hydrogen*, National Renewable Energy Laboratory, CO, 2007; Igarashi, R.Y. and Seefeldt, L.C., *Crit. Rev. Biochem. Mol. Biol.*, 38, 351–384, 2003.)

and anaerobic conditions. Organic biophotolysis gives a high yield of hydrogen gas along with the by-product CO_2 that makes this technology less environmentally friendly compared with water biophotolysis. However, water biophotolysis is a "cleaner" way to produce hydrogen compared with organic biophotolysis. But it still suffers from problems that are yet to be resolved, including low yield of hydrogen, enzyme poisoning in the presence of excess of oxygen (generated simultaneously during biophotolysis), and the difficulty faced in designing and scaling up the process of the bioreactor.

Under water biophotolysis, photosynthesis and hydrogen production are the sister reactions that proceed in the presence of green algae. Both reactions are initiated with the following reaction that involves the production of oxygen, electrons, and protons through solar energy supported water cleavage:

Step 1 (solar energy–activated splitting of water)

$$H_2O + h\nu(\text{sunlight}) \rightarrow O_2 + e^- + 2H^+ \tag{3.15}$$

Protons and electrons generated in step 1 are consumed in the enzymatic reactions of step 2. If the excess energy were not somehow dissipated, the electrons produced by the water splitting would harm or destroy the organism (plants, algae, cyanobacteria, etc.). These organisms used the following enzymatic reactions, as shown by Equations 3.16 through 3.18, to dissipate the excess of the energy liberated as shown in Equation 3.16 and form hydrogen or sugars, as shown in Figure 3.5.

Step 2 (enzymatic reactions)

1. Oxygenic photosynthesis (chlorophyll):

$$nCO_2 + 4ne^- + 4nH^+ \rightarrow (CH_2O)_n + nH_2$$

(3.16)

2. N_2 fixation (nitrogenase):

$$N_2 + 8H^+ + 8e^- + 16ATP \quad 2NH_3 + H_2 + 16ADP + 16P_i$$

(3.17)

3. Hydrogen production (green algal/cynobacterial hydrogenase)

$$2H^+ + 2e^- \rightarrow H_2$$

(3.18)

During the solar energy–activated splitting of water, oxygen, electrons, and protons, protons and electrons are produced in the first step, then the process goes to the second step of enzymatic reactions. The excess energy produced in the form of electrons as a result of the water splitting (Equation 3.18) would harm or destroy the organism if this excess energy were not dispersed. The enzymatic reactions, which involves carbon dioxide fixation to produce sugar, nitrogen fixation to produce hydrogen and ammonia, or other alternative reactions to produce hydrogen molecules, help the organism to get rid of this excess energy. All of these

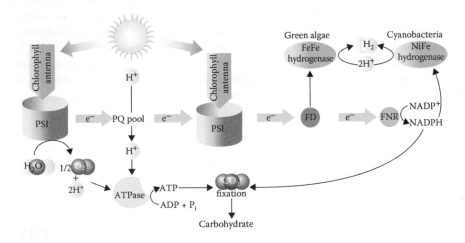

FIGURE 3.5 Photosynthetic pathway of H_2 production in green algae (employing FeFe hydrogenase) and in cyanobacteria (employing NiFe hydrogenase). PS, photosystem; PQ, plastoquinone; FD, ferredoxin; FNR, ferredoxin-nicotinamide adenine dinucleotide phosphate oxidoreductase. (From Maness, P.C. et al., *Microbe*, 4(6), 275–280, 2009.)

are temporary alternative processes, which normally only last for a few minutes. Furthermore, whenever the concentration of oxygen produced from the water-splitting reaction reaches its climax, the algae or cyanobacteria are forced to shut down the hydrogen production and fix the carbon. The oxygen physically diffuses into the enzyme's catalytic center and is irreversibly bound to it, resulting in stopping the catalytic activity of the enzyme. To overcome that inhibition, enzyme engineering has to be done on [Fe-Fe]-hydrogenase (algal)/[Ni-Fe]-hydrogenase (cyanobacteria, less sensitive to oxygen than green algae) with some of the mutations. In addition, algal hydrogen production can be sustained for a long period by inactivating the normal photosynthetic process by nutrient (sulfate) deprivation. Major challenges of both direct and indirect biophotolysis routes are that these methods involve commercially nonviable processes, low solar energy to hydrogen conversion efficiency, and difficulty in designing the perfect bioreactor.

3.4 THERMOCHEMICAL WATER SPLITTING

Thermochemical water-splitting cycles (TCWSCs) employing solar energy as a heat source can be attractive due to their relatively higher efficiency. It has been shown that TCWSCs have the potential to deliver overall system efficiencies more than 40%. TCWSC is a pure thermochemical process that has been under exploration for 25 years. Thermochemical water splitting requires a very high temperature (500°C–2500°C) to drive a series of chemical reactions to produce the hydrogen and oxygen gases by consuming water only. The chemicals (autocatalyst) produced/used in this series of chemical reactions can be reused within each cycle and creates a closed loop (cycles) to produce hydrogen and oxygen from water and heat without utilizing electricity (Chen et al. 2007; Kolb et al. 2007, Perret et al. 2007). Most of the "high-temperature" water-splitting cycles engage in thermal reduction of a metal oxide to the lower valence state/states, which is the solar energy motivated step. In a subsequent step (dark/off-sun reactions), the reduced oxide reacts with the steam to generate hydrogen and rejuvenate the original oxide. Usually, in this approach the cycles based on the redox pairs used to split water have two/three steps that lead to a low potential for energy losses between cycle steps and during separations. The necessary high temperatures can be generated by means of concentrated solar power or from the waste heat of nuclear power reactions. At present a large number of catalysts are available to be used in more than 352 thermochemical cycles at comparatively feasible temperatures. A few of the well-known thermochemical cycles are the iron oxide (FeO/Fe_3O_4) cycle, cerium(IV) oxide–cerium(III) oxide cycle, zinc/zinc-oxide cycle, sulfur–iodine cycle, copper–chlorine cycle, and hybrid sulfur cycle, which are actively being researched or in trial phases. Among the most studied TCWSCs are sulfur–halogen cycles. These thermochemical cycles can be divided into three parts on the basis of the number of steps: one step, two steps, or three steps or more cycles. Or they can be divided into "direct" or "hybrid" thermal cycles corresponding to exclusive thermochemical or thermochemical associated with electrochemical or photochemical reactions, respectively.

3.4.1 Thermodynamics of the Thermochemical Water Splitting

At ambient conditions, the thermodynamic equilibrium for water splitting can be considered by following equation:

$$H_2O(l) \rightleftharpoons H_2(g) + 1/2 O_2(g) \tag{3.19}$$

This equilibrium will be shifted to the right only if energy (ΔH) in the form of electricity/heat is provided to the system:

$$\Delta H = \Delta G + T\Delta S \tag{3.20}$$

In order to proceed with the water-splitting process at ambient temperature 298 K and 1 atm pressure, 237 kJ/mol Gibbs free energy (ΔG^0), 163 J/mol/K entropy (ΔS^0), and more than 80% of the required energy ΔH must be provided as work done. If phase transitions are neglected for the sake of simplicity, one can assume that $\Delta H/\Delta S$ does not vary significantly for a given temperature change. Consequently, the work done required at temperature T is

$$\Delta H = \Delta G^0 - (T - T^0)\Delta S^0 \tag{3.21}$$

If the T was high enough in Equation 3.21, ΔH could be nullified, meaning that thermolysis of water would occur even without work. In the aforementioned case, when the steam of water is taken in place of liquid water, tremendously high temperatures would be required ($\Delta H^\circ = 242$ kJ/mol; $\Delta S^\circ = 44$ J/mol/K), that is, above 3000 K, which would make reactor design and operation conditions extremely challenging. Therefore, single-step reactions offer only one degree of freedom (T) for the generation of hydrogen and oxygen using heat energy. Although, according to Le Chatelier's principle, a slight decrease in the thermolysis temperature would be needed, the additional work for extracting the gaseous products from the system. On the contrary, Funk and Reinstrom show that the k-step thermolysis of water allows spontaneous water splitting without work due to different entropy changes S_i^0 involved in each ith reaction. An extra bonus associated with these reactions is that the oxygen and hydrogen are produced separately, which may avoid the complex separation of these gases at high temperatures (Kogan 1998).

3.4.2 Single-Step Cycle

In these types of reactions, single-step reactions are used to produce hydrogen. Thermal decomposition of methane without using catalyst produces hydrogen-rich gas and high-grade carbon black from concentrated solar energy and methane. A solar concentrator (2 m diameter) with 1500 K mean temperature at the nozzle has shown a thermal-to-chemical conversion efficiency in the range of 2.6%–98%:

$$CH_4 \xrightarrow{\text{1500K}} C \text{ (carbon black)} + 2H_2 \quad \eta = 2.6\% - 98.0\% \tag{3.22}$$

3.4.3 Two-Step Cycle

This cycle involves two steps for hydrogen production. The first step involves solar energy–induced thermal reduction of a metal oxide to lower valence state/states, which is followed by the subsequent step (dark/off-sun reactions), where the reduced oxide reacts with the steam to generate hydrogen and rejuvenate the original oxide. A few of the popular systems of this category along with their efficiency and cost are discussed as follows: Zn/ZnO (η = 17.2%–20.7%. cost = \$4.14–5.58 kg^{-1} H_2), Cd/CdO_2 (η = 48.3%, cost = \$4.5 kg^{-1} H_2), CeO_2/Ce_2O_3 (η = NA, cost = NA) (Stéphane and Gilles 2006) (Figure 3.6a), iron oxide (Fe_3O_4/FeO) (η = \$7.4%, cost \$7.86–14.75 kg^{-1} H_2), and so on.

Iron oxide (Fe_3O_4/FeO) cycle

1. $Fe_3O_4 \rightarrow 3FeO + O_2(g)$ (1900 K) (3.23)

2. $3FeO + H_2O(g) \rightarrow Fe_3O_4 + H_2$ (850 K) (3.24)

Zn/ZnO cycle

1. $ZnO \rightarrow Zn + 1/2\,O_2$ (2300 K) (3.25)

2. $Zn + H_2O \rightarrow ZnO + H_2$ (700 K) (3.26)

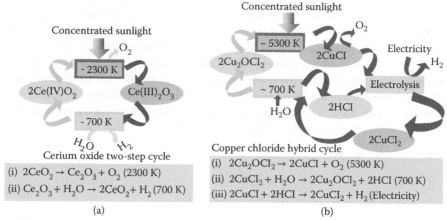

Cerium oxide two-step cycle
(i) $2CeO_2 \rightarrow Ce_2O_3 + O_2$ (2300 K)
(ii) $Ce_2O_3 + H_2O \rightarrow 2CeO_2 + H_2$ (700 K)

(a)

Copper chloride hybrid cycle
(i) $2Cu_2OCl_2 \rightarrow 2CuCl + O_2$ (5300 K)
(ii) $2CuCl_2 + H_2O \rightarrow 2Cu_2OCl_2 + 2HCl$ (700 K)
(iii) $2CuCl + 2HCl \rightarrow 2CuCl_2 + H_2$ (Electricity)

(b)

FIGURE 3.6 (a) Cerium oxide two-step thermochemical cycle and (b) copper chloride hybrid thermochemical cycle for water splitting. (From Stéphane, A. and Gilles, F., *Sol. Energy*, 80(12), 1611–1623, 2006; Ferrandon, M.S. et al., *Int. J. Hydrogen Energy*, 35(3), 992–1000, 2010.)

3.4.4 Three-Step Cycle

These cycles utilize solar concentrated energy to reduce manganese, cobalt, iron, and so on higher oxides into their lower valence metal oxides, MnO, CoO, Fe_3O_4, and FeO and free oxygen, at a temperature in the range of 1573–1873 K. Thereafter, reduced oxides react with bases NaOH and KOH to produce hydrogen and alkali metal oxides, which will regenerate the original metal oxide on reacting with water. The mechanism of the typical Mn_2O_3/MnO cycle is illustrated by Equations 3.27 through 3.29:

Mn_2O_3/MnO cycle

1. $Mn_2O_3 \rightarrow 2MnO + 1/2\, O_2$ (1873 K) $\qquad\qquad$ (3.27)

2. $2MnO + 2NaOH \rightarrow H_2 + 2NaMnO_2$ (900 K) $\qquad\qquad$ (3.28)

3. $2NaMnO_2 + H_2O \rightarrow Mn_2O_3 + 2NaOH$ (373 K) $\qquad\qquad$ (3.29)

3.4.5 k-Step Cycle

High covalence is the main advantage of using sulfur in thermochemical cycles. Hence, it can form a maximum of six chemical bonds with other elements with various oxidation states and show several redox reactions existing in sulfur compounds. This freedom allows numerous chemical steps with different entropy changes, and thus offers more complications to meet the criteria required for a thermochemical cycle. Sulfur-based cycles are beneficial, because they are economical and flexible, if they are operated in an open cycle. The sulfer–ammonia thermochemical cycle (T-Raissi et al. 2006) is the most popular among k-step sulfur cycles used for decomposing water into hydrogen and oxygen. The cycle consists of three steps: (1) photocatalytic oxidation of ammonium sulfite and reduction of water into ammonium sulfate and hydrogen, respectively (Equation 3.30), (2) ammonium sulfate decomposition into ammonia and sulfuric acid (Equation 3.31), and (3) chemical coadsorption of ammonia and sulfur dioxide to produce ammonium sulfite, which is then recycled to produce H_2 and ammonium sulfate (Equations 3.32 through 3.34).

Sulfur-ammonia cycle

1. $(NH_4)_2SO_3 + H_2O + h\nu(\text{UV light}) \rightarrow$
    $(NH_4)_2SO_4 + H_2$ (80°C) (photocatalytic step) $\qquad\qquad$ (3.30)

2. $(NH_4)_2SO_4 \rightarrow 2NH_3 + H_2SO_4$ (350°C) (thermochemical step) $\qquad\qquad$ (3.31)

3. $H_2SO_4 \rightarrow SO_3 + H_2O$ (400°C) (thermochemical step) $\qquad\qquad$ (3.32)

4. $SO_3 \rightarrow SO_2 + 1/2 O_2$ (850°C) (thermochemical step) (3.33)

5. $SO_2 + 2NH_3 + H_2O \rightarrow (NH_4)_2 SO_3$ (25°C) (chemical adsorption) (3.34)

3.4.6 HYBRID CYCLE

This process consists of thermochemical steps followed by electrochemical/photochemical reactions. Cu–Cl TCWSC is the best example of this type of cycle including two thermal and one electrochemical reaction, as illustrated in Figure 3.6b (Ferrandon et al. 2010). High efficiency and moderate temperature requirements (Rosen 2010) are the merits of the Cu–Cl TCWSC. Moreover, the process does not give any contribution to greenhouse gas emission and has implication with a series of close loop chemical reactions (Orhan et al. 2012). Furthermore, high-temperature electrolyzers are economically less advantageous than thermal-assisted electrolysis of cuprous chloride ($2.08/kg) because the Cu–Cl TCWS cycle cuts down the price of hydrogen production. Photoelectrochemical (PEC) methods cost less than $0.70/kg; photosynthetic are at $0.40/kg (cheaper than present natural gas) (Bockris 1975). Electrical energy used in this method for one of the reaction processes is meant to overcome the positive free-energy barrier inevitably occurring in a pure thermochemical cycle. Besides the copper chloride hybrid cycle, the Mark 11 cycle, also known as the Westinghouse sulfur cycle, and the Mark 13 cycle developed at Ispra, Italy are a few other examples of this class. In the Mark 11 cycle, use of SO_2 as an anodic depolarizer was originally suggested by Juda and Moulton (1972) and the first step of this reaction needs very low voltage, that is, at 0.17 V, which is much less than the 1.23 V required for the electrolysis of water. The steps involved in the Mark 11 process are the same as the Westinghouse sulfur cycle (Equations 3.35 and 3.36):

Mark 11 Cycle

$$SO_2 + 2H_2O \rightarrow H_2 + H_2SO_4 \text{ (353 K) (electrochemical)} \quad (3.35)$$

$$H_2SO_4 \rightarrow H_2O + SO_2 + 1/2 O_2 \text{ (723–1393 K) (thermochemical)} \quad (3.36)$$

The Mark 13 process developed by Ispra (Velzen et al. 1978) for producing 100 L/h of hydrogen by desulfurization of the flue gas, that is, SO_2 (Velzen and Langenkamp 1980), consists of the following steps (Equations 3.37 through 3.39):

Mark 13 Cycle

$$SO_2 + Br_2 + 2H_2O \rightarrow 2HBr + H_2SO_4 \text{ (373 – 413 K) (thermochemical)} \quad (3.37)$$

$$2HBr \rightarrow H_2 + Br_2 \text{ (electrochemical)} \tag{3.38}$$

$$H_2SO_4 \rightarrow H_2O + 5O_2 + 1/2\,O_2 \text{ (723 − 1393 K) (thermochemical)} \tag{3.39}$$

The TCWSC technique is one of the most promising techniques because it can use the best of thermal energy (harvested by renewable concentrated solar energy) and can produce carbon-free hydrogen. The main disadvantage of this technique is the high-temperature sensitive/durable reactor design that can resist the thermal shock (due to thermal shock concerns, alumina and zirconia cannot be used) and stress, stable catalytic material that can sustain the chemical integrity of the material at high temperatures and tolerate the high oxidizing atmosphere, low efficiency, and overall high capital investment. The aforementioned are the main obstacles associated with this method. Moreover, the list of materials used in TCWSC is extremely limited. To combat these challenges, an efficient and compatible material with high-temperature resistance and reactors need to be developed. Similarly, the cost of the mirrors used in the solar concentrator need to be reduced. Despite the unbeatable progress in research activities in this field, a long technological pathway and strong political will go hand in hand with potentially low or no greenhouse gas emissions, for implementation of this technique at industrial scale.

3.5   MECHANOCATALYTIC WATER SPLITTING

Mechanocatalytic splitting of water is the simplest and least studied (1998–2000) method; it is used to decompose water into H_2 and O_2 using mechanical energy that is directly converted into chemical energy (Ikeda et al. 1998). In this method, the powdered metal oxides (catalyst: NiO, Co_3O_4, Cu_2O, Fe_3O_4, AWO_4 [A = Fe, Co, Ni, Cu], and $CuMO_2$ [M = Al, Fe, Ga]) (Grimes et al. 2008) are well dispersed in distilled water by agitation through a Teflon-coated magnetic stirrer rod of different shapes, that is, triangle, round, elliptical, and cylindrical, for hydrogen generation by water splitting. During this process, the particles of metal oxide are rubbing between the stirring rod and the bottom wall of the reaction vessel. Here, the electric charge generated due to the frictional forces caused through rubbing that may be accompanied by the catalytic activity of the oxide powder plays a key role in the water-splitting process. That is why the phenomenon is known as mechanocatalytic water splitting, which is based on triboelectricity. In this process, the mechanical energy is converted into electrical energy and ultimately to chemical energy.

Domen's group resolved that the rate of the evolved gases increases with stirring speed and gets saturated above 1500 rpm. The stirring parameters and, size, and material of the rod as well as the reaction vessel material have been considered to influence the rates of hydrogen and oxygen evolution. Among the available reaction vessel materials, such as quartz, sapphire, polytetrafluorethylene (PTFE), and alumina, pyrex glass was found superior. And among the available dielectric materials such as polyester, polystyrene, polycarbonate, or polypropylene, Teflon was found highly appropriate for coating of the stirring rod. According to the Domen group, the friction generated between the catalytic particles and bottom of the vessel results in charge separation, where catalytic powder and the reaction vessel acquire negative and positive charges, respectively, that may drive the redox reactions and leads to water splitting. However, aforementioned findings and conclusions were challenged by various researchers (Ross 2004). The alternate explanations given by them included some kind of autoredox reactions between water and catalyst powder, electrical charges on the powders due to friction followed by local discharge as well as thermal decomposition that may induce increase in temperature (up to about 5000 K). To date there has been no consensus on the actual mechanism (Hara and Domen 2004). Tokio Ohta also proposed a theory of mechanocatalytic water splitting on the basis of the experimental findings (Ohta 2000 a–d). Due to the frictional rubbing of the solid, catalyst particles between the Teflon-coated stirring rod (R) and bottom of the glass vessel are subjected to develop a number of microcrevices on the surface of G (Ohta 2001) that creates separated islands of positive and negative charges on R and G, respectively. The insulated coating on stirring rod R gets positively charged due to the loss of electrons by friction, which will leave the glass vessel surface negatively charged and generate a frictional electrical capacitor between them. When the oxide particles trapped in the microcrevices are contacted by R they are subject to high electric fields. Furthermore, the excess of oxygen trapped in the interstitials of the p-type oxide semiconductor become negatively charged O^- (or O^{2-}) by receiving electrons from neighboring metal ions and leave a positive hole on this metal ion. Subsequently, the electrons and holes created in this manner are responsible for water splitting, where R and G perform a corresponding role of the anode and cathode. When R comes into contact with the O^- ion, present at the glass surface, it transfers electrons to the positively charged surface of R. These electrons are ultimately transferred to G due to rubbing action, through a hopping mechanism. Afterward, the protons move toward the bottom of the glass cell G, where they combine with electrons to generate hydrogen gas. The hole reaches the surface of the particle, where it encounters the potential barrier due to the work–function difference. If the barrier height is not sufficiently low, tunneling of the holes could take place. This happens when the potential barrier is steep and the barrier height at the Fermi level is on the order of the wavelength associated with the carrier. These holes oxidize water and generate oxygen and protons at the anode. These holes oxidize water and oxygen is evolved. Similarly, hydrogen is evolved at the cathode or G (Takata et al. 2000):

$$2h^+ + H_2O \rightarrow 2H^+ + 1/2\, O_2 \,(\text{at R}) \qquad (3.40)$$

$$2H^+ + 2e^- \rightarrow H_2 \,(\text{at G}) \qquad (3.41)$$

One other possible mechanism involves the transfer of electrons from the catalyst particle surfaces to water, so that hydrogen is evolved and the generated OH^- ions react with the positive charge at the stirring rod R to evolve oxygen, according to the following reactions (Ohta 2001):

$$2e^- + 2H_2O \rightarrow 2OH^- + H_2 \,(\text{at G}) \qquad (3.42)$$

$$2OH^- + c^{n+} \rightarrow H_2O + 1/2\,O_2 + R^{+(n-2)}\,(\text{at R}) \tag{3.43}$$

The efficiency of mechanocatalytic water splitting was calculated using the following equation (Ikeda et al. 1999; Ohta 2000a–d):

$$\eta = (E_c/E_i) \times 100 \tag{3.44}$$

where E_c is the maximum useful work that can be achieved from evolved hydrogen and E_i the mechanical energy input.

It was clearly revealed that some simple and mixed oxide powders such as NiO, Co_3O_4, Cu_2O, Fe_3O_4, and $CuMO_2$ (M = Al, Fe, Ga), when suspended in distilled water by magnetic stirring, decompose water into H_2 and O_2 catalytically. The observed mechanocatalytic overall water splitting was cycled by the conversion of mechanical energy supplied by rubbing these oxide powders against the bottom wall of the reaction vessel with the stirring rod. Typical efficiency of the mechanical-to-chemical energy conversion was estimated to be 4.3% (for NiO) (Ikeda et al. 1999).

3.6 PLASMOLYTIC WATER SPLITTING

Thermolysis of water required quite a high temperature (3000°C) and a very specific material that can sustain its integrity. To get rid of the problems associated with the thermolysis of water, electric discharge of plasma is a highly useful technique in carrying out water splitting without acquiring high temperature, and is a thermodynamically nonspontaneous reaction ($\Delta G > 0$) (Bockris et al. 1985). The plasma state involves the transformation of the energy from an electric field (microwave, radio frequency, or direct current [d.c.]) into kinetic energy of the electrons, which is further transmitted into molecular excitations and to the kinetic energy of the heavy particles. Discharge of plasmas may be broadly divided into two categories: hot (arc or thermal) discharges and cold (low temperature or glow) discharges. Electron temperature can range from thousands to tens of thousands degrees in both types of discharges. Interactions between the electron and molecule are responsible for dissociation and ionization in the plasma; the electron–molecule reaction rate constants depend on the average electron energy and the electron energy distribution function, that is, $f_e = f_{LM} \exp\left(\dfrac{e\phi}{K_B T_e}\right)$. Their state can be described by a Maxwellian velocity distribution function, which can be specified by an electron temperature, T_e. And energy per degree of freedom is equal to $\frac{1}{2}\,mv^2 = \frac{1}{2}\,K_B\,T_e$. Electric fields can arise spontaneously in the plasma due to differences in the mobility or transport of the two species (electron and molecule/ions). The electric field in turn affects the spatial distribution of particle velocities through the Lorentz force (we here ignore complications due to magnetic fields) that implies much greater mobility for the electrons. The degree of ionization, the number of free electrons divided by the gas density, nJN (where n is the electron density, J is the current density, and N is an Avogadro number) is a descriptive parameter of particular concern in the plasma chemistry of water (Givotov et al. 1981). During the water vapor decomposition in a nonequilibrium plasma state, the main portion of the energy in the discharge is expended in vibrational excitation and dissociative attachment. Chemical aspects of thermal discharges have been treated by assuming local thermodynamic equilibrium in the arc while chemical and transport kinetics control the quench of the high temperature species. There are two methods for hydrogen production: through water splitting by water plasmolysis and CO_2 plasma/$CO + H_2O$ plasma, as shown in Figures 3.7a and b, respectively (Hensen and Sanden 2014). Bochin et al. (1978) has given the reaction sequence involved in decomposition of water through vibrational excited states, shown by Equations 3.45 through 3.49:

$$H_2O* + H_2O* \rightarrow H + OH + H_2O \tag{3.45}$$

$$H + H_2O* \rightarrow H_2 + OH \tag{3.46}$$

$$OH + H_2O* \rightarrow H + H_2O_2 \tag{3.47}$$

$$OH + H_2O \rightarrow H_2 + HO_2 \tag{3.48}$$

$$HO_2 + H_2O \rightarrow H_2O_2 + OH \tag{3.49}$$

The mechanism for plasmolysis of water via dissociative attachment route is shown as follows:

$$H_2O + e^- \rightarrow H^- + OH \tag{3.50}$$

$$H^- + e^- \rightarrow H + 2e^- \tag{3.51}$$

$$H_2O + H^- \rightarrow H_2 + OH \tag{3.52}$$

$$H_2O + H^- + M \rightarrow H + H_2O + M^- \tag{3.53}$$

$$OH + H_2 \rightarrow H + H_2O \tag{3.54}$$

where the latter reaction (Equation 3.54) decreases the yield of H_2. Irrespective of the actual reaction sequence, the problem with this approach appears to be the requirement for a high ionization degree, which is difficult to attain. However, CO has been found to be a catalyst in the direct production of hydrogen from water by plasmas since it mitigates the need for high degrees of ionization and reduces the OH free radical concentration, which is a main problem in the second reaction sequence given earlier. The following reaction sequence is proposed by Bochin (Bochin et al. 1978):

$$COs^* + CO2'' \rightarrow CO + O + COs \tag{3.55}$$

$$O + COs^* \rightarrow CO + O_5 \tag{3.56}$$

$$O + H_2O^* \rightarrow {}^*OH + OH \tag{3.57}$$

$$OH + CO \rightarrow H + CO_2 \tag{3.58}$$

$$H + H_2O^* \rightarrow H_2 + OH \tag{3.59}$$

The efficiency has been projected at up to 50% for H_2O plasmolysis without the CO_2 catalysts and at about 80% with the CO_2 catalyst. (Figure 3.7b) The latter is the more attractive of the two because of the lower degree of ionization required. Givotov et al. (1981) have attempted to produce hydrogen by plasmolysis of water in one stage and in two stages with CO_2 reduction. Although they have not succeeded very well with the one-stage plasmolysis of water, the two-stage process is projected for 4 kWh per cubic meter of hydrogen production. But at present, the experimental verification of this value is needed. Plasma technology can be applied directly to CO_2, an abundantly available compound, to produce CO and O~, which upon separation could be doubly useful because it not only reacts with water to form hydrogen and CO_2 but also could form methanol and eventually other substances under different conditions. Thus, satisfactory and economic plasmolysis of CO_2 could be on the way toward the foundation of an organic chemistry in place of natural products. This process has the following limitations: it utilizes electric power, cost-effective upon a large-scale hydrogen production, and to date hydrogen yield and efficiencies are not high enough for practical application.

Chen et al. (1998, 2001) reported on water splitting conducted in a tubular reactor at atmospheric pressure using plasma and catalyst integrated technologies (PACT). The tubular PACT reactor consisted of a quartz tube fitted with an outer electrode and an inner electrode passing through the center of the tube. The inner electrode was coated with a catalytic material like Ni, Pd, Rh, or Au, and the outer electrode was made of aluminum foil. Plasma was created in the quartz tube using a low frequency (8.1 kHz) electric field applied between the electrodes. Argon bubbled through water was used as the feed gas, containing 2.3 mol% water. Maximum hydrogen production was achieved using a gold catalyst. At a voltage of about 2.5 kV, 14.2% of the water in the feed was converted into hydrogen (0.32 mol% H_2).

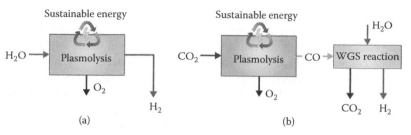

FIGURE 3.7 Production of hydrogen by plasmolysis of (a) water and (b) CO_2 and water. (From Hensen, E. and van de Sanden, R., *Solar Fuels: The Scientific Challenges*, University of Technology, Eindhoven, the Netherlands, 2014, https://static.tue.nl/.../2014_04_17_Workshop1_Sanden_Richard.)

3.7 MAGNETOLYSIS OF WATER

In magnetolysis, magnetic induction is used to create the required voltage inside the electrolyzer, using a homopolar generator. In 1985, Bockris et al. suggested this technique to eliminate the technical problems in supplying the low voltage, high current needed for electrolyzers. An electrolyzer unit operated at voltages between 1.5 and 2.0 V d.c., requiring currents of several hundred amperes. To meet these electrical requirements, transformers and rectifiers were needed, which raised the capital expenditures and resulted in losses in electrical energy. That made the electrolyzers increasingly commercially viable. Ghoroghchian and Bockris (1985) used a stainless steel disc of 30.48 cm diameter and 0.31 cm thickness that was connected to a bearing isolated shaft; the disk was mounted vertically in an electrolytic cell containing a 35% potassium hydroxide electrolyte. The whole assembly was placed within a magnetic field that was generated by an electromagnet, and the disk was rotated using an electric motor. Hence, a potential difference was created across the disk with the center acting as the anode and the rim as the cathode. Under a magnetic field of 0.86 T and at 2100 rpm, a potential difference of 2 V was created that resulted in hydrogen generation via electrolysis. The disk in such a configuration suffers viscous drag. To overcome this difficulty, the magnetic field strength should be as large as possible, or the electrolyte should also be rotated so as to avoid any relative movement between the disk and the electrolyte. It was estimated that for producing hydrogen at minimum power, a magnetic field consumption of greater than 11 T was required. The advances made over the past several decades to improve the performance of electronic circuitry, for example, rectifiers, have made electrolyzers increasingly commercially viable. Furthermore, in advanced electrolyzers a series cell configuration is used (bipolar filter press, solid polymer electrolyte [SPE], etc.) and hence there is no need to work in low-voltage, high-current mode with its inherent $I^2 R$ electrical losses. Consequently, the dormant magnetolysis field has a good chance of remaining dormant.

3.8 RADIOLYSIS OF WATER

Radiolysis of water is the dissociation of the water molecules by nuclear radiation. It is the cleavage of one or several chemical bonds resulting from exposure to high-energy flux in the form of high-energy radioactive γ rays or neutrons or charged particles like α and β. The type of radiation decided the fate of the products of radiolysis of water. For example, water dissociates under alpha radiation into a hydrogen radical and a hydroxyl radical, unlike ionization of water, which produces a hydrogen ion and a hydroxide ion. The chemistry of concentrated solutions under ionizing radiation is extremely complex. Water splitting using radioactive rays came into notice in the 1950s, when water was used as a coolant in a nuclear reactor, where radiation from the reactor core can decompose coolant water into hydrogen, hydrogen peroxide, and oxygen, in addition to giving rise to corrosion problems. When water is irradiated by emissions from radioactive materials, a number of reactions leading to the production of a variety of species take place depending upon the nature and energy of the radiation. As the high-energy particles/radiations traverse the water, they lose energy and eject electrons from the atomic shells. These high-energy electrons create low-energy secondary electrons and help to initiate further reactions. Primary products of the radiolysis of water form the following chemical species: e_{aq}^{-}, $HO^{\cdot}$, $H^{\cdot}$, $HO_2^{\cdot}$, H_3O^{+}, OH^{-}, H_2O_2, and H_2 (Spinks and Woods 1990; Le Caër 2011). The effectiveness of interaction between a given type of radiation and water in yielding a particular species is represented by parameter G, the radiochemical or radiological yield. It is defined as the number of species created per unit of deposited energy (~100 eV as the unit). It depends upon the energy transfer to the medium, called linear energy transfer (LET), from a given type of radiation. Water decomposition is observed mainly by the alpha particles, which can be entirely absorbed by very thin layers of water. The yield of hydrogen resulting from the irradiation of water with β and γ radiation is low (G values < 1 molecule per 100 eV of absorbed energy) but this is largely due to the rapid reassociation of the species arising during the initial radiolysis. If impurities are present or if physical conditions are created that prevent the establishment of a chemical equilibrium, the net production of hydrogen can be greatly enhanced. The yield is higher when energetic particles are used, with a value close to 1 when pure water is irradiated at room temperature (Katsumura et al. 1998). The radiolysis of pure steam performed using Cm-244 alpha-particles gave a yield of eight molecules per 100 eV at 300°C, whereas the yield was only two molecules per 100 eV at 250°C (Cecal et al. 2001). The hydrogen yield is limited due to the back reactions between H_2 and OH. Although the G-value for hydrogen when pure water is irradiated by γ radiation is low, studies indicate that irradiation of water in the presence of solid materials improves the yield (Sawasaki et al. 2003). Although γ radiation has a lower G-value than α radiation, the high activity of [137m]Ba and long half-distances cause γ decay to be the principal cause of radiolytic H_2 and H_2O_2 production (Figure 3.8a–d) (Dzaugis et al. 2015).

LaVerne and Tandon (2002) reported on irradiating water adsorbed on micron-sized particles of CeO_2 and ZrO_2 with γ radiation (dose rate 202 Gy/min) or 5 MeV α particles. The yield was found to dramatically increase to 150 molecules/100 eV for ZrO_2, irradiated with γ rays, when the number of adsorbed layers of water was limited to one or two. However, this yield was calculated using the energy adsorbed by the water layers alone and not using the total energy absorbed by both oxide and water. Cecal et al. (2001), from their study using oxides such as ZrO_2, TiO_2, BeO_2, or SiO_2 dispersed in water, found that yield was highest for ZrO_2. Caer et al. (2005)

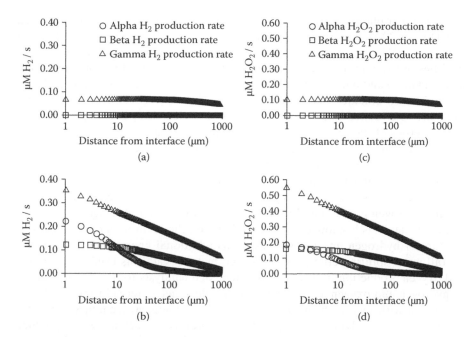

FIGURE 3.8 Radiolytic production rates for H_2 (a) before barrier failure and (b) after barrier failure. Production rates for H_2O_2 (c) before barrier failure and (d) after barrier failure. All radiolytic production rates are for 20-year-old radioactive fuel. Distance from pellet is plotted on a log scale. Production rates from alpha and beta radiation in (a) and (c) are both 0 μM/s. (From Dzaugis, M.E. et al., *Rad. Phys. Chem.*, 115, 127–134, 2015.)

also studied the effect of pore size on yield by irradiating nanoporous SiO_2 with 1 MeV electrons; an increase in yield with a reduction in pore size was observed. In general, electrons and holes formed inside the oxide that then migrate to the surface are believed responsible for the water splitting. The of water is not common due to the possibility of the radioactive species contaminating products, and at this stage the technique does not have the potential to compete.

3.9 PHOTOCATALYTIC WATER SPLITTING

Natural photosynthesis in a plant was suggested as a cost-effective indispensable route for commercialization of solar water splitting. Researchers have devoted efforts to mimic these energy reactions since 1972 (Fujishima and Honda 1972). We discussed the aspects of photocatalytic water splitting in Chapter 2 and in this chapter we brief the technique. When a semiconductor is suspended into water, oxidation and reduction reactions of water are performed at the interfacial region without any electrical work done, because the oxidation and reduction reactions are short circuited. In addition, both the conduction and valence band positions of the particulate material should be well matched with the hydrogen and oxygen evolution potential to split water (i.e., overall water splitting). Here, these microscale particles (photochemical diodes as Schottky-type devices) can catalyze water photolysis when their suspensions are exposed to the light (Maeda and Domen 2007; Rajeshwar 2007; Osterloh 2008; Maeda, 2011; Osterloh and Parkinson 2011). Besides the stability of the suspended particles, two additional qualities are required by the particulate, that is, catalysis (for water reduction and oxidation) and photovoltaic (PV) (light absorption, charge separation, charge transport) that enhance the complexity of the device (Figure 3.9a). These functions are interrelated and jointly depend on the electrical, chemical, and optical properties of the semiconductors. The water-splitting reaction is significantly more endergonic, mechanistically complex, kinetically slower, and chemically more corrosive than the electron transfer-"only" processes. Although the PV electrolysis of water is highly efficient, the cost of hydrogen made by this route is quite high. Presently, efficiency and cost are the main challenges in the path scaling hydrogen production. In comparison with PV water cleavage, the particulate photocatalysis is a low-cost system. However, the separation of the explosive mixture of hydrogen and oxygen gas needs additional energy that reduces the overall water-splitting efficiency. Moreover, metal cocatalysts used for hydrogen and oxygen production from water are often good catalysts for the recombination of hydrogen and oxygen as well, which can reduce net water-splitting efficiency. Compared with thermochemical, PEC, and photobiological techniques of water splitting, photocatalytic water splitting is a promising technology to produce "clean" hydrogen. However, if this process is assisted by suspended photocatalysts directly in water instead of using PV and an electrolytic system, the reaction is in just one step; therefore, it can be more efficient. It has the following advantages: (1) reasonable solar-to-hydrogen (STH) efficiency; (2) low process cost; (3) the ability to achieve separate hydrogen and oxygen evolution during the reaction;

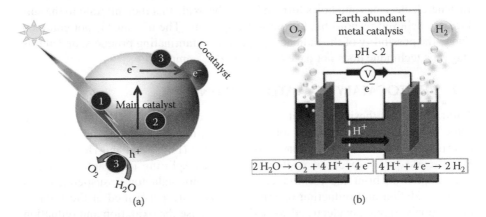

(a)   (b)

FIGURE 3.9 Illustration of the operating principle of the (a) photocatalytic (1, light absorption; 2, charge separation; and 3, charge transport) and (b) photoelectrochemical (PEC) production of hydrogen and oxygen by water splitting. (From Bloor, L.G. et al. *J. Am. Chem. Soc.*, 136(8), 3304–3311, 2014.)

and (4) small reactor systems suitable for household applications, thus providing for huge market potential.

3.10 PHOTOELECTROCATALYTIC WATER SPLITTING

The first recorded observations of a PEC phenomenon were made by Henri Becquerel (Becquerel 1839), who reported the photocurrent and photovoltage produced by sunlight acting on silver chloride-coated platinum electrodes in various electrolytes. Pure water absorbs only the infrared region of the solar radiation. In this region, photon energies are too low to simulate photochemical reactions. Therefore, water splitting is derived using solar and electrical energy support in the presence of semiconductors with small band gaps (1.6–2.5 eV) with apt band positions or wide band gap semiconductors with a sensitizer or cocatalysts or both; the molecule or semiconductor is able to absorb sunlight and stimulate the photochemical reactions needed to drive the decomposition of water and lead to the generation of hydrogen and oxygen gases. During the photoelectrolysis of water, sunlight is absorbed by the isolated fabricated molecules (semiconductors) or biogenic materials (chloroplast or algae in a configuration coupled to a hydrogen-generating enzyme) or a hybrid of previous systems, in an aqueous electrolyte. These molecules are semiconductors or enzymes, either as a suspended particle in the water or as a macroscopic electrode unit in a PV cell or an electrochemical cell that is dipped in the aqueous electrolytes, used as an electron–hole relay. If we are in search of the apt photocatalytic material then we are really looking for characteristics such as high efficiency, good durability, or stability against photocorrosion, apt band gap, good lifetime of charge carriers, low concentration of defects, free path for charge carriers, low cost, and so on. It is really a challenge to find this in the single material. Besides an integrated device, high-tech systems are required for efficient photocatalytic water splitting or hydrogen production. Research and development activities on PEC materials, devices, and systems have made important strides in developing a strong synergy with contemporary research efforts in PVs, nanotechnologies, and computational materials. For commercial hydrogen production, the Nuclear Energy Research Initiative (NERI) set the PEC hydrogen production goals as follows: 15% conversion efficiency, 900-hour replacement lifetime (1/2 year at 20% capacity factor), and ultimately synthesis via high-volume manufacturing techniques with a final hydrogen production cost of $2.10/kg and $300/m² PEC electrode cost (Deutsch et al. 2013).

Photoelectrocatalytic water splitting involves the conversion of light energy into a more useful energy product: electrical or chemical or both. The topic is well described in Chapter 4. Photoelectrocatalytic water-splitting device on absorption of solar light by one or more of the electrodes (photoanode and photocathode; at least one of the electrodes is a semiconductor) dipped in aqueous electrolyte that produce hydrogen and oxygen splitting of water. In principle, an internal electric field generated at the semiconductor–electrolyte interface uses the efficient separation of the photogenerated electron–hole pairs (excitons). Subsequently, photogenerated holes and electrons are responsible for oxidation and reduction of water, at the anode and cathode, respectively, for O_2 and H_2 production, as shown in Figure 3.9b.

3.10.1 Types of PEC Devices

There are six main types of PEC devices.

3.10.1.1 Direct PEC or Photosynthetic Cells

Photosynthetic cells (light absorption, charge separation, water electrolysis) operate on a similar principle except that there are two redox systems: one reacting with the holes at the surface of the semiconductor photoanode to oxidize the water to oxygen and the second reacting with the electrons entering the cathode for reduction of the water to hydrogen. The overall reaction is the cleavage of water by sunlight. Although cheap basic materials, low processing cost, and simple systems are the main advantages of this system, its three major challenges to increase STH efficiency are as follows:

1. Direct absorber band alignment with appropriate potential to both half reactions is required. Although such an alignment is difficult to achieve in a single material initially, any change in band alignment caused by changing surface conditions can result in efficiency degradation. This makes it difficult to design devices that maintain robust, high efficiencies in actual operation.
2. The wide band gap of a stable absorber (>1.23 eV; typically >1.6 eV) cannot drive the water-splitting reaction at the optimum visible solar spectrum, resulting in a maximum solar to fuel efficiency (SFE) of only 7% (Bolton et al. 1985; Weber and Dignam 1986).
3. The absorbers are not good catalysts and are incompetent to perform four proton-coupled electron transfer chemistry (Cukier and Nocera 1998; Concepcion et al. 2009) for water splitting.
4. Expensive with respect to the reactor design due to the use of noble metal cocatalysts and/or low efficiency of the devices.

These deficiencies can be overcome by substituting a PEC device with a multiple-junction PV and an electrochemical catalyst (EC), PV–EC tandem (Miller et al. 2003) materials, and use of an apt combination of homogeneous and heterogeneous catalysis is required to create the desired synergy. Here, the electrical field is generated at an internal junction within the semiconductor and coupled with water-splitting catalysts through ohmic contacts.

3.10.1.2 Biased PEC Devices

The bias-based cells can serve external or internal drives to perform electrolytic reaction, which is not possible in the presence of photon energy alone.

1. Electrical or grid-biased PEC devices
2. pH-biased PEC devices

3. Chemical-biased PEC devices
4. Dye-sensitized bias PEC devices
5. PV-biased PEC devices

3.10.1.3 PV Cell

This cell involves a solid-state p–n or Schottky junction to produce the required internal electric field for efficient charge separation and the production of a photovoltage sufficient to decompose water in a commercial-type water electrolyzer.

3.10.1.3 PV Electrolysis Cell or Regenerative Cell

An alternative system to PV cells that involves the semiconductor PV cell configured as a monolithic structure and immersed directly in the aqueous solution. Regenerative cells have focused on electron-doped (n-type) II/VI or III/V semiconductors using electrolytes based on sulfide/polysulfide, vanadium(II)/vanadium(III) or I_2/I^- redox couples. Conversion efficiencies of these systems go up to 19.6% (Licht 2001).

3.10.1.4 Photogalvanic/Concentration Cells

Photogalvanic cells are used to convert solar energy into electrical energy. In this type of PEC system, the solar energy is converted to both electrical and electrochemically stored energy. And the decrease in solar illumination of the cell switches the spur-of-the-moment regeneration of electrochemical energy to electrical energy, providing nearly constant (light-insensitive) electrical power for an external applied load. Here, for instance, two inert electrodes are dipped into an electrolyte with a dye solution and the light is absorbed by the electrolyte. In this case, an electron is transferred between the excited dye molecules and electron donor or acceptor molecules added to the electrolyte. Under exposure of light, the storage compartment spontaneously delivers power by metal oxidation ($Sn \rightarrow Sn^{2+}$) (Licht and Manassen 1987). The typical example of this category is a cell containing a thin-film Cd(Se,Te) photoelectrode and a thin-film CoS counter electrode immersed in an aqueous polysulfide electrolyte separated by cation-selective membrane from a tin electrode immersed in an aqueous sulfide electrolyte (Sn/S⁻). Cyclic voltammetry study indicates that unlike tin/alkali systems, the Sn/S⁻ half-cell does not undergo passivation at storage potentials. Measured over-potentials in Sn/S⁻ indicate large relative surface area tin electrodes lead to minimized polarization losses. Detailed description of a variety of membranes indicates that a sulfonated polyethylene membrane minimizes problems of polysulfide passage into the tin/aqueous sulfide half-cell. A degree of compatibility of photo and storage cells can be achieved by electrolyte modification and employing two or more photocells in series per tin half-cell. More than 90% efficiency can be achieved for these systems that can be limited by photo and not storage conversion efficiency; their 2-week outdoor operational solar to electrical efficiency (conversion plus storage) is 2.7%.

The basic difference between PEC solar cells and PEC water splitting is that in the former case electrical energy (as free energy) is produced but the net gain is zero, whereas in PEC water splitting there is a net gain (from hydrogen) with electrical energy generation. In both cases, free energy appears as the photovoltage between the electrodes that derives the electric charges through the circuit or supplies carriers under illumination. The fundamental problem in hydrogen production by water electrolysis is that today the electricity used to drive the process is primarily generated by the burning of fossil fuels.

In PEC cells, a light-absorbing semiconductor is used as either the anode or the cathode (or both) in an electrochemical cell. Fujishima and Honda (1972) used a single crystal of titanium dioxide as the photoanode and electrons released from the anode were directed through a wire to a Pt electrode, from which hydrogen evolved. Photocatalyst activity of the PEC is influenced by several factors including structure, particle size, surface properties, preparation, spectral activation, and resistance to mechanical stresses. The most efficient cells (~13%) are reported to be those involving a p-InP photocathode, onto which tiny pieces of Pt have been deposited. State-of-the-art PEC solar water-splitting cells are reviewed in light of their STH conversion efficiency (<1%–18%). These devices have been reported with PV components based on III–V semiconductors (by varying the number and type of PV junctions) and demonstrated >10% STH efficiency by using potentially less costly materials. The highest reported efficiencies were 12.4% and 18% for cells that, respectively, do and do not contain a semiconductor–liquid junction (Ager et al. 2015) (Figure 3.10). Device stability and cost are two major battlefields, which can be conquered by evolving a dynamic and collaborative work force to solve the puzzle of the fundamental efficient material for hydrogen production from water, by using cutting-edge experimental and theoretical methods and technoeconomic and life cycle assessments. There is an immediate need for adopting a globally accepted protocol for evaluating and certifying STH efficiencies and lifetimes, which exists to compare the efficiency on equal grounds. It is our recommendation that a protocol similar to that used by the PV community be adopted so that future demonstrations of solar PEC water splitting can be compared on equal ground. The whole exercise is done to achieve the DOE dream PEC device with the target of 1000 h life, 10%–25% STH efficiency, and 2–3 kg/\$ H_2 production (Deutsch et al. 2013).

Heterostructures consisting of a wide range of metal oxide combinations have been reported to facilitate charge carrier separations at their heterojunctions for efficient PEC water oxidation. A few with moderate efficiency are In_2O_3–TiO_2 and WSe_2 (Prasad et al. 1989; Licht et al. 2001), RuO_2-catalyzed AlGaAs/Si (Licht et al. 2000; Peharz et al. 2007), TiO_2/α-Fe_2O_3 (Kuang et al. 2009), α-$Fe_2O_3/SrTiO_3$ (Wang et al. 2007), $Fe_2O_3/ZnFe_2O_4$ (McDonald and Choi 2011), $TiO_2/ZnFe_2O_4$ (Yin et al. 2007), TiO_2/WO_3 (Smith et al. 2011; Wang et al. 2011), $WO_3/BiVO_4$ (Hong et al. 2011; Su et al. 2011), and $TaON/CaFe_2O_4$ (Kim et al. 2013). The metal oxide combination with proper positioning of conduction and valence band edges for potential gradients at the interfaces may lead to enhanced charge separation and inhibit recombination, and their overall PEC efficiencies increase compared with the individual single component case. Going beyond inorganic oxide PEC materials, Katakis et al. reported a molecular photochemical system in 1992 and 1994, which used to complete dissociate the water, where tris-[1-(4-methoxyphenyl)-2-phenyl-1,2-ethyleno-dithiolenic-S,S] tungsten was used as a sensitizer and MV^{2+} as an electron relay.

Finally, we can say photocatalytic water splitting is a cross-discipline technology that requires the involvement of experts from different fields (e.g., chemists, biologist, electrical engineers, material scientists, and physicists). A joint effort is needed to explore potential semiconductor materials and reactor systems that will

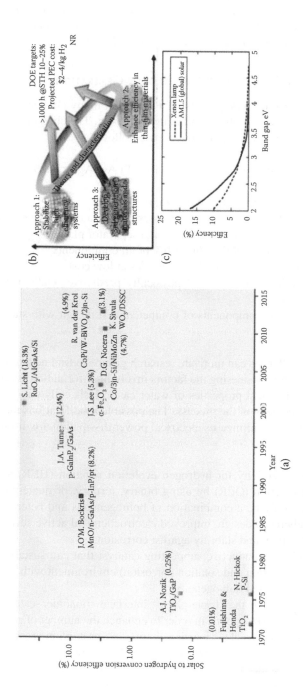

FIGURE 3.10 (a) Yearwise development in solar to hydrogen conversion efficiency vector since the discovery of Fujishima and Honda, (b) three approaches to achieve efficiency of 10%–25% at projected cost 2-4Kg H2 and (c) efficiency of PEC hydrogen generation, depending upon semiconductor band gap, under xenon arc lamp, and AM1.5 solar illuminations. (From Ager, J.W. et al. *Energy Environ. Sci.*, 8, 2811–2824, 2015; Park, S.-Y., Photoelectrochemical Water Splitting Optimizing Interfaces and Light Absorption, Ph.D. Thesis, University of Twente, the Netherlands, 2015, ISBN: 978-94-6233-082-5, doi:10.3990/1.9789462330825.)

generate the highest STH efficiency. The development of new technologies requires collaboration with a strong theoretical background for a better understanding of the hydrogen production mechanism in order to come up with a low-cost and environmentally friendly water-splitting process for hydrogen production.

3.10.2 CHALLENGES AND FUTURE OF PEC HYDROGEN GENERATION

In the past 50 years of sincere investigations on PEC cells, many challenges were recognized and only a few were resolved, and many still exist. The chief challenges observed with PEM electrolysis can be summarized as follows:

- Reduction and/or substitution of noble (expensive Pt Au, Ag, Ir, etc.) metals used in catalysis
- Increased recyclability of the catalyst utilization
- Development of the low-cost and corrosion-resistant material for current collectors, electrode materials, electrolytes, and separator plates
- Improvement of long-term stability/durability of all components
- Improvement of overall membrane characteristics at low cost
- Development of both empirically and physically predictive relations for operating parameters
- Development of various components of competent cell design with stacks concepts

Nevertheless, these challenges can motivate research groups to find new research directions or perspectives by considering the factors involving information about the physical, chemical, and electrical properties of water electrolysis cells with a view on their effects on the efficiency of the process. The possible modifications that can be done on a cell in order to minimize its electrical power dissipation are discussed in the following:

- Improvised catalytic activity for hydrogen evolution reaction (HER) and oxygen evolution reaction (OER) by using binary, ternary, or quaternary alloys, molecular catalysts, a combination of homogeneous and heterogeneous catalysts, advanced design, improved electrochemical active surface area, catalyst utilization, and stability against corrosion.
- Development of highly conductive supporting catalyst that can sustain in a corrosive (photocorrosion and solution corrosion) environment with high nanoparticle dispersion and homogeneity.
- Better understanding of the triple-phase interface (ionomer–catalyst, ionomer–support, catalyst–support) in order to enhance the number of active sites by improving nanoparticle dispersion, improving proton/ion crossover through the catalytic layer, decreasing nanoparticle hindering, and diminishing the electronic resistance provided by the ionomer. Also it is important to understand water transport across the triple-phase boundary.
- For the anode, finding catalyst alternatives to replace the short supply of iridium or unstable ruthenium will be considered a great achievement. New

catalyst configurations with engineered structures (e.g., core–shells, Bulk metallic glasses (BMGs), Nano tube- structured films (NTSFs), nanostructures, tuned alloys) and morphology could provide the necessary conditions to decrease the use of iridium or stabilize ruthenium dissolution over time.

- For the cathode, improving catalyst stability (especially when supported on nonmetal/carbon materials), exploring alternative supports other than carbon, and investigating metal-free high-surface area carbon materials (carbon blacks, carbon nanotubes [CNTs], graphenes) by controlling their pore size, functional groups, grafted polymers, and electrical conductivities for the purpose of achieving higher activity and stability.

- The utility of innovative synthesis methods (supercritical liquid, ionic liquids, phase transfer catalyst, and so on) to produce new support materials, catalysts, and electrode systems.

- Use of new advanced physical and electrochemical characterization tools, together with theoretical calculations (molecular/electronic level) to deeply understand the mechanisms for water-splitting reactions (both OER and HER) and their cozy relationship with the cell components and structures to enhance adaptability and raise performance levels.

- Development of membrane alternatives to Nafion with advanced membrane synthesis methods, resulting in electrolytes with higher proton transport but at the same time providing lower gas crossover and higher durability. This could be done using membrane composites or blends, adding inorganic or organic fillers, or introducing molecular barriers to the electrolyte.

- Development of low-cost current collectors with tuned porous structure, high corrosion resistance, low ohmic resistance, and optimized mass transport. The interface between the catalyst layer and the current collector must also be improved by developing a microporous layer that could better integrate the two layers in terms of mass and electron transport. This could be accomplished using nanostructured fillers or dopers with high electron conductivity and high corrosion resistance.

- Ti separator plates could be replaced using lower cost materials (e.g., copper, graphite, stainless steel) coated with high electron transport and high corrosion resistance materials.

- Modeling the multiphase transport of the species through the current collectors and separator plates. This could prove to be very beneficial in designing the current collectors, especially for larger-scale electrolyzers.

- Development of a predictive model for the exchange current density for various catalysts. Although this herculean task is very demanding, if successfully accomplished, modeling would prove to be a much more useful designing tool, capable of aiding in the design of all the individual components.

- High temperature and pressure may enhance the efficiency of the system but these parameters can be raised in looking to the physical strength of the apparatus.

- Utilization of highly concentrated electrolytes will result in lower impedance values. On the other hand, the use of contaminated solutions causes side reactions to take place in the cell, which will reduce the lifespan of the apparatus.

*   Placing the electrodes in an absolutely vertical position at a certain distance from each other will reduce the ohmic resistance between them. Moreover, the utilization of metals with high degrees of activity and porous materials will enhance electric and ionic conductivity of the system.
*   Compelling the gas (H_2 and O_2) bubbles to detach from the electrode and separator surfaces and to leave the system will decrease void fraction formation. It also lowers the impedance and increases efficiency levels.
*   Selecting the electrodes of larger sizes is an important factor to enlarge their overlap surface area and therefore the current path. However, since gas bubbles will accumulate in higher altitudes of the cell, certain considerations are recommended to limit the height of the electrode plates.
*   Since the equivalent circuit of a water electrolysis cell contains nonlinear elements, more research is required to study the best method of power application to a cell.

On the basis of earlier discussions, we can say that the field is full of numerous possibilities. The focus is on PEM electrolysis as a whole and we need in-depth knowledge of catalysts, membranes, electrolytes, current collectors, separator plates, effects of poisoning, fabrication methods for various components, and modeling approaches that can direct the future goal-oriented research in the direction of developing the ultimate PEM electrolyzers as a reliable, cost-effective solution to solve the issues related with the carbon-free renewable energy generation.

3.11 SUMMARY

Unlike downhill reactions, oxidation of organic compounds into CO_2 and H_2O, with large negative change in the Gibbs free energy ($\Delta G < 0$), photosynthesis ($\Delta G > 480.0$ kJ/mol) or artificial photosynthesis, that is, water splitting ($\Delta G > 237.2$ kJ/mol) are energy uphill reactions because they produce reactive materials: hydrogen and oxygen, which are set to form water by utilizing less energy from outside. Therefore, the efficiency of the water cleavage process would decrease because of these ready to recombine products. In this chapter, we discuss the basic concepts of the water cleavage process and various techniques that are used for water splitting: electrolytic, biophotocatalytic, thermochemical, mechanocatalytic, magnetolytic, plasmolytic, radiolytic, photocatalytic, and photoelectrocatalytic processes. All methods were discussed in detail with their own merits and demerits. Among them, the electrolysis of water is the most common practice for hydrogen generation and is available at a very competitive cost with gasoline. But it needs electricity, which is often created from coal. The need of electricity is further reduced by the supply of solar energy and an apt photocatalytic material to the water-splitting device. Biophotocatalytic cleavage of water involves solar energy–supported biocatalytic materials. Thermochemical water splitting requires a very a high-temperature heat (~2500°C) to empower the series of chemical reactions (autocatalytic) that creates a closed loop (cycles) for producing hydrogen and oxygen gases by consuming water and heat without utilizing electricity. If we utilize solar energy as a heat source, then we can deliver overall system efficiencies of more than 40%. Mechanocatalytic splitting of water is based on the principle of triboelectricity, where mechanical energy is directly converted into chemical energy in the presence of powdered metal oxides and agitated Teflon-coated magnetic stirrer rods of different shapes. To solve the technical problems of low voltage, high-current consumption of water electrolyzers, magnetic induction is used to create the required voltage inside the electrolyzer. Plasmolysis and radiolysis of water need plasma (by transformation of the energy from an electric to kinetic energy of the electrons, which is finally transmitted into molecular excitations) and radiation (α, β, and γ rays) energy, respectively, to break water. Photocatalytic splitting of water needs one smart photocatalyst along with solar irradiation. Techniques for the PEC splitting of water are also discussed in detail. This is an energy-saving and environmentally friendly technique with a lot of potential for energy production. Future challenges and perspectives on all the aforementioned concepts for the whole concept of the water splitting are the major concerns of this chapter.

REFERENCES

Ager, J. W., M. R. Shaner, K. A. Walczak, I. D. Sharp, and S. Ardo. 2015. Experimental demonstrations of spontaneous, solar-driven photoelectrochemical water splitting. *Energy Environ. Sci.* 8: 2811–24.

Bi, L., S. P. Shafi, and E. Traversa. 2015. Zirconia (ScSZ), Y-doped $BaZrO_3$ as a chemically stable electrolyte for proton-conducting solid oxide electrolysis cells (SOECs). *J. Mater. Chem. A* 3: 5815–19.

Bloor, L. G., P. I. Molina, M. D. Symes, and L. Cronin. 2014. Low pH electrolytic water splitting using earth-abundant metastable catalysts that self-assemble in situ. *J. Am. Chem. Soc.* 136 (8): 3304–11.

Bocanegra-Bernal, M. H. and S. D. De la Torre. 2002. Phase transitions in zirconium dioxide and related materials for high performance engineering ceramics. *J. Mater. Sci.* 37: 4947–71.

Bochin, V. P., A. A. Fridman, V. A. Legasov, V. D. Rusanov, and G. V. Sholin. 1978. *Hydrogen Energy Systems* (eds T. N. Veziroglu and W. Seifritz), Vol. 3. London, UK: Pergamon Press, p. 1183.

Bockris, J. O'M. 1975. *Methods for the Large-Scale Production of Hydrogen from Water of Hydrogen Energy*. New York, NY: Springer US, Plenum Press, pp. 371–443.

Bockris, J. O. M., B. Dandapani, D. Cocke, and J. Ghoroghchian. 1985. On the splitting of water. *Int. J. Hydrogen Energy* 10: 179–201.

Bolton, J. R., S. J. Strickler, and J. S. Connolly. 1985. Limiting and realizable efficiencies of solar photolysis of water. *Nature* 316: 495–500.

Caer, S. L., P. Rotureau, F. Brunet, et al. 2005. Radiolysis of confined water: Hydrogen production at a high dose rate. *Chem. Phys. Chem.* 6: 2585–96.

Carmo, M., D. Fritz, J. Mergel, and D. Stolten. 2013. A comprehensive review on PEM water electrolysis. *Int. J. Hydrogen Energy* 38(12): 4901–34.

Cecal, A., M. Goanta, M. Palamaru, et al. 2001. Use of some oxides in radiolytical decomposition of water. *Rad. Phys. Chem.* 62: 333–36.

Chauveau, F., J. Mougin, J. M. Bassat, F. Mauvy, and J. C. Grenier. 2010. A new anode material for solid oxide electrolyser: The neodymium nickelate. *J. Power Sources* 195: 744–49.

Chen, H., Y. Chen, H.-T. Hsieh, and N. Siegel. 2007. CFD modeling of gas particle flow within a solid particle solar receiver. *ASME J. Sol. Energy Eng.* 129: 160–70.

Chen, K., N. Ai, and S. P. Jiang. 2010. Development of (Gd, Ce) O_2-impregnated (La, Sr) MnO_3 anodes of high temperature solid oxide electrolysis cells. *J. Electrochem. Soc.* 157: P89–94.

Chen, S., K. Xie, D. Dong, et al. 2015. A composite cathode based on scandium-doped chromate for direct high-temperature steam electrolysis in a symmetric solid oxide electrolyzer. *J. Power Sources* 274: 718–29.

Chen, X., M. Marquez, J. Rozak, et al. 1998. H_2O splitting in tubular plasma reactors. *J. Catal.* 178: 372–77.

Chen, X., S. L. Suib, Y. Hayashi, and H. Matsumoto. 2001. H_2O splitting in tubular PACT (plasma and catalyst integrated technologies) reactors. *J. Catal.* 201: 198–205.

Chettiar, M. 2004. Co-production of Hydrogen and Sulfuric Acid by Electrolysis. Ph.D. Thesis and Dissertations, University of South Florida, FL.

Concepcion, J. J., J. W. Jurss, M. K. Brennaman, et al. 2009. Making oxygen with ruthenium complexes. *Acc. Chem. Res.* 42(12): 1954–65.

Cukier, R. I. and D. G. Nocera. 1998. Proton-coupled electron transfer. *Annu. Rev. Phys. Chem.* 49: 337–69.

De Silva Munoz, L., A. Bergel, D. Féron, and R. Basséguy. 2010. Hydrogen production by electrolysis of a phosphate solution on a stainless steel cathode. *Int. J. Hydrogen Energy* 35: 8561–68.

Deniele, S., I. Lavagnini, M. A. Baldo, and F. Magno. 1996. Steady state voltammetry at microelectrodes for the hydrogen evolution from strong and weak acids under pseudo-first and second order kinetic conditions. *J. Electroanal. Chem.* 404: 105–11.

Deutsch, T., J. Turner, H. Wang, and H. Dinh. 2013. 2013-Annual Progress Report, National Renewable Energy Laboratory (NREL), accessed 9 July, 2013 pp. II-70–II-74, Denver West Parkway Golden, CO. Available at: https://www.hydrogen.energy.gov/pdfs/progress13/ii_c_1_deutsch_2013.pdf.

Dzaugis, M. E., A. J. Spivack, and S. D'Hondt. 2015. A quantitative model of water radiolysis and chemical production rates near radionuclide-containing solids. *Radiat. Phys. Chem.* 115: 127–34.

Ferrandon, M. S., M. A. Lewis, and D. F. Tatterson. 2010. Hydrogen production by the Cu–Cl thermochemical cycle: Investigation of the key step of hydrolysing $CuCl_2$ to Cu_2OCl_2 and HCl using a spray reactor. *Int. J. Hydrogen Energy* 35(3): 992–1000.

Fujishima, A. and K. Honda. 1972. Electrochemical photolysis of water at a semiconductor electrode. *Nature* 238: 37–38.

Ghoroghchian, J. and J. O. M. Bockris. 1985. Use of a homopolar generator in hydrogen production from water. *Int. J. Hydrogen Energy* 10: 101–12.

Givotov, V. K., A. A. Fridman, M. F. Frotov, et al. 1981. Plasmochemical methods of hydrogen production. *Int. J. Hydrogen Energy* 6: 441–49.

Gregory, D. P. 1975. Chapter: *Electrochemistry and the Hydrogen Economy*. Volume 10 of the series *Modern Aspects of Electrochemistry*. New York, NY: Springer US, Plenum Press, pp. 239–88.

Grimes, C. A., O. K. Varghese, and S. Ranjan. 2008. *Light, Water, Hydrogen: The Solar Generation of Hydrogen by Water Photoelectrolysis*. New York, NY: Springer, pp. 84–8.

Hara, M. and K. Domen. 2004. A study of mechano-catalysts for overall water splitting. *J Phys. Chem. B* 108: 19078.

Hensen, E. and R. van de Sanden. 2014. Solar Fuels: The Scientific Challenges. Eindhoven, the Netherlands: University of Technology. Available at: https://static.tue.nl/.../2014_04_17_Workshop1_Sanden_Richard.

Hong, S. J., S. Lee, J. S. Jang, and J. S. Lee, 2011. Heterojunction $BiVO_4/WO_3$ electrodes for enhanced photoactivity of water oxidation. *Energy Environ. Sci.* 4(5): 1781–87.

Igarashi, R. Y. and L. C. Seefeldt. 2003. Nitrogen fixation: The mechanism of the Mo-dependent nitrogenase. *Crit. Rev. Biochem. Mol. Biol.* 38: 351–84.

Ikeda, S., T. Takata, M. Komoda, et al. 1999. Mechanocatalysis–A novel method for overall water splitting. *Phys. Chem. Chem. Phys.* 1: 4485–91.

Ikeda, S., T. Takata, T. Kondo, et al. 1998. Mechano-catalytic overall water splitting. *Chem. Commun.* 20: 2185–86.

Ishihara, T., N. Jirathiwathanakul, and H. Zhong. 2010. Intermediate temperature solid oxide electrolysis cell using $LaGaO_3$ based perovskite electrolyte. *Energy Environ. Sci.* 3: 665–72.

Juda, W. and D. McL. Moulton. 1972. *Chem. Eng. Syrup. Ser.*: 59.

Katakis, D. F., C. Mitsopoulou, and E. Vrachnou. 1994. Photocatalytic splitting of water: Increase in conversion and energy storage efficiency. *J. Photochem. Photobiol. A Chem.* 81: 103–06.

Katakis, D. F., C. Mitsopoulou, J. Konstantatos, E. Vrachnou, and P. Falaras. 1992. Photocatalytic splitting of water. *J. Photochem. Photobiol. A Chem.* 68: 375–88.

Katsumura, Y., G. Sunaryo, D. Hiroishi, and K. Ishigure. 1998. Fast neutron radiolysis of water at elevated temperatures relevant to water chemistry. *Prog. Nucl. Energy* 32: 113–21.

Kim, E. S., N. Nishimura, G. Magesh, J. Y. Kim, et al. 2013. Fabrication of $CaFe_2O_4$/TaON heterojunction photoanode for photoelectrochemical water oxidation. *J. Am. Chem. Soc.* 135: 5375–83.

Kogan, A. 1998. Direct solar thermal splitting of water and on-site separation of the products–II. Experimental feasibility study. *Int. J. Hydrogen Energy* 23(9): 89–98.

Kolb, G. J., N. P. Siegel, and R. B. Diver. 2007. Central-station solar hydrogen power plant, *ASME J. Sol. Energy Eng.* 129: 179–83.

Kreuter, W and H. Hofmann. 1998. Electrolysis: The important energy transformer in a world of sustainable energy. *Int. J. Hydrogen Energy* 23(8): 661–66.

Kuang, S. Y., L. X. Yang, S. L. Luo, and Q. Y. Cai. 2009. Fabrication, characterization and photoelectrochemical properties of Fe_2O_3 modified TiO_2 nanotube arrays. *Appl. Surf. Sci.* 255(16): 7385.

Laguna-Bercero, M. A. 2012. Recent advances in high temperature electrolysis using solid oxide fuel cells: A review. *J. Power Sources* 203: 4–16.

LaVerne, J. A. and L. Tandon. 2002. H_2 produced in the radiolysis of water on CeO_2 and ZrO_2. *J. Phys. Chem. B* 106: 380–86.

Le Caër, S. 2011. Water radiolysis: Influence of oxide surfaces on H_2 production under ionizing radiation. *Water* 3: 235–53.

LeRoy, R. L., M. B. Janjua, R. Renaud, and U. Leuenberger. 1979. Analysis of time-variation effects in water electrolyzers. *J. Electrochemical Soc.* 126(10): 1674–82.

Licht, S. 2001. Multiple band gap semiconductor/electrolyte solar energy conversion. *J. Phys. Chem.* 105: 6281–94.

Licht, S., B. Wang, S. Mukerji, T. Soga, M. Umeno, and H. Tributsch, 2001. Over 18% solar energy conversion to generation of hydrogen fuel; theory and experiment for efficient solar water splitting. *Int. J. Hydrog. Energy* 26: 653–59.

Licht, S., B. Wang, S. Mukerji, T. Soga, M. Umeno, and H. Tributsch. 2000. Efficient solar water splitting, exemplified by RuO_2-catalyzed AlGaAs/Si photoelectrolysis. *J. Phys. Chem. B* 104: 8920–24.

Licht, S. and J. Manassen. 1987. Thin film cadmium chalcogenide/aqueous polysulfide photoelectrochemical solar cells with in-situ tin storage. *J. Electrochem. Soc.* 134(5): 1064–70.

Liu, X. 2005. Production of Butyric Acid and Hydrogen by Metabolically Engineered Mutants of *Clostridium tyrobutyricum*. Ph.D Thesis, The Ohio State University, OH.

Maeda, K. 2011. Photocatalytic water splitting using semiconductor particles: History and recent developments. *J. Photochem. Photobiol. C* 12: 237–68.

Maeda, K. and K. Domen. 2007. New non-oxide photocatalysts designed for overall water splitting under visible light. *J. Phys. Chem. C* 111: 7851–61.

Maness, P.-C., J. Yu, C. Eckert, and M. L. Ghirardi. 2009. Photobiological hydrogen production—Prospects and challenges. *Microbe* 4 (6): 275–80.

Marinovic, V. and A. R. Despic. 1999. Pyrophsophoric acid as a source of hydrogen in cathodic hydrogen evolution on silver. *Electrochim. Acta* 44: 4073–77.

Marinovic, V. and A. R. Despic. 1997. Hydrogen evolution from solutions of citric acids. *J Electroanal. Chem.* 431:127–32.

McDonald, K. J., and K. S. Choi. 2011. Synthesis and photoelectrochemical properties of $Fe_2O_3/ZnFe_2O_4$Composite photoanodes for use in solar water oxidation. *Chem. Mat.* 23 (21): 4863–69.

Miller, E. L., R. E. Rocheleau, and X. M. Deng. 2003. Design considerations for a hybrid amorphous silicon/photoelectrochemical multijunction cell for hydrogen production. *Int. J. Hydrogen Energy* 28: 615–23.

Mohilner, D. M., R. N. Adams, and W. J. Argersinger. 1962. Investigation of the kinetics and mechanism of the anodic oxidation of aniline in aqueous sulfuric acid solution at a platinum electrode. *J Am. Chem. Soc.* 84: 3618–22.

Murray, J. N. 1982. Alkaline solution electrolysis advancement. *Hydrogen Energy Prog.* IV(2): 583–92.

Neagu, C., H. Jansen, H. Gardeniers, and M. Elwenspoek. 2000. The electrolysis of water: An actuation principle for MEMS with a big opportunity. *Mechatronics* 10(4–5): 571–81.

Ni, M., M. K. H. Leung, and D. Y. C. Leung. 2008. Technological development of hydrogen production by solid oxide electrolyzer cell (SOEC). *Int. J. Hydrogen Energy* 33: 2337–54.

Ni, M., M. K. H. Leung, D.Y.C. Leung, and K. Sumathy. 2007. A review and recent developments in photocatalytic water-splitting using TiO_2 for hydrogen production. *Renew. Sustainable Energy Rev.* 11(3): 401–25.

Ohta, T. 2000a. Efficiency of mechano-catalytic watersplitting system. *Int. J Hydrogen Energy* 25: 1151–56.

Ohta, T. 2000b. Mechano-catalytic water-splitting. *Appl. Energy* 67: 181–93.

Ohta, T. 2000c. On the theory of mechano-catalytic watersplitting system. *Int. J. Hydrogen Energy* 25: 911–17.

Ohta, T. 2000d. Preliminary theory of mechano-catalytic water-splitting. *Int. J. Hydrogen Energy* 25: 287–93.

Ohta, T. 2001. A note on the gas-evolution of mechanocatalytic water splitting system. *Int. J. Hydrogen Energy* 26: 401–02.

Orhan, M. F., I. Dincer, and M. A. Rosen. 2012. Efficiency comparison of various design schemes for copper–chlorine (Cu–Cl) hydrogen production processes using Aspen Plus Software. *Energy Conversion and Management* 63: 70–86.

Osterloh, F. E. 2008. Inorganic materials as catalysts for photochemical splitting of water. *Chem. Mater.* 20: 35–54.

Osterloh, F. E. and B. A. Parkinson. 2011. Recent developments in solar water-splitting photocatalysis. *MRS Bull.* 36: 17–22.

Park, S. Y. 2015. Photoelectrochemical Water Splitting Optimizing Interfaces and Light Absorption. Ph.D. Thesis, University of Twente, the Netherlands. ISBN: 978-94-6233-082-5, DOI:10.3990/1.9789462330825

Peharz, G., F. Dimroth, and U. Wittstadt. 2007. Solar hydrogen production by water splitting with a conversion efficiency of 18%. *Int. J. Hydrogen Energy* 32: 3248–52.

Perret, R., W. Alan. G. Besenbruch, et al. 2007. Development of Solar Powered Thermochemical Production of Hydrogen from Water. DOE Hydrogen Program FY Annual Progress Report, pp. 128–35.

Prasad, G., K. S. Chandra Babu, and O. N. Srivastava. 1989. Structural and photoelectrochemical studies of In_2O_3-TiO_2 and WSe_2 photoelectrodes for photoelectrochemical production of hydrogen. *Int. J. Hydrogen Energy* 14: 537–44.

Rajeshwar, K. 2007. Hydrogen generation at irradiated oxide semiconductor–solution interfaces. *J. Appl. Electrochem.* 37: 765–87.

Rosen, M. A. 2010. Advances in hydrogen production by thermochemical water decomposition: A review. *Energy* 35:1068–76.

Ross, D. S. 2004. A study of mechano-catalysis for overall water splitting. *J Phys. Chem. B* 108: 19076–77.

Sawasaki, T., T. Tanabe, T. Yoshida, and R. Ishida 2003. Application of gamma radiolysis of water for H_2 production. *J. Radioanal. Nucl. Chem.* 255: 271–74.

Schalenbach, M., M. Carmo, D. L. Fritz, J. Mergel, and D. Stolten. 2013. Pressurized PEM water electrolysis: Efficiency and gas crossover. *Int. J. Hydrogen Energy* 8: 14921–33.

Schröder, V., B. Emonts, H. Janßen, and H. P. Schulze. 2004. Explosion limits of hydrogen/oxygen mixtures at initial pressures up to 200 bar. *Chem. Eng. Technol.* 27(8): 819–920.

Slade, S., S. A. Campbell, T. R. Ralph, and F. C. Walsh. 2002. Ionic conductivity of an extruded Nafion 1100 EW series of membranes. *J. Electrochem. Soc.* 149(12): A1556–64.

Smith, W., A. Wolcott, R. C. Fitzmorris, J. Z. Zhang, and Y. P. Zhao. 2011. Quasi-core-shell TiO_2/WO_3 and WO_3/TiO_2 nanorod arrays fabricated by glancing angle deposition for solar water splitting *J. Mater. Chem.* 21(29): 10792–800.

Spinks, J. W. T. and R. J. Woods. 1990. *An Introduction to Radiation Chemistry*, 3rd Edn. New York, NY: John Wiley and Sons Inc.

Stéphane, A. and F. Gilles. 2006. Thermochemical hydrogen production from a two-step solar-driven water-splitting cycle based on cerium oxides. *Sol. Energy* 80(12): 1611–23.

Su, J. Z., L. J. Guo, N. Z. Bao, and C. A. Grimes. 2011 Nanostructured WO_3/$BiVO_4$ heterojunction films for efficient photoelectrochemical water splitting. *Nano Lett.* 11(5): 1928–33.

Takata, T., S. Ikeda, A. Tanaka, M. Hara, J. N. Kondo, and K. Domen. 2000. Mechano-catalytic overall water splitting on some oxides (II). *Appl. Catal. A Gen.* 200(1–2): 255–62.

T-Raissi, A., N. Muradov, C. Huang, and O. Adebiyi. 2006. Hydrogen from solar via light-assisted high-temperature water splitting cycles. *J Sol. Energy Eng.* 129: 184–89.

Turner, J., G. Sverdrup, M. K. Mann, et al. 2008. Renewable hydrogen production. *Int. J. Energy Res.* 32(5): 379–407.

Velzen, D. van, H. Langenkamp, G. Schuetz, D. Lalonde, J. Flamm, and P. Fiebelmann. 1978. *Hydrogen Energy System* (eds. T. N. Veziroglu and W. Siefretz), Vol. 2. Oxford, UK: Pergamon Press, p. 649.

Velzen, D. van and H. Langenkamp. 1980. The oxidation of sulphur dioxide by bromine and water. *Int. J. Hydrogen Energy* 5(1): 85–96.

Wan, W. and S. P. Jiang. 2010. A mechanistic study on the activation process of (La, Sr)MnO_3 electrodes of solid oxide fuel cells. *Solid State Ionics* 177: 1361–69.

Wang, J., Y. H. Han, M. Z. Feng, J. Z. Chen, X. J. Li, and S. Q. Zhang. 2011. Preparation and photoelectrochemical characterization of WO_3/TiO_2 nanotube array electrode. *J. Mater. Sci.* 46(2): 416.

Wang, Y., T. Yu, X. Y. Chen, H. T. Zhang, S. X. Ouyang, Z. S. Li, J. H. Ye, and Z. G. Zou. 2007. Enhancement of photoelectric conversion properties of $SrTiO_3$/α-Fe_2O_3 heterojunction photoanode. *J. Phys. D Appl. Phys.* 40(13): 3925.

Weaver, F. P. 2006. Low Voltage Electrochemical Hydrogen Production. Ph.D. Thesis, University of South Florida, FL.

Weber, M. F. and M. J. Dignam. 1986. Splitting water with semiconducting photoelectrodes-Efficiency considerations. *Int. J. Hydrogen Energy* 11: 225–32.

Williams, T., M. Ghirardi, and R. Remick. 2007. *Hydrogen Technologies: Photobiocatalytic production of hydrogen*. Golden, CO: National Renewable Energy Laboratory.

Yang, X. and J. T. S. Irvine. 2008. $(La_{0.75}Sr_{0.25})_{0.95}Mn_{0.5}Cr_{0.5}O_3$ as the cathode of solidoxide electrolysis cells for high temperature hydrogen production from steam. *J. Mater. Chem.* 18: 2349–54.

Yin, J., L. J. Bie, and Z. H. Yuan. 2007. Synthesis and photoelectrochemical properties of $Fe_2O_3/ZnFe_2O_4$ composite photoanodes for use in solar water oxidation. *Mater. Res. Bull.* 42(8): 1402.

Yue, X., A. Yan, M. Zhang, L. Liu, Y. Dong, and M. Cheng. 2008. Investigation on scandium-doped manganate $La_{0.8}Sr_{0.2}Mn_{1-x}Sc_xO_3$-cathode for intermediate temperature solid oxide fuel cells. *J. Power Sources* 185: 691–97.

Zhang, H., G. Lin, and J. Chen. 2010. Evaluation and calculation on the efficiency of a water electrolysis for hydrogen production. *Int. J. Hydrogen Energy* 35: 10851–58.

Zhang, H., S. Su, X. Chen, G. Lin, and J. Chen. 2012. Performance evaluation and optimum design strategies of an acid water electrolyzer system for hydrogen production. *Int. J. Hydrogen Energy* 37(24): 18615–21.

4 Electrochemical Water Splitting

4.1 INTRODUCTION TO PHOTOELECTROCHEMICAL WATER SPLITTING

Humanity is currently facing a problem of soaring energy product prices (gasoline, diesel, natural gas) due to increasing energy demand and depleting sources of conventional energy (gasoline oil will be exhausted in less than a century). Increasing human population highly contributed to the consumption of conventional energy sources resulting in increasing greenhouse gases (1 kg wood releases mainly 403 g CO_2, 86 g CH_4 131 g CO, 69 g hydrocarbon, 4.7 g NO, other gases), which is not good for the environment. The global population is expected to increase dramatically to 10.4 billion people in 2100 followed by an increase in energy demand of up to 30 terawatt (TW)/year, resulting in the release of 13.3 gigatons of carbon (GtC)/year of CO_2 in the atmosphere as shown in Figure 4.1 (Wigley et al. 1996; Hoffert et al. 1998; Lewis and Nocera 2006).

4.1.1 PHOTOELECTROCHEMICAL (PEC) WATER SPLITTING

The Sun provides us an enormous amount of energy (4.3×10^{20} J) per hour that can satisfy our global energy need for a year (Annual Energy Outlook, 2005). This means only 0.8% of the surface of the land on Earth needs to be covered with 10% efficient solar cells to generate 30 TW of power, which will be more than sufficient to satisfy our current energy needs, that is, 17.7 TW per annum. Therefore, to satisfy the long-term energy needs of the world in a cost-effective, eco-friendly, and energy economic way, solar energy is the only solution. The sunlight coming to us in a diffuse and intermittent (the Sun is diurnal with a charging and discharging cycles of ~12 hours) manner hinders the energy storage capacity of the system (Lewis and Nocera 2006; Gust et al. 2009). Due to the high usage of sunlight in stimulating electrochemical and chemical conversion (light-related chemical phenomena, electricity, energy/fuel production, pollution eradication), the current century is named an "Age of Light." Out of limitless possible chemical reactions, we are interested in hydrogen production. There are a number of methods of hydrogen generation, depicted in Figure 4.2a and b. To match the eco-friendly energy source, that is, sunlight, we also got interested in carbon-less fuel for hydrogen production, and only one source comes to our mind and that is water. Water splitting is one of the Holy Grails of chemistry. We looked for an efficient and long-lived system for splitting water to generate H_2 and O_2 in the presence of light in the terrestrial (air mass [AM]1.5)

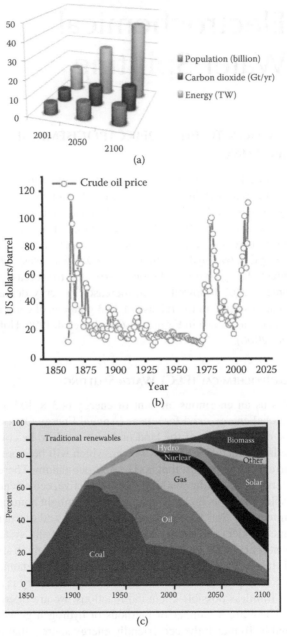

FIGURE 4.1 (a) Current and projected Earth population (blue, billion), CO_2 emission rates (red, gigatons of carbon [GtC] per year), and energy consumption (green, terawatt [TW]). (b) Average crude oil price (in US$/barrel) as a function of calendar year. (c) Current primary energy shares and future energy scenarios. (From Khnayzer, R. S., Photocatalytic Water Splitting: Materials Design and High-Throughput Screening of Molecular Compositions, Ph.D. Thesis, Graduate College of Bowling Green State University, Bowling Green, OH, 2013; (c) Nakicenovic, N. et al., *Global Energy Perspectives*, Cambridge University Press, Cambridge, 1998.)

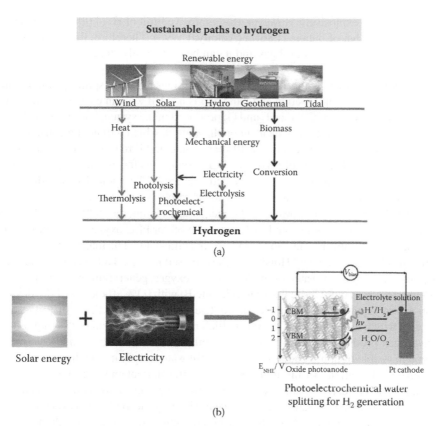

FIGURE 4.2 (a) Sustainable water-splitting methods for hydrogen generation; (b) depiction of photoelectrochemical water splitting. (From Liao, P., and Carter, E. A., *Chem. Soc. Rev.*, 2, 2401–2422, 2013.)

solar spectrum at an intensity of one Sun, as shown in Equations 4.1 through 4.3 (Youngblood et al. 2009; Nocera 2010; Reece et al. 2011):

$$O_2 + 4H^+ + 4e^- \rightarrow 2H_2O \quad E_{ox} = 1.23 - 0.059 \ (pH = 0) \ \text{or} \ E^\circ \ (pH = 7) = +0.82 \ V \quad (4.1)$$

$$4H^+ + 4e^- \rightarrow 2H_2 \quad E_{red} = 0 - 0.059 \ (pH = 0) \ \text{or} \ E^\circ \ (pH = 7) = -0.41 \ V$$
$$[V \ \text{vs. NHE}] \tag{4.2}$$

$$2H_2O \rightarrow 2H_2 + O_2 \quad E_{overall \ reaction} = -1.23 \ V \tag{4.3}$$

The overall voltage stored in hydrogen (H_2) by the reduction of two electrons and protons is 2.46 V. Hydrogen has a very high-energy storage capacity of ~119,000 J/g, which is three times higher than the capacity of oil (40,000 J/g) (Amouyal 1995). The main drawback of H_2 as a fuel is its reactivity with O_2. When it comes in contact with oxygen, it reacts explosively. However, this explosive limit of H_2 (4.00%) in air is two times less than that of butane (1.86%) (Amouyal 1995). Therefore, the utilization of hydrogen is as easy as a natural gas and it can be quite easily stored and transported like natural gas.

For a practical usage of the aforementioned reactions for water splitting, a semiconductor with the apt band position and band gap is needed with an energy efficiency of at least 10%. This means that the H_2 and O_2 produced in the system have a fuel value of at least 10% of the solar energy incident on the system, which will not be consumed or degraded under irradiation for at least 10 years (Bard and Fox 1995; Bard et al. 1995).

A photoelectrochemical (PEC) device is a good means for the water-splitting process, which converts light energy into electric energy. The photoelectric effect discovered by Becquerel first gave the idea to researchers and chemical engineers to convert light into electric power or chemical fuels (Bequerel 1939). The idea was first experimentally demonstrated by Boddy in 1968 for PEC oxygen generation from water, under the illuminating anode made from rutile-TiO_2. The major landmark was established by Fujishima and Honda (1972) who used n-rutile-TiO_2 as anode material for photoelectrolysis of water for hydrogen and oxygen generation by using pH bias (anode side A was more basic than cathode side B) with 0.1% efficiency of the device, as shown in Figure 4.3e. In Chapter 3, we discussed various methods of PEC water splitting, which were proven as the most efficient, practical, and energy saving modes of hydrogen production from water, using sunlight, semiconductors, and electricity. Although visible light has enough energy to split water (H_2O) into hydrogen (H_2) and oxygen (O_2) (Equations 4.1 through 4.3), water is transparent and does not absorb visible light. Therefore, intermediates (light-harvesting system: semiconducting catalyst) are needed to achieve water photocleavage via a cyclic pathway (Equation 4.3). A large number of intermediates have been attempted to generate simultaneous H_2 and O_2 that belongs to homogeneous and microheterogeneous systems (multimolecular, supramolecular, constrained, or confined systems) in which the illumination of a colored compound acting as a photosensitizer (dyes or quantum dots [QDs]) gives rise to photoinduced redox processes. H_2 and/or O_2 production from water by visible light requires one or more intermediates with the following main functions:

1. Visible light absorption
2. Conversion of the excited light energy ($h\nu$) into redox energy (photocarriers: photoelectron and photoholes)
3. Rigorous transfer of several electrons to water, directed to the H_2 formation, known as energy-storage compound and/or photoholes tends to the formation of O_2 (equivalent to photosynthesis)

PEC splitting of water imitates the photosynthesis (which includes oxidation of water and reduction of carbon dioxide) reaction and is popular as an artificial photosynthesis. Researchers and industrialists have been working on hydrogen generation through PEC water splitting for decades. Nozik (1975) used TiO_2/GaP electrodes for PEC water splitting, which exhibits an efficiency of 0.25%. Later on the efficiency was stimulated up to 12.5% by Tunners' group by using heterojunctions of III–V materials, such as an n-GaInP$_2$ and a p-GaAs junction photoanode

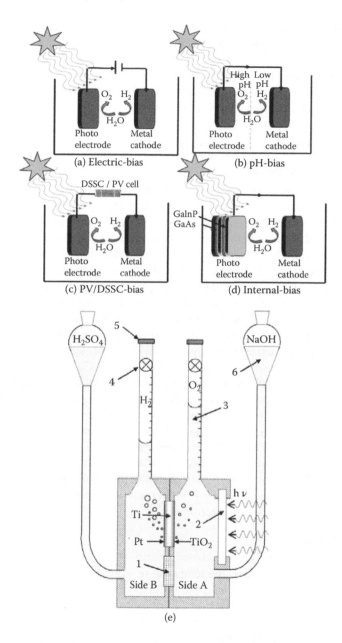

FIGURE 4.3 Methods of external/internal biasing: (a) electric/grid, (b) pH, (c) photovoltaic (PV), or (d) dye-sensitized solar cell (DSSC) used internally in photoassisted electrolysis cell, for photoelectrochemical (PEC) water splitting. (e) Schematics of the (a) Plexiglass cell for PEC water splitting, used by Fujishima and Honda for separate H_2 and O_2 evolutions. Here, 1, 2, 3, 4, 5, and 6 represent the cation exchange membrane, quartz glass filter window for light source, burette, stopcock, rubber septum, and reservoir with acidic and basic solution, respectively. (From Chouhan, N. et al., *Electrochemical Technologies for Energy Storage and Conversion*, Wiley-VCH, Weinheim, Germany, 2011.)

in series and a Pt counterelectrode. A few of the other remarkable photocatalytic materials used for water splitting are NiO-loaded $NaTaO_3$:La (56% at 270 nm) (Kato et al. 2003), Pt-loaded CdS (37% at 420 nm) (Darwent 1981), Ni-doped ZnS (1.3% at 420 nm) (Reber and Meier 1984), Pt-loaded-$(AgIn)_{0.22}Zn_{1.56}S_2$ (20% at 420 nm) (Tsuji et al. 2004), Pt-loaded $(CuIn)_{0.09}Zn_{1.82}S_2$ (13% at 420 nm) (Tsuji et al. 2006), Ru-loaded $(CuAg)_{0.15}In_{0.3}Zn_{1.4}S_2$(8% at 420 nm) (Tsuji et al. 2006), and so on.

4.1.2 FACTORS AFFECTING EFFICIENCY OF THE PEC

4.1.2.1 Electrode Material

Affordable price range, corrosion resistance, and chemical stability and appropriate band gap with apt band position suitable for water splitting are the prime criteria for the selection of electrode material (Wei et al. 2007). Modification in morphology and structure, and enhancement in active surface site leads to the facilitation of high activity for the electrodes. A spacious range of materials and their combinations are being used as electrodes. Platinum, gold, and silver are designated as the eminent choices for use as electrode material but high price limits their use at a commercial level. Metal oxides, modified metal oxides, metal chalcogenides, metal phosphides, combination of metal/metal oxides, nickel, Raney nickel, and cobalt are the most common electrode materials for use in alkaline/acidic electrolytic baths. Appleby et al. conducted a set of experiments by utilizing different electrodes such as 99.99% pure Ni, Pt, Ir, and Rh as well as Ni cloth, Ni sinter, Ni-Cd, and low impregnation nickel and cobalt molybdate catalyst on sintered nickel. Electrodes were preanodized in order to obtain stable potential characteristics. Photocatalytic material (photocarrier generator) used for electrode fabrication was modified by addition of the sensitizers (light-harvesting material) and hydrogen evolution reaction (HER)/oxygen evolution reaction (OER) cocatalysts (charge/photocarrier separator). The components of electrode material are discussed in detail in Section 4.2. Moreover, electrode material–electrolyte compatibility would decide the activity of the PEC cell. For example, platinum electrodes show higher activity in aqueous KOH solutions in comparison to the molybdenum electrodes.

4.1.2.2 Effect of Temperature

According to the thermodynamic characteristics of a water molecule, its splitting reaction potential is reduced as the temperature increases because energy demand of the reaction decreases with increase in temperature (Figure 4.4a) (Tsiplakides 2012). Nagai et al. suggested that heat can reduce the equilibrium voltage (reversible potential) of the water. This parameter also enlarges the size of the gas bubbles and reduces their rising velocities. The latter causes a larger void fraction in the electrolyte and decreases the efficiency as a result. Moreover, temperature-induced ionic conductivity and surface reaction of the electrolyte and satisfied the energy need (Udagawa et al. 2007) of the cell to reach any given current density. The report exhibits a voltage reduction up to 120 mV as the temperature was raised from 120°C to 150°C, which reduced H_2 production price. However, the increased temperature may cause some stability issues, such as container cracks and gasket leaks due to disturbance in mechanical, stability, and physical properties of such

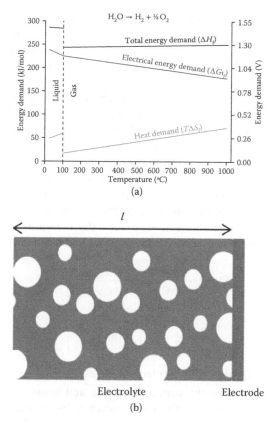

FIGURE 4.4 (a) Temperature versus energy demand graph and (b) void formation at electrolyte and electrode interface. (From Tsiplakides, D., PEM Water Electrolysis Fundamentals, http://research.ncl.ac.uk/sushgen/docs/summerschool_2012/PEM_water_electrolysis-Fundamentals_Prof._Tsiplakides.pdf, 2012; Mazloomi, K. et al., *Int. J. Electrochem. Sci.*, 7, 3314–3326, 2012.)

systems, which are a concern for manufacturers and designers of industrial and commercial hydrogen producers. Total electrolysis efficiency of the cell was controlled by individual factors, such as electrical efficiency (70%), electrolysis efficiency (22%), and thermal efficiency (8%).

4.1.2.3 Effect of Pressure

Appleby et al. (1978) conducted several pressure variation experiments (up to 40 atm) in a typical three-compartment electrolyzer cell with stable pure Pt electrode and smooth nickel plates with a 1-cm² surface area. Cell current density was kept at 1 mA/cm² with an electrolyte of either a 34 wt.% or 25 wt.% KOH solution in distilled water. Pressurization is advantageous due to the following reasons:

1. Reduction of internal cell resistance
2. Reduction of pressure on the gas bubbles reduction in the ohmic voltage drop and power dissipation

3. Production of compressed hydrogen facilitates storage
4. Reduction of volume of gas bubbles facilitates water transport
5. Electrolytic decomposition of water consumes less power to lower the costs of hydrogen production by creating higher current density conditions

Overall voltage drop of up to 100 mV was observed when the conversion process was conducted at a pressure level of under 30 atm. The cell voltage versus pressure graph has its highest reduction slope when the pressure is raised from 1 to 10 atm. At almost similar conditions, Onda et al. and LeRoy et al. estimated the ideal temperature and pressure conditions for electrolytic hydrogen production to be around 70 MPa and 250°C. High temperature and intense pressure will change both Gibbs energy and enthalpy levels of an electrolysis process. Consecutively, the electrolysis efficiency raised by 5% and 50% of energy saved. If we assume equal overall pressure at both electrodes, then the increase in the equilibrium potential E with pressure is given by Equations 4.4 and 4.5:

$$E = E_o - \frac{RT}{(nF)} \times \ln \frac{P_{H_2O}}{\left(P_{H_2} \times P_{O_2}^{1/2}\right)}$$
(4.4)

$$\Delta E = -\left(\frac{RT}{(nF)}\right) \times \left(\ln \frac{1}{(P_{O_2}^{1/2})}\right)$$
(4.5)

4.1.2.4 Electrolyte Quality and Electrolyte Resistance

Pure water is nonconductive by nature but bases and acids are known as game changers, because they improve the ionic conductivity of the aqueous electrolyte compounds by reducing the overvoltage value of an electrolyzer (Badwal et al. 2006). However, the highly corrosive behavior of such materials restricted the concentration level of acidic and alkali solutions in practice, such as the 25%–30% aqueous KOH widely used in electrolyzers (Guttman and Murphy 1983). The low concentration limited the electrocatalytic performance of the water electrolysis and contributed to the overall electrical resistance of a cell to rise and will cause the cell efficiency to drop. Therefore, substitute electrolytes such as ionic liquids [1-butyl-3-methylimidazolium tetrafluoroborate (BMI.BF$_4$, MBI.MF$_4$)] have been introduced to improve the conductivity and stability factors of electrolytic baths at ambient temperature (De Souza et al. 2003, 2006, 2008). The material of electrode plates is selected from an easily found metal such as carbon steel (CS), nickel (Ni), nickel-molybdenum (Ni-Mo) alloy, molybdenum (Mo), and so on. Magnesium, chloride, and calcium ions are the notorious names of some impurities that may cause unwanted side reactions to block and passivate the electrode plates and/or separator surfaces (Belmont et al. 1998) and interrupt the interelectrode mass and electron transfer (excess ohmic resistance).

4.1.2.5 Size, Alignment, and Space between the Electrodes

Larger surface areas ground to the less resistive current path in the electrochemical process. Nagai et al. (2003) carried out a series of experiments on ambient pressure and temperature with Ni–Cr–Fe alloy electrodes in a 10 wt.% aqueous potassium hydroxide to observe the effect of the optimum space between electrodes, their size, wet ability, inclination, and so on, the process efficiency. Among the electrodes of same width but different heights, the electrodes with larger height will cause additional power dissipation in a cell due to the formation of a larger volume of void fractions. Similarly, higher efficiency levels were obtained for the vertically placed electrodes due to the optimum bubble departure rate by reduction in ohmic resistance. Reduction in the distance between electrodes lowers electrical resistance of the PEC cell. The results clearly depict that electrodes placed too close to each other will increase the value of the void fractures and will lead to a less efficient process.

4.1.2.6 Forcing the Bubbles to Leave

Gas bubbles accumulated on the electrode surface will reduce its conductivity and cause a higher level of ohmic voltage drop (Qian et al. 1998). Therefore, forcing the bubbles to disengage from the electrodes, membrane, and electrolyte surface is required to improve the local mass and heat transfer as well as the process efficiency. Conversely, the bubble diameter depends on the current density, temperature, and pressure (Figure 4.4b) (Mazloomi et al. 2012). Pressure value has an inverse correlation with bubble size. However, current density and temperature have an opposite relationship (De Jonge et al. 1982). Moreover, the disengagement rate of gas bubbles from the surfaces and their departure velocity plays a significant role in the identification of the value of the electrical resistance of an electrolytic bath. Many efforts have been made to force the bubbles to leave the cell environment by using an ultrasonic field (Li et al. 2009a) high-gravity acceleration environment and so on. Wang et al. (2010) reported remarkably lower voltage levels for the cases with higher gravity values.

4.1.2.7 Separator Material

A separator plate (membrane/diaphragm) in a cell is used to block the free movement of mass and ions to some extent and separate the gaseous products released from different electrode sides. The electrical resistance of a separator plate depends on different variables such as corrosion, temperature, and pressure of the cell (Renaud and LeRoy 1982.) The effective electrical resistance of a separator plate should be kept three to five times higher than that of the electrolyte solutions. Due to highly wettability and porous structure, asbestos can be used as a separator plate but it is a toxic and hazardous material (Vermeiren et al. 1996.) Therefore, researchers are looking for substitute materials. Nowadays, advance materials and technologies are available to reduce the negative electrical effect of separators, that is, bipolar membranes or BMs (polyvinylidene fluoride-grafted 2-methacrylic acid 3-(bis-carboxymethylamino)-2-hydroxyl-propyl ester [PVDF-grafted-GMA-IDA], polyethersulfone-grafted-GMA-IDA [PES-grafted-GMA-IDA]) (Li et al. 2009b), highly loaded palladium nanoparticles and catalytically active asymmetric membranes, and so on (Villalobos et al. 2016). Porous composite separator material Zirfon® is composed of a polysulfone matrix and ZrO_2 (Vermeiren et al. 1998).

4.2 SEMICONDUCTING PHOTOELECTRODE MATERIALS

4.2.1 ELECTRON TRANSFER PHENOMENON

Minority carriers reaching the edge of the space charge region are transferred via interface to solution. Redox species without loss by recombination, the generation-collection problem can be formulated in terms of characteristic lengths. These are the width of the space charge region (WSC), penetration depth $(1/\lambda \cdot \alpha)$ of the light incident on the surface of the electrode, and minority carriers' diffusion length L. Here, α is the absorption coefficient of the semiconductor at wavelength λ.

The minority carrier (holes in n-type semiconductor) diffusion layer is $L_s = \dfrac{K_B T \tau_p p \mu_p}{q}$, where μ = mobility and τ = bulk lifetime.

The solution of the charge carrier's generation and collection problem comes in the form of the Gartner equation (Equation 4.6), which gives us the value of the external quantum efficiency EQE as follows:

$$\text{EQE} = \frac{J_{\text{photo}}}{q I_0} = \frac{g}{I_0} = 1 - \exp-\left(\frac{\alpha W_{SC}}{1+\tau}\right) \tag{4.6}$$

I_0 is the intensity of the incident photons, q is the charge, g is the minority carrier flux, J_{photo} is the photocurrent density in mA/cm^2, and ideal photocurrent–voltage behavior would be predicted by the Gartner equation (only seen in semiconductor/electrolyte systems involving a fast outsphere electron transfer process). The electron transfer is slow in the case of the multiple step reactions that are involved in oxygen/hydrogen evolution during water splitting. The concentration of minority carriers builds up when the carriers queue up close to the electrode–electrolyte interface. As a consequence, the recombination of electron–hole pairs occurs at the surface or in the surface charge region that competes with charge transfer reactions and reduces the external quantum efficiency. The low degree of band bending is the major limiting factor for the water-splitting reactions observed at the interface of the semiconductor electrodes. Concentration of the photogenerated minority carrier has been a factor considered for interfacial electron transfer reactions. Moreover, the lifetime of the minority carrier in bulk semiconductor is also contributing to enhancing the efficiency of the systems. For example, the lifetime is recorded in milliseconds for very low doped Si wafers and in picoseconds for impure and polycrystalline material. Recombination of the carriers in bulk sample is a pseudo-first-order process because the concentration of the majority charge carriers is much higher than the concentration of the minority carrier. This means that we can define a minority carrier lifetime. Practically, the low values of τ_{min} (lifetime for minority carrier) are observed for water-splitting photoelectrode methods where the minority carriers move only a short distance before they recombine with the majority carriers.

4.2.2 MATERIAL AND ENERGETIC REQUIREMENTS

PEC is one of the most powerful but complex ways to split water. Following are the characteristics of the electrode material for efficient water splitting:

- The semiconductor may generate sufficient voltage (>1.23 eV) that can split water on light exposure. This means that the photoelectrode must be irradiated with the light having energy higher than the band gap of the semiconductor.
- The band gap must be sufficiently small so that it can absorb an apt % portion of light. Moreover, the band edges of the band gap must straddle between the redox potentials of hydrogen (0.0 eV) and oxygen (1.23 eV). For reasonable solar efficiencies, the band gap must be less than 2.2 eV. Unfortunately, the most useful semiconductors with band gaps in this range are photochemically unstable in water.
- Long-lived material against the corrosion in aqueous electrolyte and light exposure. The most photochemically stable semiconductors in aqueous solution are oxides but their band gaps are either too large for efficient light absorption (~3 eV) or their semiconductor characteristics are poor.
- Electron transfer facilitated from the electrode surface to solution with minimum energy losses due to kinetic overpotential.
- In aqueous electrolyte H_2SO_4 of pH = 4.5, the photocatalyst is used to split overall water into H_2 and O_2 but in the presence of a sacrificial electrolyte (hole or electron), it will go for selective splitting of either H_2 or O_2.
- Low cost of the material.
- A high quantum yield (>80%) is needed to reach the efficiency necessary for a commercially viable device.

To date no material has been discovered or synthesized which can pass all of the above conditions, simultaneously. The main driving force behind water splitting is photovoltage resulting in low entropic losses (nonradioactive recombination, spontaneous emission, incomplete light trapping, nonideal band structure alignment, etc.), for the reasonably low band gap semiconductor. When a photovoltaic system (light harvesting) combines with an electrolyzer (water splitting) and is converted into a single monolithic device, then the cost of the system is reduced because it reduces semiconductor processing, the current density, electrolysis area, and sees an increase in efficiency of 30% compared to the separated system.

4.2.3 SENSITIZERS AND PHOTOCATALYST

Water splitting is an endothermic and redox process that involves an overall two-electron transfer, which requires an energy of 1.23 eV per electron and corresponds to a wavelength of $\lambda = 1008$ nm. This means water absorbs solar radiation in the infrared region, which has photon energies too low to drive photochemical water splitting. Although sunlight has an efficient amount of energy that is requisite for water splitting, water needs a visible light-harvesting surrogate system, that is, the photosensitive semiconducting material to split water. Usually, these catalytic units are either not stable in water or they have a wide band gap that can compel harvesting of the ultraviolet (UV) region of the light (only 4% of the full sunlight). Consequently, the catalytic units also need the support of light-harvesting materials (dyes/quantum dots) that can utilize the wider portion of light. These supporting units are known as sensitizers and should be allowed for rapid electron transfer. Sensitizers are light-harvesting antennas, which (photocatalyst) absorb more visible sunlight than is needed for a relay of electrons and facilitate the initiation of the basic catalytic reactions. QDs, the nanosemiconductors, are promising "sensitizers" (Chang and Lee 2007; Diguna et al. 2007; Jin et al. 2013):

1. Flexible size and shape
2. Tunability of the band gap over the UV, visible, and infrared (IR) regions
3. Broad absorption of light with high molar absorptivity
4. Robustness against photobleaching
5. Multiple excitation generation (Nozik 2002)
6. Energy transfer–assisted charge collection (Ruland et al. 2011)
7. Quantum confinement

The shape and size of the sensitizer are the deciding parameters for optimizing the electron transfer process. The quantum confinement in QDs leads to partition and separation of the conduction and valance band of the semiconductor. Band edge potential, as a function of the particle size, can be effectively selected to provide optimum overpotential. Orientation of these sensitizers should be arranged to minimize the energy transfer barrier. Finally, hierarchical preorganized photocatalyst within the reactor may be required to produce separate streams of the O_2 and H_2. Based on such promising characteristics, QDs have been used as sensitizers for PEC water-splitting applications in particular with large band gap n-type metal oxides photoanodes. Efficient electron injection from a visible light-active QD to a metal oxide requires that the conduction band edge of the QD is located above that of the metal oxide (Kamat 2007). The electron transfer between the QD and the metal oxide can facilitate charge separation and inhibit recombination through a potential gradient at the interface (Zhang 2011). Attachments of QDs on metal oxide surfaces can be achieved (1) *in situ* and (2) postsynthesis. *In situ* growth generally facilitates charge carrier transfer across the interfaces between QDs and metal oxides due to their direct contact at the atomic scale. On the other hand, the postsynthesis assembly enables charge carrier transfer to be maximized by deciding the size, size distribution, and surface functionalization of QDs (Watson 2010). Therefore, their band gaps, density and energy of their trap states are the driving forces for controlling the interfacial charge-transfer processes and the distance and electronic coupling between QDs and metal oxides. *In situ* cadmium chalcogenide QD formations have been reported for PEC photoanodes, by using hydrothermal, sol–gel, electrochemical deposition, chemical bath deposition or successive ionic layer adsorption, and reaction route techniques on various nanostructural arrays (e.g., nanotubes, nanowires, nanorods, and inverse opal) of oxides (TiO_2, Ga_2O_3, In_2O_3, ZnO, Fe_2O_3, WO_3, TiO_2/ZnO, core/shell). Visible light absorption of the QDs and type-II band edge configuration (quantum confinement) promotes the spatial separation of photogenerated carriers that could significantly enhance PEC water oxidation efficiencies under simulated sunlight. Zinc chalcogenide QDs are less toxic than cadmium chalcogenide QDs and can also be used as sensitizers. Cho et al. (2011) reported that the ZnSe-decorated ZnO nanowire photoanodes were synthesized by a dissolution recrystallization process. The ZnO/ZnSe photoanodes exhibited absorption in the visible spectrum (<550 nm in wavelength) and photocurrent. Similarly, the use of ruthenium compounds, that is, black dye (Nazeeruddin et al. 1997), N_3-dye (Nazeeruddin et al. 1993), or organic coumarin dyes (Rehm et al. 1996), along with I/I^{3-} electrolyte is commonly seen in actual applications and QDs (CdSe, CdS, noble metals, InP, PbS, etc.) and can be used as sensitizers of TiO_2 surfaces, as they show significant improvement in efficiency.

Semiconductor quantum dot molecules possess a small number of confined electrons. In the Coulomb blockade regime, conduction is dominated by cotunneling processes. These can be either elastic or inelastic, depending on whether they leave the dot in its ground state or drive it into an excited state, respectively. Due to the high photostability and their tunable absorption spectrum, semiconductor QDs have the potential of being used as photosensitizers in photocatalysis. A QD can be used to transfer a hole/electron upon photoexcitation to a water oxidation/reduction catalyst so that the catalyst can drive the water-splitting reaction. The ultimate goal is to have multiple charges transferred from one QD to one electron or hole acceptor. There are three interfacial charge transfer schemes using a model Hamiltonian approach, given by Raul, Dolzhnikov, and Edme et al., for the electron transfer from high density of states for quasi continuum of the polycrystalline catalytic unit to a low-density single, multiple, and bridge multiple electronic states of the QDs sensitizers (Figure 4.5a–c). The rate of charge transfer across the interface of semiconductor nanoparticles is influenced by the local environment at the surface. (Aruda et al., Rasmussen et al., and Weinberg et al., of Weiss Lab of Northwestern University, Chicago, IL,). The CdS QDs have a selenium-terminated surface (Figure 4.6a) that acts as a tunneling barrier for electrons by delocalization through a network of cadmium orbitals and are used to release hydrogen by reduction of water; the hole oxidized the water to release oxygen. Alternatively, a layer of cadmium surface may act as a barrier for hole transfer and can produce hydrogen by reduction of water (Figure 4.6b). The charge transfer and recombination rates can be engineered to optimize the desired performance of the PEC device by changing the surface qualities used in solar energy applications and other fields.

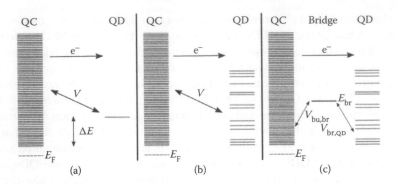

FIGURE 4.5 Electron transfer from a high density of states for quasi continuum of polycrystalline catalytic unit to a low density of states of the quantum dots (QDs) sensitizers: (a) single, (b) multiple, and (c) bridge multiple electronic state of QD (http://sites.northwestern.edu/weiss-lab/interfacial-charge-transfer/).

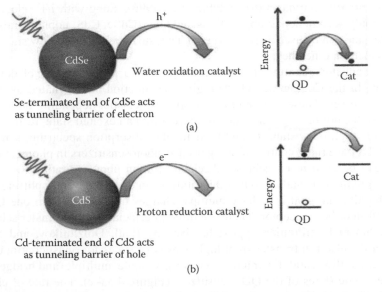

FIGURE 4.6 (a) Selenium terminating surface, which acts as a tunneling barrier for electrons by delocalization through a network of cadmium orbital and the hole, oxidizes the water to release oxygen and (b) a layer of cadmium may act as a barrier for hole transfer and can produce hydrogen by the reduction of water. (From Aruda et al., Rasmussen et al., and Weinberg et al., of Weiss Lab of Northwestern University, Chicago, IL (http://sites.northwestern.edu/weiss-lab/welcome/quantum-dot-sensitized-photocatalysis/; De Franceschi, S. et al., *Condens. Matter*, 27, 0007448, 2000.)

4.2.4 PEC Components in Action for the Water-Splitting Process

4.2.4.1 Amouyal Model

Several models were proposed by Amouyal in Figure 4.7a–e to describe the reaction steps of the water-splitting process using three-component (PS/A/Cat and PS/D/Cat), four-component [PS/R/D/Cat: reductive quenching (Figure 4.7c) and oxidative quenching mechanism (Figure 4.7d)], and five-component [PS/R_{en}/R/D/Cat: oxidative quenching mechanism (Figure 4.7e)] model systems (Amouyal 1995). PS is the

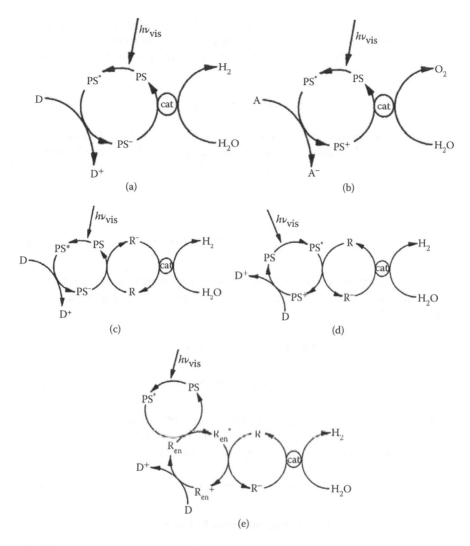

FIGURE 4.7 Schematic representation of the redox catalytic cycles for the photoreduction or photooxidation of water to hydrogen and oxygen by visible-light irradiation (a and b) three-component (PS/A/Cat and PS/D/Cat), four-component (PS/R/D/Cat: (c) reductive quenching and (d) oxidative quenching mechanism), and (e) five-component (PS/R$_{en}$/R/D/Cat: oxidative quenching mechanism) model systems for water splitting. (From Amouyal, E., *Sol. Energy Mater. Sol. Cells*, 38, 249–76, 1995.)

photosensitizer, Cat is the photocatalyst, D is the electron donor, A is the electron acceptor, R$_{en}$ is the receptor molecule, and $h\nu_{vis}$ is the visible light. In these multimolecular systems, the role of every specific molecule/component is exhibited by the following reaction steps:

1. Initially, the photosensitizer PS generates excited species PS* by absorbing visible light, which is useful in redox reactions (Equation 4.7):

$$PS + h\nu_{vis} \rightarrow PS*$$

(4.7)

2. Second compound electron relay R can be reduced or oxidized by quenching of the excited species PS* in electron transfer reactions leading to
the formation of charge pairs, such as PS$^+$/R$^-$ in the case of the oxidative
quenching of PS (Equation 4.8):

$$PS* + R \rightarrow PS^+ + R^-$$ (4.8)

3. The third compound, that is, redox catalyst Ca, is able to collect several
electrons to facilitate the exchange of two (Equation 4.9) or four electrons
with water and react with R (Equation 4.10; Figure 4.7c–e):

$$2R^- + 2H^+ + Cat \rightarrow 2R + H_2$$ (4.9)

$$4PS^+ + 2H_2O + Cat \rightarrow 4PS + O_2 + 4H^+$$ (4.10)

4. Electron acceptor A acts as a quencher to PS$^+$, which once reduced to
A (Equation 4.11) finally decomposes irreversibly as in Equation 4.12
(Figure 4.7b):

$$PS^+ + A + e \rightarrow PS^+ + A^-$$ (4.11)

$$A^- \rightarrow Decomposition\ Products$$ (4.12)

5. Reduction of the excited state photosensitizer PS* by electron donor D
yields the reduced photosensitizer PS$^-$ and the oxidized donor D$^+$ (Equation
4.13), which decomposes irreversibly (Equation 4.14):

$$PS* + D \rightarrow PS^- + D^+$$ (4.13)

$$D^+ \rightarrow Decomposition\ Products$$ (4.14)

The reduction of water to H$_2$ can be achieved in the presence of a sacrificial
electron donor D and a suitable catalyst (Equation 4.15):

$$2D + 2H^+ + Cat \rightarrow 2D^+ + H_2 + Cat$$ (4.15)

6. In this way, PS$^-$ accumulates and reacts with an electron relay R to regenerate PS and yield R$^-$ (Equation 4.16) (Figure 4.7c–e):

$$PS^- + R \rightarrow PS + R$$ (4.16)

$$2PS^- + 2H^+ + Cat \rightarrow H_2 + PS + Cat$$ (4.17)

In the presence of a suitable catalyst, R^- can lead to the formation of hydrogen as in the first scheme (Equation 4.9). It should be remarked that PS^- is a more powerful reducing species than R^- (Equation 4.17). Hence, the reduction of water to H_2 can be achieved directly by PS^- in the presence of a suitable catalyst.

As a consequence, this scheme involves only three components (PS, D, Cat) and the mechanism becomes simplified as illustrated by Figure 4.7a.

7. The five-component system (Figure 4.4e) consists of the photosensitizer PS as an antenna that transfers the excitation energy to a receptor molecule R_{en} (Equation 4.18):

$$PS^* + R_{en} \rightarrow PS + R_n^* \tag{4.18}$$

The receptor can subsequently react with the electron relay R via electron transfer (Equation 4.19) to give a charge pair (R_n^*/R^-):

$$R_n^* + R \rightarrow R_n^+ + R^- \tag{4.19}$$

In this five-component system, $PS/R_{en}/R/D/Cat$, the energy-transfer photosensitizer PS is not involved in any redox processes as are the antenna molecules in natural photosynthesis, and the receptor R_{en} acts as an energy-electron relay. It should be noted that this multimolecular approach is the simplest manner to achieve cyclic photochemical water cleavage. Several difficulties must be necessarily overcome. Indeed, the different components of such systems must fulfill the required spectral, photophysical, thermodynamic, and kinetic conditions.

4.2.4.2 Kostov et al.'s Model

Kostov et al. introduced a model for cleavage of water on a defective carbon surface of graphene or carbon nanotubes, using first principle calculations. They investigated many possible reaction pathways (Figure 4.8a and b) of water splitting over the defective sites in which the activation barriers lower down to less than half the value for the dissociation of bulk water. This reduction is caused by spin selection rules that allow the system to remain on the same spin surface throughout the reaction. Reaction route includes three basic steps: initial physisorbed state, spin-oriented intermediate stages, and final desorption of hydrogen from the catalytic surface. figures 4.8 should come here before Ulleberg Model's description and figure.

4.2.4.3 Ulleberg Model

Ulleberg established that a mathematical model for an advanced alkaline electrolyzer has been developed for water splitting. The model is based on a combination of fundamental thermodynamics, heat transfer theory, and empirical electrochemical relationships (Ulleberg 2003). A dynamic thermal model (Figure 4.9) has also been developed by considering the charge state of the battery, pressure in hydrogen storage run time of the photovoltaic hydrogen generation system, and utilization of

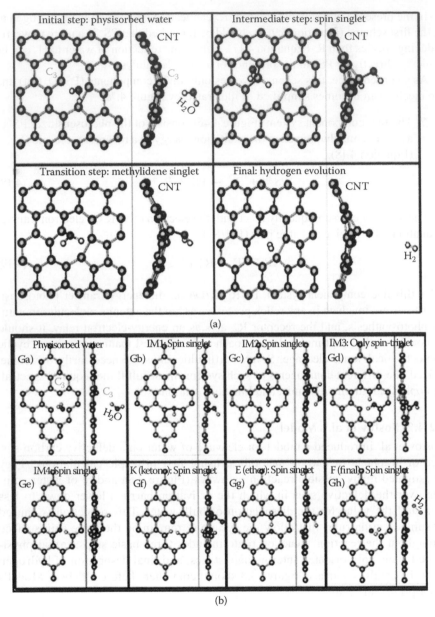

FIGURE 4.8 (a) Top and side views of carbon nanotube (CNT) for the initial (N-I), intermediate (N-IM2), transition state (N-TST), and final (N-F) conformations for a favorable dissociation pathway over a vacancy in a (10,10) nanotube. (b) Top and side views of various intermediate states in the water-splitting reaction over a vacancy site in graphene (G). (Ga) Initial physisorbed state; (Gb) IM1: spin-singlet state with the highest energy; (Gc) IM2: spin-singlet state; (Gd) IM3: the only spin-triplet state (S = 1); (Ge) IM4: spin-singlet state; (Gf) K (for "ketone"): spin-singlet state, (Gg) E (for "ether"): spin-singlet state, which is the global minimum of the potential energy surface; and (Gh) F (for "final"): spin-singlet state, the oxygen occupies the vacancy site on the surface and hydrogen is liberated from it. (From Kostov, M. K. et al., *Phys. Rev. Lett.*, 95, 13, 136105, 2005.)

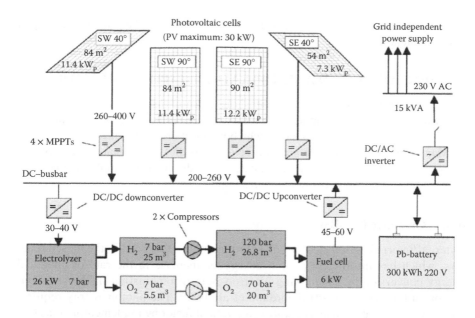

FIGURE 4.9 Stand-alone photovoltaic-hydrogen energy plant as a reference system (PHOEBUS, Julich Germany), representing the application of the Ulleberg model. (From Ulleberg, Ø., *Int. J. Hydrogen Energy*, 28, 21–33, 2003.)

the gaseous product O_2 and H_2 in a fuel cell. The model includes a photovoltaic (PV) generator, direct current (DC)–DC convertor, DC–alternative current (AC) inverter, electrolyzer, compressor, hydrogen/oxygen fuel storage section, fuel cell, secondary battery, and so on. Comparisons between predicted and measured data show that the model can be used to predict the cell voltage, hydrogen production, efficiencies, and operating temperature. The reference system used was the stand-alone photovoltaic-hydrogen energy plant in Jülich, Germany. A stand-alone power system (SAPS) is defined as an autonomous electricity generating system that is not connected with the electric grid. Climate, available energy resources, needs and requirement of the user, and the physical laws of nature and the environment are the basic factors for construction of the ultimate SAPS. Comprehensive study of SAPS is done by initially investigating all factors, separately. Afterward, all factors are connected in looking to the comfort zone or compatibility. The number of required parameters has been reduced to a minimum to make the model suitable for use in integrated hydrogen energy system simulations. Moreover, the model should be compatible with a transient system simulation program, which makes it possible to integrate hydrogen energy component models with a standard library of thermal and electrical renewable energy components. Hence, the model can be particularly useful for (1) system design (or redesign) and (2) optimization of control strategies. To illustrate the applicability of the model, a 1-year simulation of a photovoltaic-hydrogen system was performed. It was concluded that the model can work accurately for short-term as well as long-term stimulations. These results also show that improved electrolyzer operating strategies can be identified with the help of the developed system simulation model.

4.3 REACTOR DESIGN AND OPERATION (EXPERIMENT SETUP)

Photochemical water splitting is advantageous because there is no need for an additional film deposition or coating equipment. Larger surface area per unit weight is available for photocatalytic reactions, which means more active sites.

Many types of PEC device are in practice, and they can be categorized on the basis of gradient, number, and types of catalytic beds, number of multiple junctions, solar system used, design of reactor, and so on.

4.3.1 GRADIENT/BIAS-BASED REACTOR

External/internal-biased electrolytic reactions were associated in the biased reactors, which has insufficient photon energy or increase in the rate of chemical energy conversion by suppressing electron hole recombination in the semiconductor bulk (Figure 4.3a–d) (Giordano et al. 1982). The photon energy requirements would be satisfied by generating a gradient factor by implementing bias that might be generated due to the difference in concentration, photovoltaic energy, pH, electricity, sensitization, and so on, parameters of the cell. The advantage of a PEC cell is that an internal bias can be easily achieved by the photoanode with a combination of different materials. The bias formed will facilitate electron-hole separation that results in higher photocatalytic activity. Other than the internal bias, an external bias can also be applied between the electrodes for further enhancement. The PEC reactor is the system which converts solar energy into chemical energy (H_2 and O_2), where photocatalyst receives solar photons and transforms them into electrons and holes of sufficient voltage to electrolyze water and carry out electrolysis of water. These reactors are either single-junction or multijunction photosystems. A single-junction photosystem configuration (n-type or p-type) is not very effective due to the insufficient band gap or inadequate band edge positions of semiconductors. Thus, additional voltage or bias is required to meet the sufficient energetic conditions when the band edges do not overlap water-splitting redox potentials. Therefore, according to the type of the bias used for water-splitting reaction in reactors, these are further divided. In this situation, photoassisted electrolysis cells are coupled with an external or internal bias and divided into four parts:

1. Electrical bias (grid), fossil based
2. Chemical bias, pH based
3. PV cell bias
4. Dye-sensitized solar cell (DSSC) bias or internal bias

The internal-biased photoelectrode (PE) structure can be of several types such as PV/PEC (PV~a-SiGe, PEC~WO_3), PV/PV (PV1~GaInP, PV2~GaAs), and PEC/PEC (PEC1~DSSC, PEC2~WO_3) (Chouhan et al. 2011).

4.3.2 Reactors Based on Suspension and Electrode Type

On the basis of particle suspension and electrode type in photocatalysis of water, there are four basic PEC system configurations. The first two utilize aqueous reactor beds containing colloidal suspensions of PV-active nanoparticles, each nanoparticle being composed of the appropriate layered PV materials to achieve sufficient band gap voltage to carry out the electrolysis reaction. The next two configurations use planar PV cells, which are positioned in arrays facing the Sun and are immersed in a small water reservoir such that oxygen gas is produced on the anode face of the PV material and hydrogen gas on the cathode face. The specific system types are evaluated in the following.

4.3.2.1 Type 1

These types of reactor include electrode-based PEC designed systems that include the deposition of ~5-nm-thick anodic and cathodic photoactive coatings on 40-nm-sized conductive substrate. This results in a multilayer PV/PEC unit with a multiphoton response to achieve the requisite electrolysis voltage either from single above-threshold photons or from multiple below-threshold energy photons. Flat plate or tabular reactors are exhibited by Figure 4.10a and b. There is a single water bed on an electrode, with a KOH electrolyte, and a flexible clear plastic thin-film envelope, or baggie (consisting of a transparent high density polyethylene), to contain the

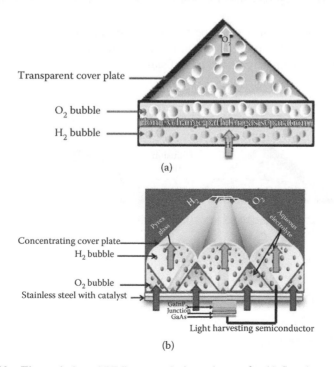

FIGURE 4.10 Electrode-based PEC reactor design schemes for (a) flat plate and (b) tabular reactor providing moderate solar concentration onto one electrode strip, for PEC water splitting.

electrode and capture. A single baggie/bed is 1060 ft long × 40 ft wide. The assumed baseline solar-to-hydrogen (STH) conversion efficiency is 10%. The system for 1 ton per day (TPD) H_2 yearly average production consists of 18 baggies. This Type 1 (Figure 4.10a and b) reactor is the simplest PEC embodiment and has the lowest capital cost ($1.63/kg H_2).

4.3.2.2 Type 2

This reactor consists of plastic baggie–covered dual bed (Figure 4.11), each having different colloidal suspension of photocatalytic material; the separate beds are for the O_2 gas and H_2 gas production reactions. The O_2 and H_2 beds are linked together with diffusion bridges to allow the transport of ions but prevent gas and particle mixing. A 0.1 M KOH electrolyte is common to both beds and facilitates transport of ionic species. These beds also contain an intermediary reactant denoted "X," which participates in the reactions, but is not consumed. "X" can be iodine, bromine, iron, or other elements. A typical set of equations (Equations 4.20 and 4.23) describe the nanoparticle photoreactions and are mentioned as follows:

Bed I (O_2 evolution bed): $4\ h\nu + 4\ X + 2\ H_2O \rightarrow O_2 + 4H^+ + 4\ X^-$ (oxidation) (4.20)

Bed II (H_2 evolution bed): $4\ h\nu + 4H^+ + 4\ X^- \rightarrow 2\ H_2 + 4\ X$ (reduction)   (4.21)

Although to some extent the Type II system is similar to the Type I system, the anodic particles would differ somewhat from the cathodic particles and the cost of the system is approximately $3.19/kg H_2. Dual water bed colloidal suspension of PV

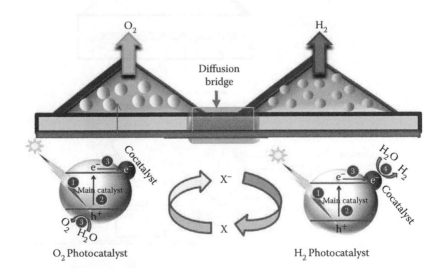

FIGURE 4.11 Type II PEC reactor design schemes for a plastic "baggie" covered dual bed particle reactor with side-by-side photocatalyst slurries (http://energy.gov/eere/fuelcells/hydrogen-production-photoelectrochemical-water-splitting).

nanoparticles, with one bed carrying out the oxidation of water as a half-reaction and the other bed carrying out the another half-reaction that is reduction of H_2O. Nanoparticles are separately tailored for O_2 production and H_2 production. The performance results pivot on minimal losses due to ion transport.

4.3.2.3 Type 3

These cells (Figure 4.12a and b) include the fixed planar PEC array tilted toward the Sun at local latitude, using multijunction PV cells immersed in a water reservoir. Tilt angle, nominally the latitude angle, can be optimized to achieve the most level H_2 gas output over the year for all the options including environmental variations. Half reactions at electrodes include a mechanism for circulating the ions between beds. The high H_2 production costs ($10.36/kg H_2) are due to large areas of PEC cell component, packaging costs, and gas compression cost (because the compressor is sized for peak hourly production).

4.3.2.4 Type 4

This uses reflectors to concentrate solar flux at greater than 10:1 intensity ratio onto a multijunction PEC element receiver immersed in a water reservoir and pressurized to approximately 300 psi. It has moderately low H_2 production costs ($4.05/kg H_2) with an offset reflector array for the concentrator, higher concentration ratio, numerous possibilities in PEC development, and high temperature operation. In-cell compression of gas eliminates the need for separate compressor. December output of the system is almost 53% of June output, so the system would need to be enlarged.

A type 4 system uses a solar concentrator reflector to focus the direct solar radiation onto the PEC cell. A solar tracking system is used to maximize the direct radiation confinement. A PEC concentrator system can potentially use a concentration ratio of 10–50 suns; however, in general practice it is limited to 10:1, which has been demonstrated in lab tests. It is estimated that a doubling of concentration ratio to 20.1 could reduce the basic reactor cost by 17%.

Solar concentrators use reflectors or lenses to focus solar energy, which substantially reduces the cost impact of the PV component of the system, but adds to the costs of the concentrators and steering systems. Therefore, although the PV components comprise high efficiency costlier cell materials (i.e., GaAs/GaInAs), the overall cost

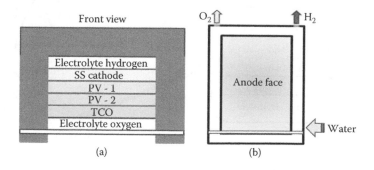

FIGURE 4.12 Type 3 for water splitting: (a) front view and (b) anode face.

of the cell is low. Moreover, in the concentrator PEC system, the inlet water pump pressurizes the collected H_2 and O_2 gases over the water reservoir, which will reduce the cost. Pressurization to 300 psi prevents the need for a separate compressor, which minimizes water vapor loss by the reactor, and reduces O_2 gas bubble size, which minimizes potential bubble scattering of incident photons at the anode face.

The solar concentrator PEC design includes a linear array of PEC cells at the offset parabolic reflector's focal point to focus the radiation on a linear PEC receiver, as shown in Figure 4.13. The system for 1 TPD H_2 average production consists of 1885 such reactors. The offset parabolic array has advantages of reduced structural weight, no aperture blockage, and location of the active receiver components, water feed, and hydrogen collection piping in the reflector base assembly. The array has two-axis steering to track solar radiation.

4.3.3 Miscellaneous Reactor Types

In addition to the above-mentioned basic unit structures of reactors, some special reactors are used by different research groups. The best example of PEC cells is the H-type reactor system proposed by Anpo et al. (Figure 4.14a). This consists of an H-type reactor with photoanode, Pt cathode, and Nafion or proton-exchange membrane. Proton-exchange membrane is used to divide the reactor into two compartments with one photoelectrode (either cathode or anode) in each chamber as shown in Figure 4.14a. The photoelectrode is made up of a Ti foil substrate that is sandwiched between the visible light-active TiO_2 photocatalyst anode and a Pt cathode, prepared by sputtering. The metal Ti foil provides the channel for electron transfer so that the external circuit can be eliminated and electrical resistance can be significantly reduced. Upon light irradiation, water oxidation occurs on TiO_2 to give oxygen gas and protons, which are then transferred to the Pt side via proton exchange membrane, while reduction of hydrogen ion occurs on Pt to give hydrogen gas. As a result, the separate evolution of H_2 and O_2 can be achieved.

In addition to the H-type reactor system, Wu and his group have developed a unique twin-reactor system that combines the advantages of both Z-scheme and

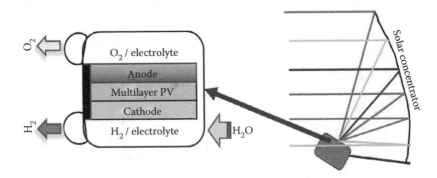

FIGURE 4.13 Type 4 system consisting of a steered solar concentrator and tracker system, focusing solar flux on PEC planar element receivers pressurized to approximately 300 psi.

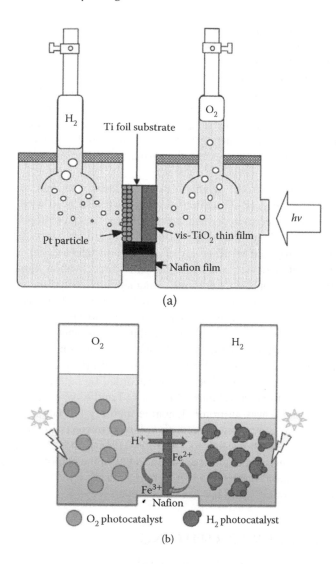

FIGURE 4.14 (a) H-type reaction system for photocatalytic water splitting. (From Matsuoka, M. et al., *Catal. Today*, 122, 51–61, 2007.) (b) Concept of a novel twin-reactor system. (From Yu, S.C. et al., *J. Membr. Sci.*, 382, 291–299, 2011.)

H-type reactor systems to carry out the water-splitting reaction (Lo et al. 2010; Yu et al. 2011). In this novel system, Pt/SrTiO$_3$:Rh and WO$_3$ are used as a H$_2$ photocatalyst and O$_2$ photocatalyst, respectively, and are discretely placed into the compartments of a connected twin reactor separated by a modified ion-exchange membrane (Figure 4.14b). This dual-layer photoelectrode has the advantage of improved light absorption efficiency in the visible region and better charge separation. The modified ion-exchange membrane used in the cell not only allows the transport of protons but also the exchange of the mediator ions (Fe^{2+}/Fe^{3+}; quantitatively measured by the colorimetric method) in solution. The major merit of this novel system is that separate hydrogen and oxygen evolution can be achieved, while using only powder photocatalyst to perform the water-splitting reaction. In a subsequent study, $Pt/SrTiO_3$:Rh and $BiVO_4$ (Liao et al. 2012) were used as the H_2-photocatalyst and the O_2-photocatalyst, respectively, to run a water-splitting reaction in the novel twin-reactor system. It was also concluded in the study that by using the novel twin-reactor system, the deactivation of $Pt/SrTiO_3$:Rh often occurring in the conventional Z-scheme system can be successfully minimized by suppressing the formation of $Fe(OH)_3$ on the photocatalyst surface.

In addition, photoelectrodes made from oxide-based material, semiconductor materials, or particularly, III–V semiconductors have also been used to prepare a high-efficiency photoelectrode for the water-splitting reaction. Figure 4.15a represents a monolithic photovoltaic-PEC device for hydrogen production via water splitting that was proposed by Khaselev et al.; it is an internal and voltage-biased integrated photovoltaic device, and can split water directly upon illumination. The reactor (Figure 4.15b) consists of a 45-mL, magnetically stirred quartz vessel containing the photoactive suspension, connected to a closed stainless steel circulation system, in which an inert gas (either helium or nitrogen) continuously flows. The analysis of the species evolved from the aqueous suspension under irradiation was performed by gas chromatography (GC, with thermal conductivity detector) or quadrupolar mass spectrometry (QMS). The signals at m/z = 2, 18, 28, 32, and 44 were monitored during the runs. The signals recorded by both detectors were calibrated by injections of known volumes of H_2 (or O_2) in the system through a six-way valve. The light source was an iron halogenide mercury arc lamp, emitting in the 315- to 400-nm wavelength range with an irradiation intensity in the reactor of 3×10^{-7} Einstein/s/cm^2. All photocatalysts (10 mg) were suspended in water, sonicated for 20 minutes, and thoroughly flushed with inert gas in the photoreactor, prior to the beginning of irradiation. In Figure 4.15b, A = reactor, B = refrigerator, C = pump, D = detector (QMS or GC), and E = six-ways valve.

4.4 EFFICIENCY OF WATER SPLITTING

Nature smartly uses sunlight and carbon dioxide in the presence of in-built photocatalysts of plants for efficiently breaking water and creating energy along with the chemicals (O_2 and carbohydrates) during the photosynthesis process. Similarly, we also want to imitate the nature (to split water for producing hydrogen and oxygen) using less complicated photocatalytic material than the plants. Hydrogen is one of the most important product of water splitting and is a well-known and promising energy carrier and fuel due to its high energy content per unit mass (142 J/g, three times that of gasoline) and high fuel efficiency, that is, 75%, and clean combustion product (only water). In principle, PEC water splitting possesses a high potential for low-cost hydrogen generation efficiency (up to >30%). PEC water splitting is one of the most attractive methods to generate hydrogen for capturing and storing solar energy in the form of chemical energy. The effiiciency of these systems are registered in the form of current

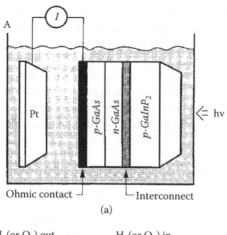

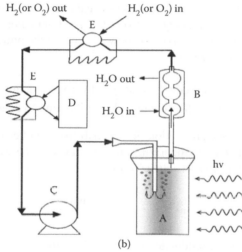

FIGURE 4.15 Schematic of the (a) monolithic PEC/PV device. (From Khaselev, O., and Turner J., *Science*, 280, 425–27, 1998.) (b) Sketch of the Plexiglas cell for photocatalytic water splitting, with separate H_2 and O_2 evolution. A = reactor, B = refrigerator, C = pump, D = detector (quadrupolar mass spectrometry [QMS] or gas chromatography [GC]), and E = six-way valves.

density (current per unit area) at applied voltage (bias) and solar to hydrogen (STH) conversion efficiency. The efficient water-reduction reaction (hydrogen generation) to produce H_2 and the efficient oxidation of water (oxygen evolution and H^+ formation) are the fundamental requirements to obtain PEC solar fuels at high STH efficiency. There are four fundamental steps to determine the overall performance for PEC water splitting (Tsang et al. 1995): (1) light absorption and charge carrier generation, (2) charge separation, (3) charge transport, (4) charge carrier extraction and electrochemical product formation. Thus, the STH efficiency (η STH) can be expressed as:

$$\eta \, STH = \eta A \times \eta CS \times \eta CT \times \eta CR \qquad (4.22)$$

where the η's represent efficiencies of light absorption (ηA), charge separation (ηCS), charge transport (ηCT), and charge collection/reaction efficiency (ηCR). The second ratio in Equation 4.22 is generic for any STH production system, while the third is derived specifically for PEC hydrogen processes. The fourth term, explicitly relating conversion efficiency to the photocurrent density, is calculated for a PEC system under AM1.5 global solar illumination. To achieve a high STH efficiency for a PEC system, the efficiency of each step should be improved and it is necessary to scrutinize the design of the efficient photoanodes/photocathodes. The quantum efficiency ($\eta\%$) of the system is measured as follows:

$$\eta(\%) = \frac{\text{No. of reacted electrons}}{\text{No. of incident photons}} \times 100 = \frac{\text{No. of evolved H}_2 \text{ molecules} \times 2}{\text{No. of incident photons}} \times 100 \quad (4.23)$$

The STH conversion efficiency of any solar-based hydrogen production system is estimated as the ratio of the exploitable chemical energy during the generation of hydrogen gas to the total solar energy delivered to the system. For steady-state operations, this is equivalent to the ratio of the power output to the power input. In words, this can be expressed as follows:

$$\frac{P_{\text{Out}}}{P_{\text{In}}} = \frac{\left(\text{Hydrogen production rate}\right) \times \left(\text{hydrogen energy density}\right)}{\text{Solar flux integrated over illuminated area}} \quad (4.24)$$

where the rate of hydrogen production (R_{H2}) is half the rate of electron flow because two electrons (or half the photocurrent) are consumed in the evolution of one H_2 molecule. This is technically written as follows:

$$R_{H2} = \frac{I_{ph}}{2e} = \frac{(J_{ph} \times \text{Area})}{2e} \quad (4.25)$$

where R_{H2} is the hydrogen production rate (s^{-1}), I_{ph} is the photocurrent (A), e is the electronic charge (C), Area is the illuminated photoelectrode area (cm^2), and J_{ph} is photocurrent density (mA/cm^2). Hydrogen production rate, mentioned in Equation 4.25, solar energy flux (P_{solar}; integrated solar flux of 1000 W/m^2 for AM1.5G solar irradiation), area, and the Gibbs free energy are used to calculate the STH efficiency for a PEC system and can be expressed as follows:

$$STH(\%) = \frac{\Delta G \times R_{H2}}{(P_{solar} \times \text{Area})} = \frac{\left(J_{ph}(V_{WS} - V_B)\right)}{eE_s} (\text{AM1.5G solar irradiation}) \quad (4.26)$$

where J_{ph} is the photocurrent produced per unit irradiated area, $V_{WS} = 1.23$ eV is the water-splitting potential per electron, E_S is the photon flux, V_B is the bias voltage applied between the working and the counterelectrodes, and "e" is the electronic charge. Efficiency can also be measured in terms of incident photon to current efficiency (IPCE). The IPCE provides a measure of the efficiency of conversion of photons to current in a PEC cell flowing between the working and counterelectrodes (Murphy et al. 2006). The IPCE can be calculated using the following equation:

$$\text{IPCE} \ (\%) = \frac{J_{ph}(\lambda)}{eE_s(\lambda)} J_{ph} \qquad (4.27)$$

$$\text{IPCE}(\%) = 1240 \times \text{photo current density (mA/cm}^2) \times (\text{light intensity (mW/cm}^2))^{-1} \times 100 \qquad (4.28)$$

where $E_s(\lambda)$ is the incident photon flux at wavelength λ. 100% IPCE indicates that one incident photon generates an electron hole pair, which is effectively separated and results in current flow in the external circuit. The plot of IPCE as a function of wavelength is known as the action spectra.

Maximum theoretical water-splitting efficiency for α-Fe$_2$O$_3$ is 12.9% within the thermodynamic limit, which cannot be exceeded (Murphy et al. 2006). These values are calculated using ideal conditions, but the achievable practical efficiency is expected to be significantly lower on account of imperfect absorption, reflection losses, and recombination; putting these losses all together can reduce the efficiency by a factor of up to three.

Quantum efficiency (η) of PEC water splitting is calculated using the following formula:

$$\eta = \frac{V_{OC} \cdot I_{SC} \cdot FF}{P_{Light}} = \frac{(IV)_{max}}{P_{Light}} \quad \text{and} \quad \eta(\%) = \frac{V_{OC} \cdot I_{SC} \cdot FF}{P_{Light}} = \frac{(IV)_{max}}{P_{Light}} \times 100 \qquad (4.29)$$

where V_{OC} is open circuit voltage, I_{SC} is saturated current, FF is fill factor, and P_{light} is power density (W/cm^2). The general shape of the characteristic curve is shown in Figure 4.16. Parameters indicate, respectively, the points of open circuit voltage, short circuit current, voltage at maximum power, and current at maximum power. The efficiency is less than 1 due to various factors. The photons of energy less than

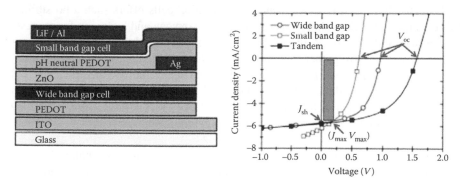

FIGURE 4.16 Characteristic current density versus voltage (J–V) plot representing open circuit voltage V_{OC}, saturated current density J_{sh}, maximum voltage V_{max}, and maximum current density J_{max} for two-terminal polymer tandem solar cells. (From Gilot, J. et al., *Organ. Electron.*, 12, 4, 660–65, 2011.)

the band gap are wasted as they cannot be used for the water-splitting purpose. Also the useful photons may contain energy above the band gap that is also wasted because of the discrete nature of the energy of photons. Another important factor is that the cell voltage needs to be some value below the maximum voltage (open circuit voltage) in order to obtain a finite current. The maximum power for an ideal cell is the theoretically achievable maximum value. Attempting to achieve anything greater than that would be akin to getting work output from an engine with efficiency greater than the Carnot efficiency. The quality of any solar cell can then be quantified as a ratio of the maximum power to this fictitious maximum power if the cell could be run at the open circuit voltage and short circuit current. Two major factors that cause the shape of the real characteristic curve to be different from the ideal curve are the series resistance and the shunt resistance. Ideally, the series resistance is 0 and the shunt resistance is infinity. Figure 4.16 shows the behavior of the curve as the series resistance becomes greater than 0. To get better efficiency for each step and consequently better overall STH efficiency, the following strategies are applied: controlled size and morphology/structure of the photocatalytic sample, metal oxide heterostructuring, doping (metal/nonmetal), quantum dot sensitization, plasmonic metal nanoparticle attachment, cocatalyst coupling, and so on.

4.5 CHALLENGES AND PERSPECTIVES

Hydrogen-based energy technologies are expected to play a considerable role in meeting future energy needs. Recently, renewable hydrogen research is getting some new life all around the world, as is evidenced by the increase in the number of recent publications, including several informative books and patents. There is an important unifying theme linking all renewable-energy research fields, including PEC hydrogen, photovoltaics, nanotechnology, and solar hydrogen production by other solar hydrogen routes. Hydrogen fuel is a silent, emission free, and more efficient than any fuel used in an internal combustion engine. However, the steam reforming of hydrocarbon fuels, the most common hydrogen production technique, does little to reduce overall carbon emissions in the system. A renewable, carbon-free source of high-quality hydrogen is water. Water-splitting PEC cells offers such a possibility, this technique absorbed photon energy to produce oxygen and hydrogen for processing, storage, and later use. Although PEC water splitting is advantageous over photochemical water splitting, there are still some technical challenges associated with it. The researchers of the world are thoughtfully working to improve the efficiency and reduce the cost of the device, which is a great challenge of the time. When we meet this goal, then we are able to commercialize hydrogen fuel at large scale. Following are the areas which need attention to help to meet the goal.

- Current cost of hydrogen generation including semiconductor material.
- Oxides are the most stable semiconducting material in aqueous solution, but their band gaps are either too large for efficient light absorption (~3 eV) or their semiconductor characteristics are poor; therefore, new materials are under investigation.
- Low light absorption.
- Reproducibility or reuse of photocatalytic material for at least 10 cycles.

- Imperfections in the crystalline structure.
- Different bulk and surface properties of the photoelectrodes.
- The material's corrosion under aqueous electrolytes and solar energy.
- Energetics of the electrochemical reaction must be harmonized with the solar radiation spectrum.
- Minimize the formation of H_2O through the backward reaction of H_2 and O_2.
- More than 10% efficiency.
- To overcome technical challenges which can hinder commercial application.
- Routes suffer from a number of challenges that currently prohibit them from becoming commercially viable processes.
- Low solar energy to hydrogen conversion efficiency.
- Reactor design.
- Light saturation effect.
- The remaining portion of the energy wasted as heat or fluorescence.
- The purity of the hydrogen produced is a critical issue for its commercial use.
- Safety is of paramount interest in designing the reactors, particularly if oxygen and hydrogen are simultaneously produced. This exothermic reaction yields 232 kJ/mol of water, resulting in a sharp explosion, which will also consume energy and additionally reduce the overall water-splitting efficiency.
- Metal cocatalysts used for hydrogen and oxygen production from water are often good catalysts for the recombination of hydrogen and oxygen as well, which can reduce net water-splitting efficiency.
- Depending upon the technique used, the produced hydrogen may contain water *vapor*, oxygen, carbon dioxide, and various species.
- 50% w/w potassium hydroxide has been used as a carbon dioxide absorbent, and an alkaline pyragallol solution for removing oxygen; the evolved gases would need to be passed through a dryer or condensation unit to remove moisture.
- Difficulties involving the transfer of hydrogen out of the reactor the hydrogen produced under water.
- Separation of and collection of the various hydrogen species is difficult.
- Effective charge separation and extended photovoltage response in the range of 300–800 nm.
- Losses due to charge recombination of trapped photogenerated holes with the accumulated electrons.
- Morphological degradation in the sample due to etching or photocurrent degradation.
- Seizure of hydrogen evolution probably due to passivation of Mg.
- The stability of elemental species on a photocatalyst's surface depends on the electrolyte pH, as well as the applied potential (As at GaAs surface). The stability of elemental arsenic on a GaAs surface depends on the electrolyte pH, as well as the applied potential. At very low pH and a potential of −0.45 V, arsenic is oxidized to stable As_2O_3 (Frese et al. 1980; Menezes and Miller 1982; Allongue and Blonkowski 1991).

To explore all solutions and aspects of the ongoing discussions on the hydrogen energy provision via PEC splitting of the water; from its generation and storage to its distribution and management, the needs and demands of the different factors, limitation, and the impacts that energy technologies and policies have on societies. Despite the above limitations, there is great hope for the development of an apt material that could be used in appropriate reactors with suitable electrolytes for efficient (more than 20%) generation of hydrogen by water splitting. For achieving the above target one should have to be opted for modification at several levels of photocatalytic semiconductor and reactor. Fundamental semiconducting materials along with their materials-interface properties hold the keys to successful PEC research and development. Thus, the PEC network is constructing a road map to clean energy by including state-of-the-art materials theory, synthesis, and characterization, to facilitate research progress, and to arouse fundamental breakthroughs. Beyond the boundaries of countries, the investigation of promising PEC technologies even becames a more valuable asset to expand collaborative activities. Many international (formal/informal) watch dogs are coordinating in fostering collaborative PEC research activities. Some are focused on the R&D activities of hard-core PEC material classes, and others focus on critical activities to advance the supporting science and technologies. These activities for sure support the development in the following aspects:

1. Standardized testing and reporting protocols for evaluating PEC materials systems.
2. Advanced characterization techniques to acquire better understanding of PEC materials and interfaces, and their optoelectronic properties.
3. Exploration of new theoretical models of PEC materials and brand new semiconductor systems.
4. Refinement of techno-economics analyses of PEC hydrogen production on the basis of the performance (efficiency) and processing cost feedback.

The above will provide a basis for evaluating the long-term feasibility of large-scale PEC production technologies in comparison with other renewable approaches. On an impressive scale, the PEC research community is applying its tool chest to investigate a broader spectrum of promising materials classes.

4.6 SUMMARY

The limited store of conventional fossil fuels and their negative impact on ecohealth accelerate the concern of the present generation on the issue of worldwide energy that must be satisfied in an eco-benign way. Additionally, there is a growing political and public recognition and acceptance on benign energy sources. Moreover, governments are putting forward careful efforts to motivate the research of transforming of energy resources from conventional to renewable/sustainable ones. Hydrogen as a fuel is the most powerful renewable source. Water is a good source of hydrogen and breaking of water can generate hydrogen without releasing the carbon content in the atmosphere. But due to the unavailability of highly efficient photocatalytic material and apt knowledge of the technology, commercial level hydrogen production is not possible. In this chapter, we discussed the theoretical and practical aspects of PEC splitting of water. We minutely discuss the components of the PEC device such as the photocatalytic electrodes, sensitizers, receptors, scavengers, or sacrificial materials (electron donor or acceptors) along with their material and energetic requirements and electron transfer phenomenon. Different kinds of scavengers or sacrificial electrolyte systems such as hole and electron scavengers were discussed in light of their mechanism in PEC cells and their respective role in hydrogen and oxygen generation. Usually, quantum dots and dyes are used as the sensitizers for the photocatalyst to facilitate the wide band gap—the main photocatalytic system for light-harvesting power. They support charge transfer and suppress recombination of photo carriers (hole and electrons) through the interface potential gradient. Their mechanisms along with suitable examples are expressed in detail. We also described how these components act together in a PEC water-splitting device with the help of the various models in light of energetic, chemical, thermodynamic, and kinetic parameters and surface chemistry. This chapter focused on the identification of the various aspects of PEC reactor design and their operational experimental setup. PEC reactors based on bias and type of photocatalytic material are well illustrated in the chapter. The typical reactor designs and their mode of operations are also elaborated. Efficiency of PEC for hydrogen generation in terms of STH, IPCE, and the apparent quantum efficiency of the reactors is also described. Strategies to improve and optimize the efficiency of the overall PEC system for hydrogen generation through water splitting were also discussed. The current challenges of PEC water splitting were highlighted with their most probable solutions and future prospectives.

REFERENCES

Allongue, P. and S. Blonkowski. 1991. Corrosion of III-V compounds: A comparative study of GaAs and InP II. Reaction scheme and influence of surface properties. *J. Electroanal. Chem.* 317: 77–99.

Amouyal, E. 1995. Photochemical production of hydrogen and oxygen from water: A review and state of the art. *Sol. Energy Mater.. Sol. Cells.* 38: 249–76.

Annual Energy Outlook. *Energy Information Administration.* Washington, DC: US Department of Energy, 2005.

Appleby, A. J., G. Crepy, and J. Jacquelin. 1978. High efficiency water electrolysis in alkaline solution. *Int. J. Hydrogen Energy* 3: 21–37.

Aruda, K. O., M. Tagliazucchi, C. M. Sweeney, D. C. Hannah, and E. A. Weiss, 2013. The role of interfacial charge transfer-type interactions in the decay of plasmon excitations in metal nanoparticles. *Phys. Chem. Chem. Phys.* 17: 7441–49.

Bard A. J. and M. A. Fox. 1995. Artificial photosynthesis: Solar splitting of water to hydrogen and oxygen. *Acc. Chem. Res.* 28(3): 141–45.

Bard, A. J., G. M. Whitesides, R. N. Zare, and F. W. Mclafferty. 1995. Holy grails of chemistry. *Acc. Chem. Res.* 28(3): 91.

Belmont, C., R. Ferrigno, O. Leclerc, and H. H. Girault. Coplanar interdigitated band electrodes for electrosynthesis. Part 4: Application to sea water electrolysis. 1998. *Electrochim. Acta* 44: 597–603.

Bequerel, E. 1939. Recherches sur les effets de la radiation chimique de la lumière solaire, au moyen des courants électriques. *C.R. Acad. Sci.* 9: 145–49.

Boddy, P. J. 1968. Oxygen evolution on semiconducting TiO₂. *J. Electrochem. Soc.* 115: 199–203.

Chang C. H. and Y. L. Lee. 2007. Chemical bath deposition of CdS quantum dots onto mesoscopic TiO₂ films for application in quantum-dot-sensitized solar cells. *Appl. Phys. Lett.* 91(5): 053503.

Cho, S., J. W. Jang, S.-H. Lim, et al. 2011. Solution-based fabrication of ZnO/ZnSe hetero-structure nanowire arrays for solar energy conversion. *J. Mater. Chem.* 21: 17816–22.

Chouhan, N., C. K. Chen, W.-S. Chang, K.-W. Cheng, and R.-S. Liu. 2011. Chapter 12: Photoelectrochemical cells for hydrogen generation. In: *Electrochemical Technologies for Energy Storage and Conversion* (eds. J. Zhang, L. Zhang, H. Liu, A. Sun, R.-S. Liu), Vol. 2. Weinheim, Germany: Wiley-VCH.

Darwent, J. R. 1981. H_2 production photosensitized by aqueous semiconductor dispersions. *J. Chem. Soc. Faraday Trans.* 277: 1703–09.

De Franceschi, S., S. Sasaki, J. M. Elzerman, W. G. V. Wiel, S. Tarucha and L. P. Kouwenhoven. 2000. Electron cotunneling in a semiconductor quantum dot. *Condens. Matter* 27: 0007448 (10 p.).

De Jonge, R. M., E. Barendrecht, L. J. J. Janssen, and S. J. D. van Stralen. 1982. Gas bubble behaviour and electrolyte resistance during water electrolysis. *Int. J. Hydrogen Energy* 7: 883–94.

De Souza, R. F., J. C. Padilha, R. S. Gonçalves, et al. 2003. Room temperature dialkylimid-azolium ionic liquid-based fuel cells. *Electrochem. Commun.* 5: 728–31.

De Souza, R. F., J. C. Padilha, R. S. Gonçalves, et al. 2006. Dialkylimidazolium ionic liquids as electrolytes for hydrogen production from water electrolysis. *Electrochem. Commun.* 8: 211–16.

De Souza, R. F., G. Loget, J. C. Padilha, et al. 2008. Molybdenum electrodes for hydrogen production by water electrolysis using ionic liquid electrolytes. *Electrochem. Commun.* 10: 1673–75.

Diguna, L. J., Q. Shen, J. Kobayashi, and T. Toyoda. 2007. High efficiency of CdSe quantum-dot-sensitized TiO_2 inverse opal solar cells. *Appl. Phys. Lett.* 91(2): 023116.

Frese Jr. K. W., M. J. Madou, and S. R. Morrison 1980. Investigation of photoelectrochemical corrosion of semiconductor. *J. Phys. Chem.* 84: 3174–78.

Fujishima, A. and K. Honda. 1972. Electrochemical photolysis of water at a semiconductor electrode. *Nature* 238(5358): 37–8.

Gilot, J., M. M. Wienk, and R. A. J. Janssen. 2011. Measuring the current density–voltage characteristics of individual subcells in two-terminal polymer tandem solar cells. *Organ. Electron.* 12(4): 660–65.

Giordano N., V. Antonucci, S. Cavallaro, R. Lembo, and J. C. J. Bart. 1982. Photoassisted decomposition of water over modified Rutile electrodes. *Int. J. Hydrogen Energy* 7(11): 867–72.

Gust, D., T. A. Moore, and A. L. Moore. Solar fuels via artificial photosynthesis. 2009. *Acc. Chem. Res.* 42: 1890–98.

Guttman, F. and O. J. Murphy. 1983. *Modern Aspects of Electrochemistry*. Plenum Press, New York.

Hoffert, M. I., K. Caldeira, A. K. Jain, et al. 1998. Energy implications of future stabilization of atmospheric CO_2 content. *Nature* 395: 881–84.

Jin, H., S. Choi, H. J. Lee, and S. Kim. 2013. Layer-by-layer assemblies of semiconductor quantum dots for nanostructured photovoltaic devices. *J. Phys. Chem. Lett.* 4(15): 2461–70.

Kamat, P. V. 2007. Meeting the clean energy demand: Nanostructure architectures for solar energy conversion. *J. Phys. Chem. C* 111(7): 2834–60.

Kato, H., K. Asakura, and A. Kudo. 2003. Highly efficient water splitting into H_2 and O_2 over lanthanum-doped $NaTaO_3$ photocatalysts with high crystallinity and surface nanostructure. *J. Am. Chem. Soc.*, 125(10): 3082–89.

Khaselev, O. and J. A. Turner. 1998. A monolithic photovoltaic-photoelectrochemical device for hydrogen production via water splitting. *Science* 280: 425–27.

Khnayzer, R. S. 2013. Photocatalytic Water Splitting: Materials Design and High-Throughput Screening of Molecular Compositions. Ph.D. Thesis, Graduate College of Bowling Green State University, Bowling Green, OH.

Kostov, M. K., E. E. Santiso, A. M. George, K. E. Gubbins, and M. B. Nardelli. 2005. Dissociation of water on defective carbon substrates. *Phys. Rev. Lett.* 95(13): 136105.

LeRoy, R. L., C. T. Bowen, and D. J. LeRoy. 1980. The thermodynamics of aqueous water electrolysis. *J. Electrochem. Soc.* 127: 1954–62.

Lewis, N. S. and D. G. Nocera. 2006. Powering the planet: Chemical challenges in solar energy utilization. *Proc. Natl. Acad. Sci. U.S.A.* 103: 15729–35.

Li, S., C. Wang, and C. Chen. 2009a. Water electrolysis in the presence of an ultrasonic field. *Electrochim. Acta.* 54: 3877–83.

Li, S.-D., C. C. Wang, and C.-Y. Chen. 2009b. Water electrolysis for H_2 production using a novel bipolar membrane in low salt concentration. *J. Membrane Sci.* 330(1–2): 334–40.

Liao, C.-H., C.-W. Huang, and J. C. S. Wu. 2012. Hydrogen production from semiconductor-based photocatalysis via water splitting. *Catalysts* 2: 490–516.

Liao, P. and E. A. Carter. 2013. New concepts and modeling strategies to design and evaluate photo-electro-catalysts based on transition metal oxides. *Chem. Soc. Rev.* 2: 2401–22.

Lo, C. C., C. W. Huang, C. H. Liao, and J. C. S. Wu. 2010. Novel twin reactor for separate evolution of hydrogen and oxygen in photocatalytic water splitting. *Int. J. Hydrogen Energy* 35(4): 1523–29.

Matsuoka, M., M. Kitano, M. Takeuchi, K. Tsujimaru, M. Anpo, and J. M. Thomas. 2007. Photocatalysis for new energy production: Recent advances in photocatalytic water splitting reactions for hydrogen production. *Catal. Today* 122: 51–61.

Mazloomi, K., N. B. Sulaiman, and H. Moayedi. 2012. Electrical efficiency of electrolytic hydrogen production. *Int. J. Electrochem. Sci.* 7: 3314–26.

Menezes, S and B. Miller. 1982. Surface and redox reactions at GaAs in various electrolytes. *J. Electrochem. Soc*. 130: 517–23.

Murphy, A. B., P. R. F. Barnes, L. K. Randeniya, et al. 2006. Efficiency of solar water splitting using semiconductor electrodes. *Int. J. of Hydrogen Energy* 31: 1999–2017.

Nagai, N., M. Takeuchi, T. Kimura, and T. Oka. 2003. Existence of optimum space between electrodes on hydrogen production by water electrolysis. *Int. J. Hydrogen Energy* 28: 35–44.

Nakicenovic, N., A. Grübler, and A. McDonald. 1998. *Global Energy Perspectives*. Cambridge, London: Cambridge University Press.

Nazeeruddin, M. K., A. Kay, I. Rodicio, et al. 1993. Conversion of light to electricity by cis X_2bis (2, 20 bipyridyl 4, 40 dicarboxylate) ruthenium (II) charge transfer sensitizers (X : Cl, Br, I, CN, and SCN) on nanocrystalline titanium dioxide electrodes. *J. Am. Chem. Soc.*, 115: 6382–90.

Nazeeruddin, M. K., P. Pechy, and M. Gratzel. 1997. Efficient panchromatic sensitization of nanocrystalline TiO_2 films by a black dye based on a trithiocyanato ruthenium complexes. *Chem. Commun.* (18): 1705–06.

Nocera, D. G. 2010. "Fast food" energy. *Energy Environ. Sci.* 3: 993–95.

Nozik, A. J. 1975. Photoelectrolysis of water using TiO_2 crystals. *Nature* 257: 383–86.

Nozik, A. J. 2002. Quantum dot solar cells. *Physica E* 14(1): 115–20.

Onda, K., T. Kyakuno, K. Hattori, and K. Ito. 2004. Prediction of production power for high-pressure hydrogen by high-pressure water electrolysis. *J. Power Sources* 132: 64–70.

Qian, K., Z. D. Chen, and J. J. J. Chen. 1998. Bubble coverage and bubble resistance using cells with horizontal electrode. *J. Appl. Electrochem.* 28: 1141–45.

Rasmussen, A. M., S. Ramakrishna, E. A. Weiss, T. Seideman. 2014. Theory of ultrafast photoinduced electron transfer from a bulk semiconductor to a quantum dot. *J. Chem. Phys.* 140: 144102.

Raul C., D. Dolzhnikov, K. Edme, et al. Available at: www.sites.northwestern.edu/weiss-lab/welcome/quantum-dot-sensitized-photocatalysis/.

Reber, J. F. and K. Meier. 1984. Photochemical production of hydrogen with zinc sulfide suspensions. *J. Phys. Chem.* 88(24): 5903–13.

Reece, S. Y., J. A. Hamel, K. Sung, et al. 2011. Wireless solar water splitting using silicon-based semiconductors and earth-abundant catalysts. *Science* 334: 645–48.

Rehm, J. M., G. L. McLendon, Y. Nagasawa, et al. 1996. Femtosecond electron transfer dynamics at a sensitizing dye semiconductor (TiO_2) interface. *J. Phys. Chem.* 100: 9577–78.

Renaud, R. and R. L. LeRoy. 1982. Separator materials for use in alkaline water electrolysers. *Int. J. Hydrogen Energy*, 7: 155–66.

Ruland, A., C. Schulz-Drost, V. Sgobba, and D. M. Guldi. 2011. Enhancing photocurrent efficiencies by resonance energy transfer in CdTe quantum dot multilayers: Towards rainbow solar cells. *Adv. Mater.* 23(39): 4573–77.

Tsang, S. C., J. B. Claridge, and M. L. H. Green. 1995. Recent advances in the conversion of methane to synthesis gas. *Catal. Today* 23(1): 3–15.

Tsiplakides, D. 2012. PEM Water Electrolysis Fundamentals. Available at: www.research. ncl.ac.uk/sushgen/docs/summerschool_2012/PEM_water_electrolysis-Fundamentals_ Prof._Tsiplakides.pdf.

Tsuji, I., H. Kato, and A. Kudo. 2006. Photocatalytic hydrogen evolution on ZnS-$CuInS_2$-$AgInS_2$ solid solution photocatalysts with wide visible light absorption bands. *Chem. Mater.* 18(7): 1969–75.

Tsuji, I., H. Kato, H. Kobayashi, and A. Kudo. 2004. Photocatalytic H_2 evolution reaction from aqueous solutions over band structure-controlled $(AgIn)_xZn_{2(1-x)}S_2$ solid solution photocatalysts with visible-light response and their surface nanostructures. *J. Am. Chem. Soc.* 126(41): 13406–13.

Udagawa, J., P. Aguiar, and N. P. Brandon. 2007. Hydrogen production through steam electrolysis: Model-based steady state performance of a cathode-supported intermediate temperature solid oxide electrolysis cell. *J. Power Sources* 166: 127–36.

Ulleberg Ø. 2003. Modeling of advanced alkaline electrolyzers: A system simulation approach. *Int. J. Hydrogen Energy* 28: 21–33.

Vermeiren, P., W. Adriansens, and R. Leysen. 1996. Zirfon: A new separator for Ni-H2 batteries and alkaline fuel cells. *Int. J Hydrogen Energy* 21: 679–84.

Vermeiren, P., W. Adriansens, J. P. Moreels, and R. Leysen. 1998. Evaluation of the Zirfon® separator for use in alkaline water electrolysis and Ni-H₂ batteries. *Int. J. Hydrogen Energy* 23(5): 321–24.

Villalobos, L. F., Y. Xie, S. P. Nunes, and K.-V. Peinemann. 2016. Polymer and membrane design for low temperature catalytic reactions. *Macromol. Rapid Commun.* 37(8): 700–06.

Wang, M., Z. Wang, and Z. Guo. 2010. Water electrolysis enhanced by super gravity field for hydrogen production. *Int. J. Hydrogen Energy* 35: 3198–205.

Watson, D. F. 2010. Linker-assisted assembly and interfacial electron-transfer reactivity of quantum dot–substrate architectures. *J. Phys. Chem. Lett.* 1(15): 2299–309.

Wei, Z. D., M. B. Ji, S. G. Chen, et al. 2007. Water electrolysis on carbon electrodes enhanced by surfactant. *Electrochim. Acta* 52: 3323–29.

Weinberg, D. J., S. M. Dyar, Z. Khademi, et al. 2014. Spin-selective charge recombination in complexes of CdS quantum dots and organic hole acceptors. *J. Am. Chem. Soc.* 136: 14513–518.

Wigley, T. M. L., R. Richels, and J. A. Edmonds. 1996. Economic and environmental choices in the stabilization of atmospheric CO_2 concentration. *Nature* 379: 240–43.

Youngblood, W. J., S.-H. A. Lee, Y. Kobayashi, et al. 2009. Photoassisted overall water splitting in a visible light-absorbing dye-sensitized photoelectrochemical cell. *J. Am. Chem. Soc.* 131: 926–7.

Yu, S. C., C. W. Huang, C. H. Liao, J. C. S. Wu, S. T. Chang, and K. H. Chen. 2011. A novel membrane reactor for separating hydrogen and oxygen in photocatalytic water splitting. *J. Memb. Sci.* 382: 291–299.

Zhang, J. Z. 2011. Metal oxide nanomaterials for solar hydrogen generation from photoelectrochemical water splitting. *MRS Bull.* 36(1): 48–55.

5 Oxide Semiconductors (ZnO, TiO$_2$, Fe$_2$O$_3$, WO$_3$, etc.) as Photocatalysts for Water Splitting

5.1 INTRODUCTION

Energy is on the top of 10 great problems of humanity for the next 50 years, that is, energy, water, food, environment, poverty, terrorism and war, disease, education, democracy, and population, which underlines the importance of energy for the survival of the human race. Unfortunately, most of our energy sources are coming from carbon-rich conventional sources, which are gradually depleted day by day and impose a negative impact on nature. Therefore, there is a great need for searching for a solution of this problem in renewable resources. In the words of Thomas Alva Edison (1931), we should not wait for the time when oil and coal run out before we tackle that situation (Smalley 2003). The splitting of water by using sunlight to generate hydrogen is one of the gentlest ways to produce energy. In this case, the importance of solar-harvesting devices may enhance the use of sustainable energy, which will essentially reduce the consumption of fossil fuels. To meet the target of 10% solar energy conversion efficiency, a considerable amount of research is required to design a catalyst which can work efficiently under solar radiation (especially the visible light). For this, a thoughtful choice of stable materials will be made, which may have a band gap (BG) around 2 eV and at the same time satisfy the band potential requirements. Sulfides meet the BG requirement, but they are not stable. Metal oxides (mixed and single) are the most studied photocatalyst because they have the greatest stability and not so good efficiency (due to their wide BG). Due to the strong power of oxidation or reduction or both of the photoexcited oxides they can also be used for overall water splitting. A large number of metal oxides such as TiO$_2$, ZnO, Fe$_2$O$_3$, and WO$_3$, BiVO$_4$, SrTiO$_3$, and so on, with various morphologies have been utilized for water splitting. In general, the valence bands (VBs) of stable oxide semiconductor photocatalysts based on metal cations with d^0 and d^{10} configurations are composed of O 2p orbitals. The energy levels are considerably more positive (~+3 eV) than the oxidation potential of H$_2$O and more negative than the reduction potential of water. Therefore, the BGs of oxide semiconductor photocatalysts for water splitting will inevitably become wider than 3 eV. However, most metal oxides have a large BG greater than 3 eV, limiting the absorption of light in the ultraviolet (UV) region and needs the improvement in overall solar energy conversion efficiency. Thus, oxide photocatalysts usually respond to only UV light. Frank and Bard first examined the possibilities of using TiO_2 to decompose cyanide in water. The research in this field was stimulated after the discovery of the Honda–Fujishima effect in 1972 (Fujishima and Honda 1972). Fujishima and Honda reported the photoelectrocatalytic decomposition of water on TiO_2 and platinum electrodes and corresponding production of oxygen and hydrogen.

5.2 DESIGN OF METAL OXIDE PHOTOCATALYSTS WITH VISIBLE LIGHT RESPONSE (EFFECT OF MORPHOLOGY OF SEMICONDUCTOR AND REACTION MECHANISM OF PHOTOELECTRODES)

Sunlight has enormous power (~600,000 nuclear reactors or 40 times more than our present energy need) to serve humanity without polluting the planet but we are utilizing only 0.6% of it. A photocatalyst is the powerful medium to convert solar energy to an energy carrier or fuel, that is, production of hydrogen by breaking water. This is a pollution-free mode of hydrogen generation. Unfortunately, to date there is no low-cost and stable photocatalyst available that can be efficiently used to generate hydrogen. Generally, oxide materials are used as the stable photocatalysts but they suffer with the problem of low light-harvesting capacity (sensitive to UV light that is <5% of the total sun light). The first photocatalyst used for hydrogen and oxygen generation in an electrochemical cell was a titania photoelectrode, with an applied external bias. Since then, a great number of researchers have explored the ways to achieve direct water dissociation with limited success. The following reactions are involved in the production of hydrogen and oxygen from water (pH = 7):

$$\text{At cathode (reduction)}: 2H_2O + 2e^- \rightarrow H_2 + 2OH^-, E^\circ_{(H_2/H_2O)} = -0.41 \text{ V (vs. NHE)}$$

(5.1)

$$\text{At anode (oxidation)}: 2H_2O \rightarrow O_2 + 4H^+ + 4e^-, E^\circ_{(H_2/H_2O)} = +0.82 \text{ V (vs. NHE)}$$

(5.2)

$$\text{Overall reaction}: H_2O \rightarrow H_2 + \frac{1}{2}O_2, E^\circ_{\text{overall}} = E^\circ_{(H_2O/O_2)} - E^\circ_{(H_2/H_2O)} = 1.23 \text{ V} \quad (5.3)$$

where NHE is the normal hydrogen electrode. The overall reaction is a two-electron redox process and an endothermic reaction that involves a change in the Gibbs free energy (ΔG° = 2.46 eV or 237.2 kJ/mol). The energy of 1.23 eV per electron transferred corresponds to the wavelength of λ = 1008 nm. Water absorbs solar radiation in the infrared (IR) region, with a very low photon energy, which is unable to drive photochemical water splitting. Thus, it needs a photochemical or semiconductor that can absorb light for water to break it. Semiconductors, under solar irradiation, are capable of decomposing water into hydrogen and oxygen. This semiconductor with an appropriate value for BG induces the oxidation and reduction reactions of water. The free energy considerations as discussed earlier for water splitting dictate that the material must have a BG of at least 1.23 eV and similar BGs for a combined CO_2 reduction/water oxidation Z-scheme. However, at least in the case of water splitting, thermodynamic losses have been estimated to be 0.4 eV (Murphy et al. 2006a), while kinetic barriers to be overcome for the surface reactions may add another 0.3–0.4 eV (De Krol et al. 2004), leading to a proposed ideal BG of ~2 eV for water splitting. Such a material would absorb photons in the visible range (wavelength of 610 nm) near the peak in the solar spectrum (500 nm). The solar intensity drops off dramatically for light with wavelengths ~400 nm, suggesting an upper limit to a useful BG of ~3 eV. The solar spectrum profile suggests aiming for a BG closer to its peak intensity at ~2.5 eV. Therefore, the favorable BGs range between 2 and 2.5 eV. The conduction band (CB) level should be more negative than the reduction potential of the hydrogen, $E^{\circ}_{(H_2/H_2O)}$, while the valance band level should be more positive than the water oxidation level, $E^{\circ}_{(H_2O/O_2)}$. When a semiconductor is irradiated by the photons of energy equal to or greater than that of its BG, excitation of an electron from the VB to the CB occurs that leaves a hole behind in the VB.

If the photogenerated electrons and holes recombine in bulk or on the semiconductor surface, they then release energy in the form of heat or a photon. Otherwise, the electrons and holes migrate to the semiconductor surface without recombination and can, respectively, reduce and oxidize water (or the reactant). The efficiency of converting light energy into hydrogen energy using suspended oxide catalysts is very low to date. The primary reasons for the low efficiencies are as follows:

1. Chemical instability under both illumination and dark.
2. A stable and low cost material of BG of ~2.0 eV to absorb maximum solar radiation is difficult to obtain.
3. Presence of the charge recombination centers to prevent recombination of the photogenerated charge carriers.
4. Moderate conductivity; too high and the series resistance degrades the efficiency, too low and the electrolyte electrically shorts to the photoelectrochemical (PEC) cell.
5. No suitable band edge position with respect to the H^+/H_2 reduction potential (0.00 eV) and O_2/OH^- oxidation potentials (1.23 eV).
6. Photochemical water splitting involves at least one exothermic reaction, which is made easy to recombine molecular hydrogen and oxygen (backward reaction).

$$H_2(g) + \frac{1}{2}O_2(g) \rightarrow H_2O(g) + 57 \text{ kcal} \qquad (5.4)$$

7. The poor visible spectrum response of corrosion-resistant oxide catalysts due to wide BG. For example, TiO_2, ZnO, and so on, which are the suitable photocatalysts with superb chemical stability and strong catalytic activity, do not respond to visible light.

A general mechanism for photocatalytic water splitting is illustrated in Figure 5.1.

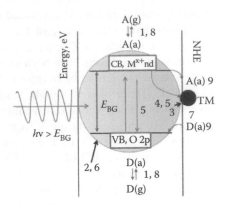

1. Adsorption
2. Surface structure
3. Interface
4. Charge diffusion
5. Charge trapping
6. Particle size
7. Surface electronic states
8. Dark reactions
9. Photoreaction

FIGURE 5.1 A schematic of the different photoreaction/photoexcitation processes relevant to photocatalytic reactions. The figure presents a schematic of the main chemical and physical reactions involved on the surfaces of a semiconductor under photon irradiation (legend at the right-hand side of the figure). Large sphere, semiconductor, small sphere, transition metals (TMs), such as Au, Ru, or Rh; VB, valence band (mainly O 2p in the case of oxides); CB, conduction band (mainly M^{x+} 3d levels in oxides); A, acceptor; D, donor; NHE, normal hydrogen electrode; E_{BG}, band gap; (a), adsorbed; (g), gas phase.

Band positions of several important oxide semiconductors in the presence of aqueous electrolyte at pH = 7 are listed in Table 5.1. Energy scales of bands are indicated in electron volts calculated with respect to either NHE or vacuum level as a reference.

5.2.1 Effect of Morphology of Semiconductor

5.2.1.1 Design of Photocatalyst at Nanoscale

To address the current challenges in PEC water splitting, chemically modified nanostructured electrodes could provide a new avenue with high-efficiency and lightweight devices. Because of the nanostructuring of the material that offers extremely large semiconductor/electrolyte interfacial area and short diffusion distance for minority carriers, which can manifest the better charge separation and reduce the loss due to electron–hole recombination. In actuality, the discovery of the microscope (350 years ago) was the turning point to tune the morphology at nanolevel of the materials dealt with by chemists, biologists, and other scientists. Nanoscience deals with the study of the ability to imagine, design, investigate, modify, model, and manipulate the substance at the molecular or atomic level, and there are properties that differ significantly from those at a larger scale. Nanotechnology could become much more important and open new avenues in the fields of optoelectronics, photonics, display systems, chemical and biosensors, catalysis, photovoltaics (PV) and fuel cells, and so on, to regulate the molecular unexplored expedition. The unique properties of nanomaterials, which make them different from large particles are the greater surface to volume ratios, optical (e.g., color, transparency), electrical (e.g., conductivity), physical (e.g., hardness, boiling point), chemical (e.g., reactivity, reaction rates), and surface (e.g., morphology, organization of molecular particle composites in one-dimensional

TABLE 5.1

Band Positions at Valance Band/Conduction Band and Band Gap Wavelength of Various Oxide Semiconductors at pH = 7

Serial No.	Oxide	E_{CB} (eV)	E_{VB} (eV)	λ_g (nm)
1.	H^+/H_2	**−4.1**	−	−
2.	OH^-/O_2	**0.82**	−	−
3.	Cu_2O	−4.22	−6.42	563.5
4.	Ce_2O_3	−4.00	−6.40	516.7
5.	WO_3	−3.88	−6.73	485.09
6.	In_2O_3	−3.88	−6.68	442.9
7.	PbO	−4.02	−6.82	442.9
8.	$MnTiO_2$	−4.04	−7.14	400
9.	ZnO	−4.19	−7.39	387.5
10.	TiO_2	−4.21	−7.41	387.5
11.	$KNbO_2$	−3.64	−6.94	375.8
12.	Cr_2O_3	−3.93	−7.43	354.3
13.	NiO	−4.0	−7.50	354.3
14.	$KTaO_2$	−3.57	−7.07	354.3
15.	MnO	−3.49	−7.09	344.4
16.	$AlTiO_2$	−3.64	−7.24	344.4
17.	$MgTiO_2$	−3.75	−7.45	335.1
18.	Tb_2O_3	−3.44	−7.74	326.3
19.	Pr_2O_3	−3.24	−7.14	317.9
20.	$LiTaO_2$	−3.55	−7.55	310.0
21.	$LaTl_2O_7$	−3.9	−7.9	310.0
22.	SnO	−3.59	−7.79	293.2
23.	Sm_2O_3	−3.07	−7.47	281.8
24.	Nd_2O_3	−2.87	−7.57	263.8
25.	Ga_2O_3	−2.95	−7.75	258.3
26.	Yb_2O_3	−3.02	−7.92	253.1
27.	ZrO_2	−3.41	−8.41	248
28.	La_2O_3	−2.53	−8.03	225.5

Note: The bold potentials indicate water reduction and oxidation half reactions.

[1D], two-dimensional [2D], and three-dimensional [3D] assemblies) properties, and so on, because microforces such as Brownian motion, van der Waals forces, chemical bonds, and viscosity became more important than macroforces. Here, the gravitational forces present in nano-objects become negligible and electromagnetic forces begin to dominate. Therefore, quantum mechanics is used to describe the motion and energetics of the nano-objects instead of classical mechanics. PEC water splitting based on oxide-semiconductors represents a green and low-cost technique to generate hydrogen and oxygen. However, their solar-to-hydrogen (STH) conversion efficiency is quite low and the desired morphology of the working photoelectrode system needs to be engineered. Nanoscale photoelectrochemistry represents a fusion of photoelectrochemistry and nanoscience. However, nanotechnology is proven to be a superb tool in modification of the nanostructured photoelectrodes to improve PEC water-splitting efficiency, light-harvesting power, charge transport, kinetics, and energetic parameters. Engineering on nanostructures cannot change their intrinsic electronic properties but increases the electrolyte accessible area along with the number of active centers on the surface, and shortens minority carrier diffusion distance. More importantly, recent developments in chemically modified nanostructured electrodes, including surface modification with catalyst and plasmonic metallic structures, element doping, and incorporation of functional heterojunctions, has directed to significant enhancements in the efficiencies of charge separation, transport, collection, and solar energy harvesting.

5.2.1.2 Unique Aspects of Nanotechnology

1. **Shortened carrier collection pathways.** Photoexcitation produces charge carriers with finite mobility and lifetime, depending on the material, the carrier type, and the light intensity. To drive water redox reactions, these carriers need to reach the material interfaces at the electrolyte and at the back contact. In the absence of an external field, charge carriers move by diffusion and their range is defined by the mean-free diffusion length L. To improve minority carrier collection at the semiconductor (sc)–electrolyte interface, the surface roughness of the film can be increased, as shown in Figure 5.4b later in the chapter. This surface nanostructuring approach is particularly useful for first-row transition metal oxides (MnO_2, Fe_2O_3), which suffer from low hole mobility and lifetimes (Cox 2010; Huda et al. 2011). Ideal electron–hole collection is possible with suspended nanoparticles, if their particle size $d < L_n$, L_p. In this situation, both carrier types need to be extracted at the sc–electrolyte interface. Thus, there is a need for selective redox agents.

2. **Improved light distribution.** The ability of a material to absorb light is determined by the Lambert–Beer law and the wavelength-dependent absorption coefficient α. The light penetration depth α^{-1} refers to the distance after which the light intensity is reduced to $1/e$ of the original value. For example, for Fe_2O_3, $\alpha^{-1} = 118$ nm at $\lambda = 550$ nm (Luo et al. 2014), for CdTe, $\alpha^{-1} = 106$ nm (550 nm), and for Si, $\alpha^{-1} = 680$ nm (510 nm) (Würfel 2005; Berger et al. 2008). To ensure greater than 90% absorption of the incident light, the film thickness must be greater than 2.3 times the value of α^{-1}. Surface structuring on the micro- or nanoscale can increase the degree of horizontal light distribution via light scattering (Figure 5.2i). This "trapped" light would otherwise be lost by direct reflection from a flat surface. Light scattering is maximal in particle suspensions because it can occur at both the front and back sides of the particles. The dimensions of nanostructured photocatalysts are usually smaller than α^{-1}, so each nanoparticle only absorbs a small fraction of the incident light. However, complete light absorption by the suspension can be achieved by adjusting the particle concentration and optical path length (diffusion length) of the reaction container. Ideal electron–hole collection is possible with the suspension of nanoparticles of particle size $d < L_e$, L_h. d: film or particle thickness; L_e: electron diffusion length; L_h: hole diffusion length.

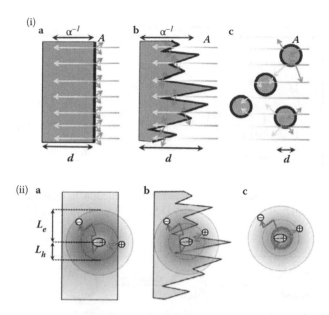

FIGURE 5.2 (i) Light distribution (a) in flat films, (b) in rough films, and (c) in particle suspensions. (ii) Charge collection (a) in flat films, (b) in nanostructured films, and (c) in particle suspensions. d, film or particle thickness; A, surface area; d, film or particle thickness; α^{-1}, optical penetration depth; L_e, electron diffusion length; L_h, hole diffusion length. Short arrows show scattered or reflected light. (From Osterloh, F. E., *Top. Curr. Chem.*, 371, 105–42, 2016.)

The concentration of the majority carriers increases with their lifetime and diffusion length, upon the doping on the metal oxide. On the contrary, the lifetime and diffusion length of the minority carriers decreases. For the optimum collection of both carriers (electron and holes) at the interface, the thickness d of the semiconductor film has to be in the same range as L_e and L_h (Figure 5.2ii). This can be achieved by increasing the surface roughness of the film by introducing nanostructuring to the metal oxides, as shown in Figure 5.2ii.

3. **Quantum size confinement.** The dependence of semiconductor energetics on particle size has been established in the mid-1980s (Yoffe 2001). The quantum size effect also depends on the type of the material used and shape of the nanocrystal (NC) (Butler and Ginley 1978). With decreasing the particle size, BG of quantum dots (QDs) increases with the shifting of their CB edge to more reducing and the VB to more oxidizing potentials. This will increase light absorption efficiency via reducing surface light reflection and increasing light scattering. Moreover, quantum confinement could modify the band structure of semiconductors, for instance, the tunable CB potential. From Marcus–Gerischer theory, it is expected that the increase in the thermodynamic driving force increases the rates for interfacial charge transfer and water electrolysis (Figure 5.3) (Salvador 2001); this phenomena can be experimentally observed for the charge transfer across the

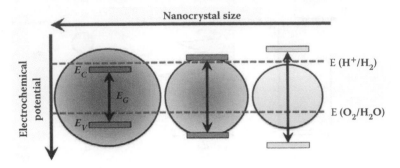

FIGURE 5.3 Impact of quantum confinement with reduction in quantum dot size of the same material. (From Butler, M. A. and Ginley D.S., *J. Electrochem. Soc.*, 125, 228–32, 1978.)

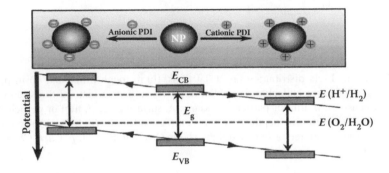

FIGURE 5.4 Effect of potential determining ions on nanocrystal energetic. (From Cowan, A. J. et al., *J. Phys. Chem. C*, 114, 4208–14, 2010.)

solid–solid interfaces (Sant and Kamat 2002; Robel et al. 2007). A logarithmic dependence of the photocatalytic proton reduction rate on the BG has also been verified with CdSe QDs (Holmes et al. 2012). These effects will likely be used more often in advanced solar water-splitting devices.

4. **Potential determining ions.** The effect of potential determining ions (PDIs) on the interfacial energetics is well known (Singh et al. 1980; Peter et al. 2003). Due to the small thickness of nanostructures, external electric fields can reach into the nanomaterial interior and modify the local energetic structure. Thus, the band edge potentials of nanomaterials and resulting functions as well as interfacial charge transfer can be controlled with PDIs, as shown in Figure 5.4. Examples are the band shift caused by hydrosulfide on nano-Bi_2S_3 in TiO_2 NC films (Chakrapani et al. 2010), and the effect of pH on interfacial charge transfer (Osterloh 2013).

5. **Surface area–enhanced charge transfer.** The larger specific surface area of nanomaterials promotes charge transfer across the material interfaces (solid–solid and solid–liquid), allowing water redox reactions to occur at relatively low current densities and, correspondingly, low overpotentials.

In other words, the increase of surface area allows better matching the photocurrents with the slow kinetics of the water redox reactions. In particular, water oxidation is known to require milliseconds to seconds to proceed at Fe_2O_3 and TiO_2, according to recent transient absorption measurements (Schaller and Kimnov 2004; Tang et al. 2008). Thus, increases of surface area reduce the need for highly active and often expensive cocatalysts, based on Ir, Rh, or Pt.

6. **Multiple exciton generation.** Multiple exciton generation (MEG) is a process whereby multiple electron–hole pairs, or excitons (Figure 5.5a and b), are produced upon absorption of a single photon in semiconductor NCs; this represents a promising route to increased solar conversion efficiencies in single-junction PV cells. The MEG process proceeds via impact ionization, coherent superposition of single and multiexciton states, and multiexciton formation through a virtual exciton state. MEG was first demonstrated in 2004 using colloidal PbSe QDs (Beard et al. 2007). Later on the phenomena was observed for QDs of other compositions including PbS, PbTe, CdS, CdSe, In As, Si (Wang et al. 2010) single-walled carbon nanotubes (SWNTs, efficiency 130%) (Tielrooij et al. 2013) graphene with external efficiency 120% and internal efficiency 150% (Bohm et al. 2015), and InP, and so on.

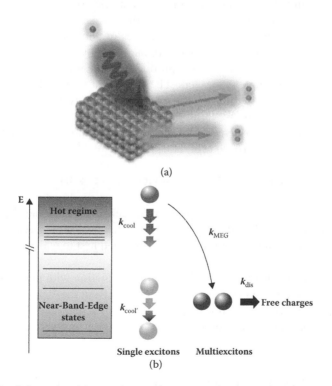

FIGURE 5.5 Schematic of (a) multiple exciton generation in semiconductor quantum dot; (b) relaxation pathways of photoexcited hot single excitons. (From Böhm, M. L. et al., *Nano Lett.*, 15, 12, 7987–93, 2015.)

Although the effect has been applied to water photoelectrolysis at a very small extent, it has enormous possibility to be utilized in future MEG-enhanced water-splitting devices that will likely be tandem or multijunction devices, because the individual QDs cannot produce a sufficient potential for overall water splitting. This is because for efficient solar energy conversion, the BGs of the relevant dots need to be a fraction of the energy of visible light photons ($E = 1.55–3.1$ eV). In Section 5.7, we provide a brief overview of the recent process in PEC water splitting of using chemically modified nanostructured photoelectrodes of metal oxides.

5.2.2   Reaction Mechanism of Typical Oxide Photoelectrodes

5.2.2.1   TiO$_2$

This possesses suitable energy levels for photocatalytic water splitting. A significant enhancement in its photocatalytic activities has been found, and if particle size is reduced to 10–15 nm in diameter, this behavior has been attributed to quantum size effects. Photochemical reactions at the semiconductor surface occur if the light-generated charge carriers find suitable reaction partners, acceptors/protons for electrons, and donor molecules for holes. In small particles, the charge carrier diffusion length is large compared with the large-sized particles and no internal electric field (depletion layer) is required to separate the photogenerated electron–hole pairs. Therefore, the probability of the charge carriers to reach the surface of the particle increases with the decrease in particle size. Moreover, rates of photocatalytic hydrogen production in presence of alcohol scavenger were increased by photocatalytic reforming of the alcohols as compared to the same produced by the water. This is because the alcohols are very active hole scavengers. The loading of titania photocatalysts with metal particles dramatically enhances the rate of hydrogen evolution, presumed to be due to trapping of CB electrons within the metal clusters. The ambient temperature reforming of methanol over a variety of metal-loaded titania photocatalysts can be described as follows (Al Mazroai et al. 2007):

$$CH_3OH + H_2O + M - TiO_2 \rightarrow CO_2 + 3H_2 \tag{5.5}$$

However, the methanol is made up of fossil fuel (CO + H$_2$) and therefore contributes to increasing carbon dioxide emissions. This method is largely considered nonefficient because it needs hydrogen to be consumed first to make methanol in the first place.

Within a semiconductor (such as TiO$_2$), upon excitation with photons of energy equal or higher than its BG (E_{BG}), electron transfers from the VB to the CB, which will consequently create electron (e$^-$) and hole (h$^+$) pairs. The detailed mechanism of water splitting in the presence of ethanol at TiO$_2$ is given by Equations 5.6 through 5.12:

$$\text{Photoexcitation: } TiO_2 + 2 \text{ UV photons} \rightarrow 2e^- + 2h^+ \tag{5.6}$$

$$\text{Another surface intermediate site of } TiO_2\text{: } CH_3CH_2OH \rightarrow CH_3CHO^\bullet + H^+ \tag{5.7}$$

Dissociative adsorption of ethanol on $Ti - O$: $CH_3CH_2OH + Ti - O \rightarrow$

$$CH_3CH_2OTi + OH^- \qquad (5.8)$$

First electron reduction at the CB: $OH^- + e^- \rightarrow O + H_2$ $\qquad (5.9)$

First hole trapping at the VB: $CH_3CH_2OTi + h^+ \rightarrow CH_3CHO^\bullet Ti \rightarrow H^+$ $\quad (5.10)$

Second hole trapping at the VB: $CH_3CHO^\bullet Ti + h^+ \rightarrow CH_3CHO \text{ (g)} + Ti$ $\quad (5.11)$

Second electron reduction at the CB: $H^+ + e^- \rightarrow \dfrac{1}{2}H_2$ $\qquad (5.12)$

5.2.2.2   ZnO

In the presence of the supporting electrolytes Na_2S and K_2SO_3 and main photocatalytic unit CdS/Co–ZnO, the major chemical reactions (Reber and Meier 1984) that are responsible for hydrogen evolution by photocatalytic cleavage of the water are given by Equations 5.13 through 5.16.

Photoexposure: $CdS / Co - ZnO + h\nu \rightarrow e^-(CdS / Co - ZnO) + h^+(CdS / Co - ZnO)$
$$(5.13)$$

Cathode: $2HO + 2e^- \rightarrow 2OH^- + H_2$ $\qquad (5.14)$

Anode: $S^{2-} + SO_3^{2-} + 5H_2O \xrightarrow{h\nu} 5H_2 + 2SO_4^{2-}$ $\qquad (5.15)$

Overall: $SO_3^{2-} + S^{2-} + 2OH^- + 1.5H_2O + 6h^+ \rightarrow 1.5S_2O_3^{2-} + SO_4^{2-} + 4H^+$ (5.16)

5.3   DOPED PHOTOCATALYSTS

The performance of the photoelectrodes can also be improved by manipulating their chemical nature. Chemical modifications including elemental doping, surface modification of electrochemical catalysts such as oxygen evolution reaction (OER) catalyst and hydrogen evolution reaction (HER) catalyst have been demonstrated via controlled addition of cocatalyst (noble metal/metal oxide/metal hydroxide) sensitizers (plasmatic material/QD/dye). Although a large number of photocatalytic classes has been reported with increased efficiency by chemical modification, wide BG semiconducting metal oxides can be easily doped and their optoelectronic properties can be modified. The oxides such as TiO_2 (Feng et al. 2011 Hwang et al. 2009; Hoang et al. 2012a), ZnO (Lin et al. 2012), α-Fe_2O_3 (Franking et al. 2013; Wang et al. 2013), $BiVO_4$ (Ng et al. 2010; Zhong et al. 2011), WO_3 (Esposito et al. 2012; Hill and Choi 2012), and Ta_2O_5 (Kronawitter et al. 2011) are able to split gaseous or liquid phase water into H_2 and O_2 under UV light irradiation (not visible light) and have especially benefited from this technique. Simultaneously, the CB potentials of the small BG metal oxides such as Fe_2O_3 and WO_3 are more positive than water reduction potential, and thus, the application of external bias is needed to split water. In addition, the poor conductivity and extremely short carrier diffusion length of these oxides often result in significant electron–hole recombination loss. Therefore, chemical modification by elemental (cationic and anionic) doping (Serpone 2006) to the metal oxides is a promising approach to modify the electronic and optical properties of the semiconductor photocatalysts/electrodes. Moreover, chemical doping can intrinsically change the electronic and optical properties of semiconductor electrodes by narrowing the BG of the semiconductor. Furthermore, the electronic and optical properties of the semiconductor can be modulated by controlled introduction of dopants as electron acceptors and donors. In Section 5.3, we will review the doping effect on a number of commonly used semiconductor photoelectrodes.

The donor densities were calculated by the equation:

$$N_d = \left(2/e_0\varepsilon\varepsilon_0\right)\left[d(1/C^2)/dV\right]^{-1} \tag{5.17}$$

where e_0 is the electron charge, ε is the dielectric constant of TiO_2, ε_0 is the permittivity of vacuum, N_d is the donor density, C is the capacitance, and V is the applied bias at the electrode.

This is mainly because when the BG energy of the metal oxide becomes smaller, the CB edge shifts more positively (Scaife 1980). Thus, the CB edge of metal oxides having a small BG energy tends to locate quite close to the redox potential of H^+/H_2 redox couple (0 eV vs. NHE), in some cases, locate more positive than the redox potential of H^+/H_2, losing their ability to reduce H^+ into H_2. However, much work has been done to develop visible light-responsive metal oxide photocatalysts such as the chemical doping of TiO_2 with transition metals ions or oxides (Hoffmann et al. 1995). Although TiO_2 chemically doped with metal ions could, in fact, induce visible light response, these catalysts showed limitations in sufficient reactivity for practical applications. When metal ions or oxides are incorporated into TiO_2 by a chemical doping method, the impurity of energy levels formed in the BG of TiO_2 may cause an increase in the recombination between the photoformed electrons and holes (Anpo et al. 1998; Yamashita et al. 2002; Anpo and Takeuchi 2003). Researchers have reported on TiO_2 photocatalysts capable of absorption of visible light up to 550 nm by applying an advanced metal ion-implantation method to modify the electronic properties of the semiconductors by bombarding them with various metal ions accelerated by high voltage. The absorption band of the TiO_2 photocatalysts implanted with metal ions, such as Cr, V, Fe, Ni, and so on, was found to shift to visible light regions, the extent of the red shift depending on the amount and type of metal ions implanted. Various dopants including metal ions such as transition metals (Cu, Co, V, Mn, Fe, Zn, Ti, etc.) (Hameed et al. 2004), Cr (Shen et al. 2013), In (Sasikala et al. 2010), Ni (Hou et al. 2015), Mo (Pilli et al. 2011), Ti (Hu et al. 2009; Wang et al. 2011a), Sn (Ling et al. 2011; Frydrych et al. 2012; Xu et al. 2012b), and Fe (Singh et al. 2008; Dholam et al. 2009) and nonmetal elements such as nitrogen (Hoang et al. 2012b), sulfur (Rockafellow et al. 2009; Xiang et al. 2011b), silicon

(Kay et al. 2006; Sivula et al. 2011), boron (Lu et al. 2007), fluorine, phosphorus, and carbon (Park et al. 2009) are shown in Figure 5.6. Transitional metal ion doping and rare-earth metal ion doping have been extensively investigated for enhancing TiO_2 photocatalytic activities (Choi et al. 1994; Litter et al. 1999; Wilke and Breuer 1999; Dvoranova et al. 2002; Paola et al. 2002; Xu et al. 2002; Hameed et al. 2004; Wang et al. 2004; Xu et al. 2004). Choi et al. (1994) carried out a systematic investigation to study the photoreactivity of 21 metal ions doped into TiO_2. It was found that doping of metal ions could expand the photoresponse of TiO_2 into the visible spectrum. As metal ions are incorporated into the TiO_2 lattice, impurity energy levels in the BG of TiO_2 are formed, indicated as follows (Equations 5.18 through 5.21):

$$M^{n+} + h\nu \rightarrow M^{(n+1)+} + e_{CB}^- \tag{5.18}$$

$$M^{n+} + h\nu \rightarrow M^{(n-1)+} + h_{VB}^+ \tag{5.19}$$

where M and M^{n+} represent the metal and the metal ion dopant, respectively. Furthermore, electron (hole) transfer between metal ions and TiO_2 can alter electron–hole recombination as

$$\text{Electron trap: } M^{n+} + e_{CB}^- \rightarrow M^{(n-1)+} \tag{5.20}$$

$$\text{Hole trap: } M^{n+} + h_{VB}^+ \rightarrow M^{(n+1)+} \tag{5.21}$$

The energy level of $M^{n+}/M^{(n-1)+}$ should be less negative than that of the CB edge of TiO_2, while the energy level of $M^{n+}/M^{(n-1)-}$ should be less positive than that of the VB edge of TiO_2. For photocatalytic reactions, carrier transferring is as important as carrier trapping. Only if the trapped electron and hole are transferred to the surface, can photocatalytic reactions occur. In cases of deep doping, metal ions likely behave as recombination centers, since electron/hole transferring to the interface is more

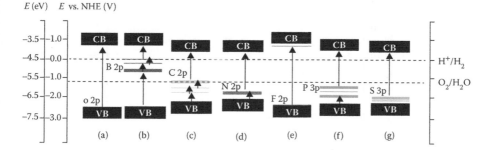

FIGURE 5.6 Effect of nonmetal element monodoping on band structure of $SrTiO_3$. (a) Undoped $SrTiO_3$, (b) one B atom doped $SrTiO_3$ at 4.167 atom%, (c) one C atom doped $SrTiO_3$ at 4.167 atom%, (d) one N atom doped $SrTiO_3$ at 4.167 atom%, (e) one F atom doped $SrTiO_3$ at 4.167 atom%, (f) one P atom doped $SrTiO_3$ at 4.167 atom%, and (g) one S atom doped $SrTiO_3$ at 4.167 atom%. The VB maximum and CB minimum values are given with respect to the NHE potential (V) and energy with respect to vacuum (eV). (From Zhang, C. et al., *Comput. Mater. Sci.*, 79, 69–74, 2013a.)

difficult. Among the 21 metal ions studied, Fe, Mo, Ru, Os, Re, V, and Rh ions can increase photocatalytic activity, while Co and Al ions cause detrimental effects (Choi et al. 1994). 4%Ni doped CdS produced hydrogen with the rate of 25.848 mmol/(h g) (QE = 26.8%, λ = 420 nm) from water splitting under aqueous $(NH_4)_2SO_3$ and the system retains its stability and activity even after 20 h. (Wang et al. 2015)

Figure 5.7 illustrats the significant increase in photocurrent and incident photo-to-electron conversion (IPCE) at 1.02 V for Ti-doped α-Fe_2O_3 films in 1.0 M NaOH solution (Wang et al. 2011b). Besides the above classes of doped semiconductors, nowadays hydrogen-treated oxides such as WO_3 (Wang et al. 2012), ZnO (Lu et al. 2012a), and $BiVO_4$ (Wang et al. 2013), TiO_2 (Wang et al. 2011a), photoanodes exhibited the enhanced electrical conductivity and better STH efficiency performance for PEC water splitting. The donor densities of H–TiO_2 were increased three times, which was supposed to be due to the formation of oxygen vacancies

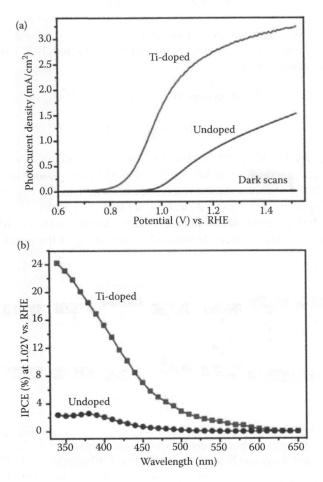

FIGURE 5.7   (a) The linear sweep voltammograms of undoped and Ti-doped α-Fe_2O_3 films. (b) Incident photo-to-electron conversion (IPCE) spectra of these two films collected at 1.02 V versus reversible hydrogen electrode (RHE), in 1.0 M NaOH solution. (From Wang, G. M. et al., *Nano Lett.*, 11, 3503–09, 2011b. With permission.)

that improve charge collection efficiency and causes. Substantial increase in pho-tocurrent densities (2.5 mA/cm^2 at 0 V vs. Ag/AgCl) compared with pristine TiO_2 electrode was observed. IPCE studies suggested that the enhanced photoactivity was due to the improved charge collection efficiency. There was no visible light photoactivity, although visible light absorption was observed for H–TiO_2 nanow-ires. By integrating the IPCE spectrum with the standard AM 1.5G solar spectrum, the H–TiO_2 sample achieved the excellent STH conversion efficiency of ∼1.1%, at −0.6 V versus Ag/AgCl. Li et al. have further demonstrated that the hydrogen treatment is a general method to introduce oxygen vacancies into metal oxides and improve their PEC performance (Hensel et al. 2010).

5.4 QD-SENSITIZED METAL OXIDE PHOTOCATALYSTS

To harvest more visible light by using oxides, numerous methods, such as the use of photosensitive dyes and semiconductor NCs, have been adopted. Semiconductor NCs (QDs) have many significant advantages over dyes. Compared with organic dyes, they show better optical stability, large extinction coefficients, and adaptabil-ity to the large range solar spectrum. The major drawback is that almost every high-quality QD comprises highly toxic elements. However, the recent arrival of less-toxic alternatives like InP or CIS seems to solve this problem. Moreover, $CuInX_2$, $CuGaX_2$, $AgInX_2$, $AgGaX_2$ (X = S, Se), and similar materials are alterna-tives to InP because they are composed of abundant, readily available, nontoxic elements (no class A and B elements of QDs) that form an effective heterojunc-tion with solid hole/electron conductors and are used to improve the matching of the solar spectrum with their absorption spectrum by controlling the particle size, which can also be used to increase the visible light absorption of large BG metal oxide-based photoelectrodes. By controlling the particle size of QDs, one can vary the energetics of the particles. For example, small BG metal sulfides or selenides such as CdS and CdSe QDs have been used to sensitize the wide BG metal oxides such as ZnO and TiO_2 for PEC hydrogen generation (Wang et al. 2010, 2012). QDs, demonstrating size-dependent quantization of the electronic energy levels as depen-dent upon particle size and shape, particle–particle spacing, and nature of the outer shell. These types of heterojunctions also allow the rapid separation and transfer of photoexcited electron–hole pairs. Now questions arise how the QDs are differ-entiated from the same sized nanomaterials, which may be answered in terms of quantum confinement and electron tunneling. Here, the Bohr diameter is a useful parameter used to determine the type of confinement that decided which particle will be the QD among the same sized nanodot of different materials. Usually, the particles having size smaller than 3 Bohr diameter (the Bohr radius of an exciton in PbSe is as large as 46 nm, possess strong confinement; $\Delta E \sim 1/\mu^*$ where μ^* is the effective mass of the excitons), are QDs. As the NC becomes more confined, the peak (Raman vibration/absorption/x-ray diffraction [XRD] spectra) will broaden and shrink. On the basis of the free movement of the particles in 3D, 2D, 1D, and 0D spaces, the nanoparticles are defined as bulk, quantum well, quantum wire, and QDs, as shown in Figure 5.8. Because of this strong confinement in QDs, sev-eral novel quantum phenomena emerge, such as phonon bottlenecks and coulomb or spin blockades. Nozik (2001) predicted a novel MEG phenomenon in QDs and

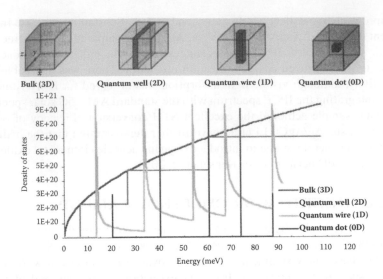

FIGURE 5.8 Schematic representation of the different sized particles of the same material. (a) Bulk; three-dimensional (3D) carriers act as free carriers in all three directions. (b) Quantum well; two-dimensional (2D) carriers are one-dimensional (1D) confined but act as free carriers in a plane. (c) Quantum wire; 1D carriers are 2D-confined but free to move along direction of the wire. (d) Quantum dot; 0D carriers are confined in all three directions (no free carriers). The interesting and sometimes unexpected nanoparticle properties are partly due to aspects of the material surface dominating those of the bulk. (From https://arxiv. org/pdf/0811.1937.pdf [cond-mat.mes-hall] last access on 12 Nov 2008.)

Schaller and Klimov (2004) provided the experimental evidence of MEG in PbSe QDs. MEG of up to seven excitons per single phonon has been reported (Schaller et al. 2006). Although the microscopic mechanism of MEG is still enigmatic, several mechanisms have been suggested to be applicable to this process, including impact ionization, an inverse Auger process, direct carrier multiplication (Califano et al. 2004a, 2004b), and MEG (Wang et al. 2003). In MEG, the excitons are generated in a decoherent process. The Auger mechanism assumes an incoherent process and requires a faster decoherence time than the creation of multiple excitons (Ellingson et al. 2005). Direct carrier multiplication relies on multiexciton coupling to virtual single-exciton states and the coherent superposition of these during an optical pulse. The discovery of the MEG process of QDs apparently became important as promising building blocks in solar devices and intensive research is going on involving the fabrication of dye-sensitized solar cells (DSSCs) or photocatalytic devices for water splitting. When irradiated with light, several excitons can form per photon in a semiconductor QD, which gives a solid assurance for high efficiency up to the Shockley–Quiesser limit (estimated maximum efficiency limit is 31% for photodevices). The photon absorbed in the VB of the semiconductor promotes an electron from VB to CB. If the difference between the photon energy and BG will be dissipated as heat, the phenomenon that cools down the hot electron is known as Auger cooling. It has been shown that QDs need to possess hydrophilic surface properties in order to adsorb onto metal oxide surfaces efficiently (QDs with hydrophobic surface ligands do adsorb to some degree, but not as readily as hydrophilic ones).

There are two major strategies to obtain hydrophilic QDs: first, synthesis of QDs in aqueous solution; second, synthesis of QDs in organic solution with subsequent ligand exchange. Both methods lead to the similar results (Murphy et al. 2006b).

QD-sensitized PEC cells have the potential to convert more solar energy than dye-sensitized cells. The sensitization of semiconductor metal oxide for PEC cells using CdS, CdSe, and CdTe QDs was recently reported (Kamat 2008). Although cadmium chalcogenide QDs have been extensively used as a sensitizer in harvesting visible light, the photostability is a major concern for CdSe- and CdS-based sensitizers due to their self-oxidation reaction; another concern is their toxicity. Furthermore, the sacrificial chemicals are certainly not desirable to generate hydrogen but it is necessary to develop small BG semiconductors to sensitize large BG metal oxides for water splitting. Hole-sacrificial reagents such as SO_3^{2-}/S^{2-}, were added to prevent the self-oxidation, while the photoanode is no longer oxidizing water. DSSCs made from TiO_2 and CdSe (or PbS) QDs have already been reported in the literature (Kamat 2008). Their efficiency is around 3% (Niitsoo et al. 2006), which is still smaller than that of conventional DSSCs.

Nann et al. (2010) utilized InP QDs as a suitable sensitizer for solar water-splitting applications, whereby they were coated with an iron catalyst forming a 3D nanophotocathode on gold material for the production of hydrogen and commonly used n-type semiconductor oxide photocatalysts, as photoanodes. Recently, Yang et al. (2014) developed a Z-scheme device, which shows spontaneous overall water splitting with efficiency of 0.17%, in a nonsacrificial environment under visible light illumination ($\lambda > 400$ nm). They reported a corrosion-resistant and photostable PEC cell system that possesses nanocomposite photoelectrodes, consisting of a CdS QD-modified TiO_2 photoanode and a CdSe QD-modified NiO photocathode, where cadmium chalcogenide QDs are protected by a ZnS passivation layer and gas evolution cocatalysts (Figure 5.9a). Another interesting example of QD-sensitized overall water splitting under visible light irradiation is nitrogen-doped graphene oxide QDs, where a nitrogen-rich site exhibited n-type and a nitrogen-deficient site exhibited p-type conductivities and served as p–n-type photochemical diodes, in which the carbon sp^2 clusters of graphene serve as the interfacial junction for H_2 and O_2 evolution from their p- and n-domains, respectively (Figure 5.9b) (Yeh et al. 2014). A mesoporous TiO_2 layer is on the top. CdSe/H–TiO_2 exhibits a maximum photocurrent density of ~16.2 mA/cm^2, which is 35% higher than that of the optimized control sample (CdSe/P25). CdSe/H–TiO_2 under filtered exposure conditions recorded a current density of ~14.2 mA/cm^2, the greatest value in the visible range (Kim et al. 2013). Enhanced PEC water splitting was demonstrated in Si QDs/TiO_2 nanotube array composite electrodes under visible light illumination and the photocurrent density was 1.6 times larger than that of pristine TiO_2 electrodes (Li et al. 2015). Photocatalytic H_2-production rate over Co_3O_4-QDs and Co_3O_4-SSR with 420-nm visible light cut off filter is estimated as 1.10 and 0.00 µmol/h, respectively (Zhang et al. 2014b). TiO_2/CdZnS/CdZnSe electrodes prepared from three different batches provide a remarkable photon-to-hydrogen efficiency of 7.3 ± 0.1% (the rate of the photocatalytically produced H_2 by water splitting is about 172.8 mmol/h/g), which is the most efficient QDs-based photocatalysts used in solar water splitting (Wang et al. 2013). The photocurrent response of the rGO–CdS–H_2W_{12} composite film was enhanced fivefold compared with CdS film (Wang

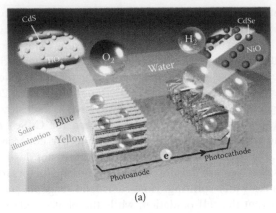

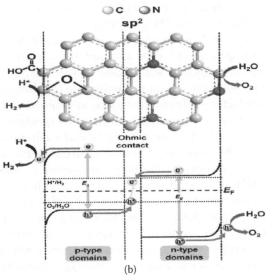

FIGURE 5.9 (a) ZnS passivation layer and protected quantum dots (QDs) assisted photo-electrodes for overall water splitting; (b) N-doped graphene oxide QDs serve as *p–n*-type photochemical diodes. (From Yang, H. B. et al., *ACS Nano*, 8, 10, 10403–10413, 2014; Yeh, T.-F. et al., *Adv. Mater.*, 26(20), 3297–303, 2014.)

et al. 2014). Branched TiO_2 nanoarrays are sensitized with CdS QDs as anodes for PEC water splitting. The remarkable photocurrent density (~4 mA/cm² at a potential of 0 V vs. Ag/AgCl) and high STH efficiency of the materials obtained were ascribed to the novel branched nanostructure and efficient electron transfer from CdS to TiO_2 (Su et al. 2013). The InN/In$_{0.54}$ Ga$_{0.46}$N-QDs-photoelectrode reveals a maximum IPCE of up to 56% at a wavelength of 600 nm with a hydrogen generation rate of 133 mmol/h/cm² at zero voltage under illumination of a 1000-W Xenon arc lamp. The bare In$_{0.54}$Ga$_{0.46}$N-layer-photoelectrode reveals a much lower IPCE of 24% with hydrogen generation rate of 59 mmol/h/cm² (Alvi et al. 2015). Trevisan et al. (2013) also demonstrated in situ growth of CdS and PbS QDs on mesoporous TiO_2 aiming at harvesting light in both visible and near-infrared (NIR) regions and their application for PEC hydrogen generation.

5.5 PLASMONIC MATERIAL–INDUCED METAL OXIDE PHOTOCATALYSTS

Recently, Ebbesen et al. (Barnes et al. 2003) introduced a novel class of optoelectronic material, that is, plasmonic photocatalysts, to the world. Their presence stimulated research in photocatalysis due to their good light-harvesting qualities along with their BG breaking effect and sensitizer effect. They are found to be useful for a wide range of applications such as wastewater treatment, air purification, water splitting, enhancing computational power of electronics, the treatment of breast cancer, manufacturing bright light-emitting diodes (LEDs), rendering a satellite object invisible, and so on (Atwater 2007). During the water-splitting process, the studied photocatalytic materials (oxides, oxynitride, nitrides, sulfides, oxysulfides, etc.) face two main problems: (1) low photocatalytic efficiency (due to the recombination of photoelectrons and photoholes) and (2) lack of satisfactory visible light responsive photocatalytic materials because most commonly available and stable oxides have a wide BG, which allow them to absorb UV light, and short BG materials did not sustain photoactivity for a long time (Hashimoto et al. 2005; Wang et al. 2012). Both of the deficiencies would be overcome by the use of plasmonic photocatalytic materials. These novel metamaterials have two prime parts: (1) localized surface plasmonic resonance (LSPR) in nanoparticles of noble metals; and (2) Schottky junction results of the contact between the noble metal and semiconductor/polar material (support/carrier). The prominent benefits associated with plasmonic material in terms of LSPR are given as follows:

1. **Tunable resonating wavelength** (UV, near UV, visible, or IR) of the metal nanoparticles can control the size, shape, and surrounding environment (Kelly et al. 2003).
2. **Enhanced light absorption effect:** The LSPR can drastically enhance the absorption of visible light (low BG photocatalysts)/UV light (Thomann et al. 2011) (large BG materials) (Awazu 2008).
3. **Reduced diffusion length:** This is compatible to the minority carrier diffusion length of ~10 nm (Zhdanov et al. 2005; Linic et al. 2011; Mubeen et al. 2011); with the strong absorption of the incident light in a thin layer, it will be beneficial to materials that present poor electron transport.
4. **Enhance local electric field:** The LSPR creates an intensive local electric field, which favors photocatalytic reactions (powers the excitation of the electrons and holes) (Wu et al. 2010; Torimoto et al. 2011) or increase the number of photoexcited electron and holes, heats up the surrounding environment to increase the redox reaction rate and the mass transfer (Sun et al. 2009; Christopher et al. 2011), and polarizes the nonpolar molecules for better adsorption (Mubeen et al. 2011).
5. **Other contributory effects:** In addition to the aforementioned enhancements in the major aspects of photocatalysis, plasmonic photocatalysis enjoys some other benefits such as higher light utilization efficiency, better temperature dependence, and enhanced molecule adsorption, polarization, the catalytic effect of the noble metal itself (e.g., Pt on hydrogen evolution), and the quantum tunneling effect (Wang et al. 2012).

6. **Benefits of Schottky junction:** Electrons and holes are compelled to move in different directions once they are created inside or near the Schottky junction (Thimsen et al. 2011) and the phenomena prevents the recombination of the photocarrier.

5.5.1 ADVERSE EFFECTS OF METAL NANOPARTICLES

1. A common problem is the shading effect, that is, the metal nanoparticles on the surface of the semiconductor reducing the light-receiving area of the semiconductor. The shading effect affects the photocatalytic systems that require irradiation of the semiconductor part.
2. Some surface area of the semiconductor may even block the pores of the semiconductor. This leads to a reduction of the specific surface area of the semiconductor and thus affects adversely the photocatalytic activity.
3. The metal nanoparticles could act as a recombination center. This is because the contact between the metal and the semiconductor could form surface states at the interface, which promotes charge trapping, recombination, and Fermi level pinning (Nakato et al. 1975; Nakato and Tsubomura 1985; Primo et al. 2011).
4. Some other problems are related to the stability of the metal/semiconductor mixture. The metal nanoparticles might undergo photocorrosion and leaching, causing a gradual loss of the photocatalytic performance over time (Herrmann et al. 2005); the encapsulated form and the isolated form can circumvent this problem, at the cost of losing the "fast lane" transfer of the charge carriers to the adsorbates.

Heterogeneous photocatalysis has five basic independent steps (Sun et al. 2009), which can be benefited by both of the aforementioned phenomena as follows:

1. *Transfer of the reactants* onto the photoreaction surface was boosted by fluid mixing.
2. *Adsorption of the reactants* was improved by polarization (Awazu et al. 2008).
3. *Redox reactions* in the adsorbed phase were facilitated by the metal's fast transfer, charge carrier trapping, and large contact surface would significantly enhance the creation and separation of active electrons/holes and the localized heating effect to raise the reaction rate (Yu et al. 2009).
4. *Desorption* rate of the production from the surface was increased by the localized heating effect that increases the reaction rate (Chen et al. 2008; Adleman et al. 2009) and boosts fluid mixing.
5. *Transfer of the products away* from the surface was benefited by the localized heating effect that increases the reaction rate (Adleman et al. 2009) and boosts fluid mixing.

In usual practice, metallic structures are known for high optical losses, which is not good for transmitting light signals. This is because the electrons oscillating in the electromagnetic field collide with electrons present in the surrounding lattice of atoms and soon dissipate the field's energy. However, plasmonic material is exposed to sunlight, free electrons of the nanoparticles of noble metal are integrated with the photon energy that produces subwaves and conduct/propagate electrons in oscillating mode, which can result in astounding optical properties and lowering the energy losses at the interface (low electron density zone) between the nanometals and a dielectric than inside the bulk of a metal because the field spreads into the nonconductive material, where there are no free electrons to collide and to dissipate the energy. Hence, plasmons can propagate at interface for several centimeters before dying. The propagation length can be maximized if the waveguide employs an asymmetric mode, which pushes a greater portion of the electromagnetic energy away from the guiding metal film and into the surrounding dielectric, thereby lowering energy losses. Thus, electrons of noble metals act as LSPR material (Figure 5.10) that must have the thermal redox reaction-active centers on the catalyst that can trap, scatter, and concentrate light, and enhance the number of active sites, and rate of electron–hole formation by providing a fast lane for charge transfer on the semiconductor surface. The material systems will be mainly grouped based on the structure of the metal and photocatalyst, for example, nanoparticles in the sole-metal form, nanocomposites in the embedded form, nanocomposites in the encapsulated form, and nanocomposites in the isolated form.

The charged chemical compounds are selectively attracted or repelled by the accumulated charges on the surfaces of the metal nanoparticles and the semiconductor. In addition, the dipole nature of the local electric field of the surface plasmon attracts the polar molecules and polarizes nonpolar molecules. $Au@ZnFe_2O_4$ is a representative example of the plasmonic photocatalysis, in which a hot electron of the plasmonic Au nanoparticle transfers its electron to the CB of the $ZnFe_2O_4$ and ZnO and facilitates the absorbance of the light to increase the efficiency of the device (Figure 5.11a) (Sheikh et al. 2013). In a second example, $Au@TiO_2$, the presence of Au nanoparticle induces a space charge region (i.e., Schottky junction) in the TiO_2

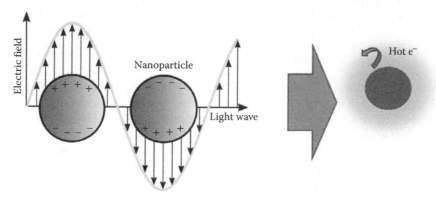

FIGURE 5.10 Schematic presentation of the localized surface plasmonic resonating electron of a noble metal. (From Sheikh, A. et al., *Small* 9(12), 2091–96, 2013.)

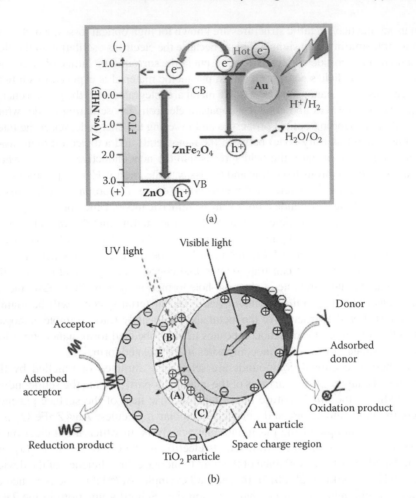

FIGURE 5.11   (a) Representative example of the enhanced adsorption in the plasmonic photocatalysis in Au@ZnFe$_2$O$_4$. A hot electron is transferred from Au to CB of the ZnFe$_2$O$_4$ and this electron further jumps into the CB of ZnO. (b) Electron transfer mechanism of Au nanoparticle to TiO$_2$. (From Sheikh, A. et al., *Small*, 9(12), 2091–96, 2013; Zhang, X. et al., *Rep. Prog. Phys.*, 76, 046401, 2013b.)

particle, which builds up an internal field E pointing from the TiO$_2$ to the Au. The LSPR was created in response to the electromagnetic field of the incident light; the Au nanoparticle drives a collective oscillation of the electrons, which excites more electrons and holes (depicted as [a]). Such internal field forces induce the separation of electrons and holes which suppresses their recombination. When an electron–hole pair is excited in or near the space charge region by an incident UV photon (depicted as [b]), the LSPR excited electrons in Au have sufficient energy to go across the space charge region and are fed into the CB of TiO$_2$ as depicted by (c) (Zhang et al. 2013b) as shown in Figure 5.11b. Figure 5.12a and b shows the photocurrent for Au–TiO$_2$ nanocomposite in the presence and absence of Au under 633-nm light irradiation; the experiment supports the presence of plasmonic effect in this compound.

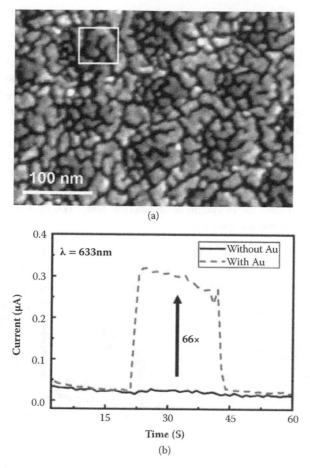

(a)

(b)

FIGURE 5.12 (a) Scanning electron microscopy (SEM) image of Au–TiO$_2$ nanocomposites for photocatalytic water splitting under visible light and (b) photocurrent of anodic TiO$_2$ with and without Au nanoparticles irradiated with 633-nm light for 22 seconds. (Reprinted with permission from Liu, Z. et al., *Nano Lett.* 11, 1111–16, 2011. Copyright [2011] American Chemical Society.)

5.6 Z-SCHEME PHOTOCATALYSTS

A number of oxide photocatalysts exhibit good photocatalytic activity for water splitting. Unfortunately, most of them function in UV light due to their wide band. However, almost half of all incident solar energy at Earth's surface falls in the visible region (400 < λ < 800 nm); the efficient utilization of visible light remains indispensable for realizing practical H$_2$ production based on photocatalytic water splitting. Therefore, the focus of the research has subsequently turned to achieving more efficient utilization of visible light by widening the band of photoabsorption and increasing the quantum efficiency of the photocatalyst. This can happen by combining the two wideband oxide photocatalysts in the redox pair of an electrolyte, one whose VB maxima exists below the oxidation potential of water, and is suitable for O$_2$ production and another with CBs that exist in the upper side of the reduction side of the water. This scheme is known as "Z-scheme or tandem photocatalysis" and achieves overall water splitting, which can show the similarity of the intrinsic process to photosynthesis in plants. This is one of the possible forms of artificial photosynthesis and was inspired by natural photosynthesis in green plants, where photosystems I and II harvest 700- and 680-nm photons, respectively, to oxidize H_2O into O_2 under sunlight, with a quantum yield (QY) close to unity. Similarly, the Z-scheme also includes two-step photoexcitation using two different semiconductor powders (one having a VB position suitable for oxidation of water and another one having a CB position suitable for reduction of water) and a reversible donor/acceptor pair (so-called shuttle redox mediator). The mechanistic elaboration of Z-scheme water splitting is shown in Equations 5.22 through 5.25 and Figure 5.13a. In an H_2 evolution system, the forward reactions, that is, the reduction of protons by CB electrons coupled with the oxidation of an electron donor (D) by VB holes, yield the corresponding electron acceptor (A):

$$2H^+ \rightarrow 2e^- + H_2 \,(\text{photoreduction of } H^+ \text{ to } H_2) \qquad (5.22)$$

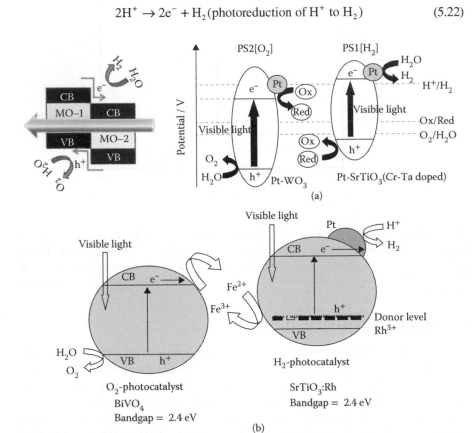

(a)

(b)

FIGURE 5.13 Z-scheme system for photoelectrochemically water splitting by a two-photon process with a visible light response for (a) Pt–WO$_3$ and Pt–SrTiO$_3$ (Cr–Ta doped) in mediator IO_3^-/I^- and (b) BViO$_4$ and SrTiO$_3$:Rh in mediator Fe^{2+} and Fe^{3+}. (From Sayama, K. et al., *Chem. Commun.*, 2416–17, 2001; Kato, H. et al., *Chem. Lett.*, 33, 1348–49, 2004.)

$$D + nh^+ \rightarrow A \quad \text{(photooxidation of D to A)} \quad (5.23)$$

On the other hand, the forward reactions on an O_2 evolution photocatalyst are as follows:

$$A + ne^- \rightarrow D \text{ (photoreduction of A to D)} \quad (5.24)$$

$$2HO^- + 4h^+ \rightarrow O_2 + 4H^+ \text{(photooxidation of } HO^- \text{ to } O_2) \quad (5.25)$$

The development of Z-scheme water-splitting systems has relied on finding a new semiconductor photocatalyst that efficiently works in the presence of a shuttle redox mediator and creates active sites to promote surface chemical reactions while suppressing backward reactions involving redox mediators. Basically, the Z-scheme photocatalytic system was classified into two categories: (1) PEC device and (2) photocatalytic device. Z-scheme PEC devices include a photoanode that is coated with oxygen evolution material coupled to a photocathode, which is coated with hydrogen evolution material into the redox electrolyte solution. Principally, it consists of a hydrogen evolution catalyst and an oxygen evolution catalyst, with an electron mediator as a link between the two oppositely charged electrodes. One major advantage that a Z-scheme device has over a single semiconductor polymerized complex (PC) is its potential to use a greater proportion of the solar spectrum via the use of two different semiconductors whose BG is suitable only for half reactions (oxidation/reduction), as in a multijunction PV device. This should result in improved overall STH conversion efficiency of the device. Another advantage is that these systems allow evolution of hydrogen and oxygen in separate sites, and thus remove the obstacle of gas separation before collection. This also leads to a safer process, as it prevents the flammable composition of hydrogen and oxygen, and at the same time prevents back reactions to form water. In a Z-scheme PEC cell, the thermodynamic constraints are that the anode's VB must be more positive, while the cathode's CB must be more negative than that shown for water. If these constraints hold, only then the photogenerated hole/electron might be capable of oxidizing water and reducing water. In addition to the above-mentioned requirements for BGs, electron–hole pair lifetime, conductivity, and band edge placement, fast and selective redox kinetics will be required as well. One of the significant challenges to acquire/synthesize a photocatalytic material is that it should be stable under operating conditions (sunlight and water) for many years and inexpensive to manufacture so that it can be deployed at a large scale. According to the early works, it is strongly suggested that suppressing the backward reactions involving redox reagents, which are thermodynamically downhill in most cases, while promoting forward reactions is the key to achieve water splitting into H_2 and O_2, according to the Z-scheme principle. Actually, for a successful Z-scheme water-splitting systems are one of the most important issues with the Z-scheme, where the redox cycle of the mediator challenges the reduction or the oxidation process of water at the hydrogen and oxygen evolution sites (Kudo and Miseki 2009; Tabata et al. 2010). Following are the notable examples of this category. Kransnovskii and Brin in 1962 reported the first case of (an O_2 evolution half-reaction of Z-scheme water splitting) water oxidation using oxide semiconductors such as TiO_2 and WO_3 in the presence of Fe^{3+}

as an electron acceptor. The apparent quantum yields (AQYs) of these three systems $(Pt/SrTiO_3:Rh)-(BiVO_4)$, $-(Bi_2MoO_6)$, and $-(WO_3)$, were 0.3%, 0.2%, and 0.2% at 440 nm, respectively (Maeda 2013). Recently, Irie et al. reported Z-scheme water-splitting systems using only $SrTiO_3$-based photocatalysts (Hara et al. 2012). Among the materials they prepared, $SrTiO_3$:Ga/Bi and $SrTiO_3$:In/V were able to reduce and oxidize water to form H_2 and O_2, respectively, in the presence of an IO^{3-}/I^- redox couple, achieving Z-scheme water splitting under UV irradiation ($\lambda < 400$ nm). The highest apparent quantum efficiency (AQE = 6.3% at 420 nm) ever reported for Z-scheme overall water splitting is using $Pt-ZrO_2/TaON$ and $PtOx-WO_3$ as a H_2-evolving photocatalyst and an O_2-evolving photocatalyst, respectively (Maeda et al. 2010). However, the current efficiency of the photocatalytic water reduction is still much lower than that of $PtOx-WO_3$ (about 20% at 420 nm). when a couple of Ir and rutile-TiO_2 modified-Ta_3N_5 powder that treated as oxygen evolution catalyst, and the hydrogen evolution catalyst $Pt/ZrO_2/TaON$, combined with an iodate/iodide shuttle redox mediator under visible light, gives. (Tabata et. al. 2010) Another exclusive example includes $Ru/SrTiO_3$ and $BiVO_4$ as hydrogen and oxygen evolution catalysts (pH = 3.5 maintained using H_2SO_4), respectively, with 1.7% apparent efficiency and without an electron mediator (Sasaki et al. 2009). The results indicate that an interparticle electron transfer occurs between the two catalysts. The system consisting of $Pt/SrTiO_3$:Cr/Ta, Pt/WO_3 and an IO^{3-}/I^- redox couple shows that the QY is 0.1% at 420.7 nm (Sayama et al. 2001). Another important Z-scheme system is considered to occur on the $Pt-BaTaO_2N/Pt-CaTaO_2N$ and $Pt-WO_3$ (Higashi et al. 2009) photocatalyst under visible (420–440 nm) light in iodide (I^-) or iodate (IO^{3-}). The AQE for overall water splitting for the aforementioned combination is ~0.1% at 420–440 nm and two-step mechanisms for these reactions are considered in Equations 5.26 through 5.28:

$$BaTaO_2N \text{ (or } CaTaO_2N) + h\nu(\lambda > 420 \text{ nm}) \rightarrow e^- + h^+ \qquad (5.26)$$

$$2H^+ + 2e^- \rightarrow H_2 \qquad (5.27)$$

$$I^- + 3H_2O + 6h^+ \rightarrow IO^{3-} + 6H^+ \qquad (5.28)$$

The H_2 production ceased after prolonged irradiation due to the backward reaction taking place over the reduction site (Pt) of the photocatalyst, where preferential reduction of IO^{3-} anion to I^- occurs instead of water reduction (Equation 5.29):

$$IO^{3-} + 6H^+ + 6e^- \rightarrow I^- + 3H_2O \qquad (5.29)$$

$Pt-WO_3$ and $Pt-SrTiO_3$ (Cr–Ta doped) were found effective in utilizing visible light for H_2 and O_2 production in a NaI solution (Sayama et al. 2001). Over an extended duration H_2 and O_2 evolution under monochromatic light 420.7 nm, 57 mW was measured at 0.21 μmol/h of H_2 and 0.11 μmol/h of O_2; the estimated quantum efficiency was 0.1%. Similarly, redox couple Fe^{3+}/Fe^{2+} and Ce^{4+}/Ce^{3+} pairs are also effective for hydrogen production by water splitting (Kato et al. 2001). $Pt/SrTiO_3$:Rh works as a photocatalyst for H_2 production using Fe^{2+} ions, while $BiVO_4$ acts as a photocatalyst for O_2 production using Fe^{3+} ions (Figure 5.13b) (Sayama et al. 2001).

A powdered Z-scheme system for the photocatalytic water-splitting reaction under visible light irradiation was also successfully demonstrated by utilizing a combination of a metal–complex catalyst, reduced graphene oxide (RGO), and semiconductor photocatalysts. For this purpose, a Ru-complex electrocatalyst [Ru (2,2′-bipyridine) (4,4′-diphosphonate-2,2′-bipyridine)(CO)$_2$]$^{2+}$ is linked with SrTiO$_3$:Rh (Rh: 4 at%) as a H$^+$ reduction photocatalyst was coupled with BiVO$_4$ as a water oxidation photocatalyst and RGO as a solid-state electron mediator, as represented in Figure 5.14a. H$_2$ and O$_2$ evolved stoichiometrically under visible light irradiation ($\lambda > 420$ nm) with the turnover number for the Ru-complex calculated to be 450 after 8.5 hours of irradiation time (Suzuki et al. 2015).

In the (ZnRh$_2$O$_4$/Ag/Ag$_{1-x}$ SbO$_{3-y}$) system, Ag acted as a solid-state electron mediator for the transfer of electrons from the CB of Ag$_{1-x}$ SbO$_{3-y}$ to the VB of ZnRh$_2$O$_4$. The system simultaneously produced H$_2$ and O$_2$ from pure water at a molar ratio of ~2:1 under irradiation with visible light at 500 nm (Figure 5.14b) (Kobayashi et al. 2014). An (oxy)nitride-based heterostructure for powdered Z-scheme overall water-splitting system, MgTa$_2$O$_{6-x}$N$_y$/TaON, was fabricated by a simple one-pot nitridation route to effectively suppress the recombination of carriers by efficient spatial charge separation and decreased defect density. By employing Pt-loaded MgTa$_2$O$_{6-x}$N$_y$/TaON as a H$_2$ evolving photocatalyst and PtOx–WO$_3$ and IO$_3^-$/I$^-$ pairs were used as an O$_2$-evolving photocatalyst and a redox mediator IO$_3^-$/I$^-$ pairs, gives an AQE of 6.8 % at 420 nm, which was the highest AQE among the powdered Z-scheme overall water-splitting systems ever reported (Chen et al. 2015).

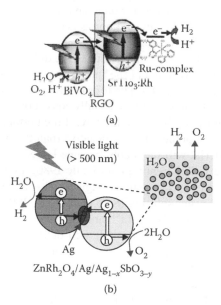

FIGURE 5.14 Schematic presentation of Z-scheme solid-state photocatalytic water splitting using (a) Ru-complex cocatalyst ([Ru (2,2′-bipyridine) (4,4′-diphosphonate-2,2′-bipyridine)(CO)$_2$]$^{2+}$) linked with SrTiO$_3$:Rh (Rh: 4 at%) as a H$^+$ reduction photocatalyst, coupled with BiVO$_4$ as a water oxidation photocatalyst and reduced graphene oxide as a solid-state electron mediator. (b) Ag-embedded solid solution ZnRh$_2$O$_4$/Ag/Ag$_{1-x}$ SbO$_{3-y}$ (Kobayashi, R. et al., *J. Phys. Chem. C*, 118, 39, 22450–456, 2014.)

5.7   METAL ION–INCORPORATED METAL OXIDE

Optimal surface structural design of the photocatalysts is a key point to efficient overall water splitting into H_2 and O_2. The surface modification method was devised for a photocatalyst to effectively promote overall water splitting. Amorphous oxyhydroxides/oxides of group IV and V transition metals (Ti, Nb, V, Ta) over a semiconductor photocatalyst are a better choice because they possess layered perovskite structures that provide reaction sites between the layers, and BG between O 2p and d^0/d^{10} is usually too big. These compounds are comparatively less soluble in water, which is also suitable for water splitting. Historical background of these compounds reveals that until the 1980s, research on photocatalytic overall water splitting had prominently been devoted to a few of the oxide semiconductors, that is, ZnO, $SrTiO_3$, and TiO_2 (Domen et al. 1980, 1982). But after the discovery of $K_4Nb_6O_{17}$, which has a layered structure and high photocatalytic activity, research on these materials became a prominent area (Domen et al. 1986a; Kudo et al. 1988). All these photocatalysts have unique structures, such as layered, tunnel, pillared, and so on (Inoue et al. 1990; Takata et al. 1997; Shimizu et al. 2002). Afterward, multicomponent semiconductors based on d^0/d^{10} transition metals have been meticulously designed, fabricated, and modified using noble metal/metal loading, metal ion doping, and composite metal oxide semiconductors, and then investigated for application to water splitting. Photocatalysts based on transition-metal cations with empty d orbitals/d^0 electronic configuration and typical metal cations with filled d orbitals/d10 electronic configuration are representative photocatalysts and also expected to be active as photocatalysts for breaking of water. Some have achieved high quantum efficiencies as high as 30%–50%, but to date no homogeneous or heterogeneous photocatalysts have been found suitable for stoichiometric water splitting with efficiency more than 30% in visible light. This is primarily due to the difficulty in oxygen formation (four-electron consuming process) and the rapid reverse reaction between products (Ikeda et al. 1998a; Kato et al. 2003). The tops of the VBs of such metal oxide photocatalysts, having d^0 or d^{10} metal cations, usually consist of O 2p orbitals, which are located at ~+3 V or higher (vs. NHE). Therefore, if the bottom of the CB of a given metal oxide is located at a potential more negative than the water reduction potential, the BG of the material will inevitably become larger than 3 eV, rendering the material inactive in the visible light region (Scaife 1980). Inoue et al. also reported typical metal oxides with d^{10} electronic configuration as a new class of photocatalysts suitable for overall water splitting (Sato et al. 2001, 2004). The metallic oxide semiconductors (Ti^{4+}, Zr^{4+}, V^{5+}, Nb^{5+}, Ta^{5+}, and W^{6+}) with their high oxidation states have the potential to break water, with their well-defined structures, apt BGs and active surface. This route of synthesis plays an important role. They can be synthesized in solid liquid and gas phase syntheses routes. The majority of perovskite or layered perovskite-type materials of formula $M_x M'_y M''zO_7$ (x, y, z = 0–4; M, M', and M″ represent three different metals) based on Nb(V), Ta(V), Ti(IV), and so on, had been prepared by a conventional solid-state reaction method, in which appropriate amounts of precursor oxides or, nitrates, carbonates are ground together and the mixture obtained is calcined at high temperatures (typically 1000°C–1300°C) for times sufficient to allow cation interdiffusion. The materials produced by this method have low surface area, uncontrolled grain growth, localized segregation of the components, and possible loss of stoichiometry due to volatilization of the constituent components, resulting in decrease in the photocatalytic activity of a given catalyst. However, some other methods, like solgel, hydrothermal, microemulsion, template, chemical vapor deposition, electric-arc, laser ablation, and so on were used for their synthesis. A Pechini-type solgel process (polymerizable complex) based on polyesterification between citric acid and ethylene glycol also provides a better alternate strategy to synthesize mixed metal oxides, which yield ultrafine nanoscale particles. Furthermore, they improvised with metal loading, ion doping, and composite formation, and redox mediators are added to the main oxide to prevent the rapid recombination of the photogenerated electron/hole pairs with high surface area and better visible light response for the semiconducting particles. Computational and analytical studies were conducted to quantitatively probe the relationships between composition, crystal structure (primarily perovskite and perovskite-related structures) and the electronic structure of oxides containing octahedrally coordinated d^0 high oxidation number transition metal ions such as Ti^{4+}, Nb^{5+}, and Ta^{5+}. For isostructural niobate, titanate, and tantalate compounds, the BG increases as the effective electronegativity ($Nb^{5+} > Ti^{4+} > Ta^{5+}$) of the transition metal ion decreases. Eng et al. (2003) also observed that the BG is sensitive to changes in the conduction bandwidth, which can be maximized for structures possessing linear M–O–M bonds, such as the cubic perovskite structure. As this bond angle decreases the CB narrows and the BG increases. This tendency indicates that photogenerated electron–hole pairs can migrate relatively easily in the corner-shared framework of TaO_6 units. Niobate photocatalysts show differences in photocatalytic properties mainly due to CB energy levels, which are lower than in their Ta counterparts, and the degree of excitation energy delocalization. The simple perovskite crystal structure has corner connected BO_6 octahedra and 12 oxygen coordinated A cations, located in between the eight BO_6 octahedra (Figure 5.15a). The perfect structure of the octahedral connection results in a cubic lattice. However, depending on the ionic radii and electronegativity of the A and B site cations, tilting of the octahedra takes place, which gives rise to lower symmetry structures. As seen from the crystal structure B site cations (Figure 5.15b) are strongly bonded with oxygen (or any other anion), while A site cations have relatively weaker interactions with oxygen. Depending on the type of the cations occupying the lattice sites, these interactions could be altered to yield different perovskite crystal geometries.

Vanadium- and niobium-based perovskites are also similar to tantalum (Ta)-based photocatalysts and they show good photocatalytic activity under UV irradiation. Domen et al. (1980) observed the effect of NiO over $SrTiO_3$ powder for water splitting into H_2 and O_2 and later they studied the mechanism and found that NiO acts as a hydrogen evolution site while oxygen evolution occurs at $SrTiO_3$ (Domen et al. 1986b).

5.7.1 TANTALATE PHOTOCATALYSTS

Tantalate perovskites are well known for efficient overall water-splitting reaction under UV irradiation as they possess both VB and CB potentials suitable for water-splitting reactions. NiO-loaded $NaTaO_3$ doped with lanthanum showed a high photocatalytic activity (56%) for water splitting into H_2 and O_2 with evolution rates of 19.8 and 9.7 mmol/h, respectively, under 270-nm UV light irradiation. NiO cocatalysts, loaded on the edge of the high crystalline nanostepped structure (Figure 5.16) of NaTaO$_3$:La photocatalysts as ultrafine particle, were responsible for high photocatalytic activity (Kato and Kudo 2001; Kato et al. 2003). To enable visible light photocatalysis, doping of various elements has been studied to achieve visible light activity.

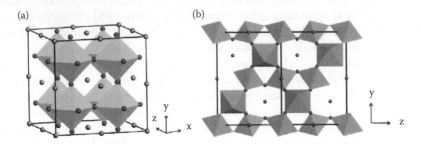

FIGURE 5.15 Crystal structure of simple perovskite (a) BaTiO$_3$ (A site cation) and (B site cation); (b) double-perovskite Na$_2$Ta$_2$O$_6$ (red: oxygen, green and purple: A and B site cation, grey and blue: BO$_6$ octahedra). (From Eng, H. W. et al., *J. Solid State Chem.*, 175, 94–109, 2003.)

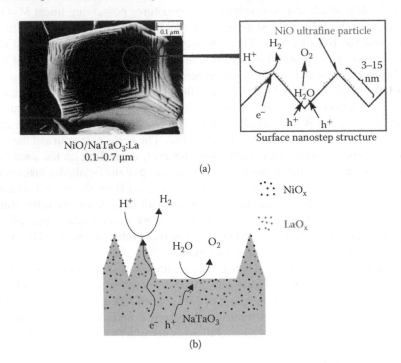

FIGURE 5.16 (a) Field emission scanning electron microscope (FESEM) of the layered structure of La-doped NiO-loaded/NaTaO$_3$ with nanostep structure sodium tantalite decorated with ultrafine NiO and grooves with holes for oxidation of water. (b) Reduction and oxidation of water at NiO/NaTaO$_3$:La's surface. (From Kato, H. et al., *J. Am. Chem. Soc.*, 125, 3082–3089, 2003.)

Kanhere et al. (2012b) reported a detailed study on visible light active Bi-doped $NaTaO_3$ for different Bi concentrations (2.5%, 5.0%, and 7.5% by moles). The optical properties of the doped unimolar samples of the Na and Ta (1.01–1.03, contain Bi ions located at both Na and Ta sites in the lattice) resulted in the highest BG narrowing. It was shown that the photocatalytic hydrogen evolution occurred from these samples under visible light irradiation ($\lambda > 390$ nm) after loading of appropriate amount of platinum cocatalyst. The Mott–Schottky plots revealed that the flat-band potential of the pristine $NaTaO_3$ is highly negative to the H_2/H_2O reduction potential (−1.19 eV vs. SCE, pH = 7) and was sufficient for hydrogen generation (Kanhere et al. 2011, 2012b). Furthermore, codoping of the mixed metals such as La–N La–Co, La–Cr, La–Ir, La–Fe in $NaTaO_3$ has shown successful visible light absorption and subsequent hydrogen evolution (Yang et al. 2010; Zhao et al. 2011). Solid solution of $Na_{1-x} La_x Ta_{1-x} Cr_x O_3$ ($x = 0$–1.00) was prepared with a conventional solid-state reaction method. The crystal structure changed from monoclinic ($x \leq 0.60$) to orthorhombic ($x \geq 0.70$) with the modulation of chemical composition. H_2 generation under visible light irradiation ($\lambda > 420$ nm) was successfully obtained with the tuning of band structures of $Na_{1-x} La_x Ta_{1-x} Co_x O_3$. The highest performance for H_2 generation (4.5 µmol/h) was observed at $x = 0.4$ with AQY of 0.5% (Yi and Ye 2007, 2009). Another successful solid solution is $LaFeO_3$–$NaTaO_3$ (Kanhere et al. 2012a). Codoping of $LaFeO_3$ in $NaTaO_3$ has been studied for hydrogen evolution by Zhao et al.; studies have indicated that both anion and cation doping in $NaTaO_3$ are useful for visible light photocatalytic applications. Among the doped $NaTaO_3$ systems, computational studies on the anionic (N, F, P, Cl, S) doping were reported by Han et al. (2009), who showed that certain anions like N and P may be useful for visible light absorption. Additionally, doping of magnetic cations such as Mn, Fe, and Co in $NaTaO_3$ has also been studied using density functional theory (DFT)-PBE (hybrid exchange correlation (XC) functionals of Perdew, Burke, and Ernzerhof) (Zhou et al. 2011). Recently, our group studied DFT calculations of a number of doped $NaTaO_3$-based photocatalysts by PBE0 hybrid calculations (Figure 5.9) (Kanhere et al. 2014). Further, anion doping was also explored in detail using (HSE06) hybrid DFT calculations, where N, P, C, and S doping at O sites were studied. The study also reports the thermodynamics and effect of coupling between N–N, C–S, and P–P on the optical and electronic properties (Wang et al. 2013). A DFT study was used to design a novel photocatalyst and to explain the properties of existing materials systems. Particularly, use of a hybrid functional such as PBE0 or HSE06 allows for the accurate definition of the VB structure and location of bands or energy states that are crucially important for visible light-driven photocatalysis. Hybrid DFT calculations could be useful in predictive modeling, where BGs and band edge potential of useful doped photocatalysts are identified (Kanhere et al. 2014). $AgTaO_3$ exhibits similar behavior to alkali tantalates; however, it has a smaller BG value of 3.4 eV. $AgTaO_3$ doped with 30% Nb absorbs visible radiation and shows a stoichiometric overall water-splitting reaction under visible light when loaded with NiO cocatalyst (Ni et al. 2013). Codoping of N–H and N–F in $AgTaO_3$ has been studied in detail. The study indicates that codoping not only balances the charges but also improves carrier mobility. N–F codoped $AgTaO_3$ has an effective BG value of 2.9 eV and shows H_2 generation under visible light (Li et al. 2013). Similarly, $KTaO_3$ ($E_g = 3.6$ cV) photocatalysts have been studied for water splitting under UV radiation. However, work on the development of visible light-driven $KTaO_3$-based photocatalysts is limited. The photocatalytic decomposition of water has been studied to achieve an artificial photosynthesis using the alkali and alkaline earth tantalates, such as $LiTaO_3$, $NaTaO_3$, $KTaO_3$, $MgTa_2O_6$, and $BaTa_2O_6$, shown with and without cocatalysts. Here, tantalates worked as a photocatalytic material, possessing reasonable activities for the decomposition of water into H_2 and O_2 in stoichiometric amounts under UV irradiation. Among them, $BaTa_2O_6$ in the orthorhombic phase was the most active. The addition of a small amount of $Ba(OH)_2$ into the water and supporting NiO drastically enhanced the photocatalytic reaction on the $BaTa_2O_6$ catalyst. On the other hand, in the transition metal-incorporated tantalates, $NiTa_2O_6$ produced both H_2 and O_2 without cocatalysts. Intercalation of nanosized Fe_2O_3 and TiO_2 particles within layered compounds such as $HNbWO_6$, $HTaWO_6$, $HTiNbO_5$, and $HTiTaO_5$ has been found to dramatically improve visible light photocatalytic activity (Wu et al. 1999; Jang et al. 2005) due to the more effective separation of the photogenerated electrons–holes due to their rapid diffusion.

5.7.2 VANADATE PHOTOCATALYSTS

Bismuth vanadate $BiVO_4$ is the most popular candidate of this class. It has a direct BG of approximate value of 2.4 eV and monoclinic structure that confirmed its good candidature for a metal oxide semiconductor (PEC catalyst), which can work under visible light. The optical properties of this monoclinic metal oxide ($BiVO_4$) are either similar to or even better than WO_3 and suitable for exploiting the OER using visible light irradiation. Additionally, the band-edge alignment allows generation of a relatively large photovoltage for water oxidation compared with other metal oxides (Sayama et al. 2006a; Walsh et al. 2009). There are several ways for technical improvements to be made in $BiVO_4$ for enhancement in its efficiency, that is, by controlled growth of $BiVO_4$ by modifying synthesis methods, more suitable interface for heterojunctions, suppression of recombination losses by decorating its surface with cocatalyst, BG engineering by metal or element doping, and so on. Berglund et al. (2011) synthesized the vanadium-rich nanostructured films of $BiVO_4$ on glass substrate (used as photoelectrode), which can exhibit several times higher OER activity than that of stoichiometric films, under visible light illumination. In this series, Luo et al. (2010) synthesized a compact polyhedron microcrystalline $BiVO_4$ film by adjusting the deposition time, temperature, pH of solution, and buffer solution, during the chemical bath deposition method, resulting in various microstructured morphologies. Their findings are quite interesting because low crystalline but relatively compact films showed high photocatalytic activities over the high crystalline films. This is because diffusion of holes to the surface might limit the performance of these films. Furthermore, the addition of the group VI metals (Mo, W, etc.) is used to enhance the PEC water oxidation activity of $BiVO_4$ films. The Bard group carried out a combinatorial study on the Bi–V–W oxide system using a scanning electrochemical microscopy (SECM) technique (Ye et al. 2010). The result in material composed of the ratios 4.5:5:0.5 of Bi, V, and W, respectively, gave the highest photoactivity (Ye et al. 2011), both in the combinatorial experiment and in bulk film studies. They incorporated overlayers of oxide cocatalysts for the OER and found that the highly active OER catalyst iridium oxide did not significantly enhance the photoactivity, whereas the less active catalyst materials Pt and Co_3O_4 did enhance the photocurrent densities by nearly an order of magnitude due to compatible heterojunction. Hong et al. (2011) also synthesized thin film with heterojunctions of high light-harvesting $BiVO_4$ and favorable charge-transfer properties WO_3 by sequential deposition of multilayers of the respective precursors.

Several other research groups have found that deposition of oxide catalysts on the surface of $BiVO_4$ photoelectrodes significantly enhances its water oxidation efficiency. Pilli et al. (2011) observed an enhancement in water oxidation activity for $BiVO_4$ upon doping with 2% Mo as well as upon deposition of a cobalt oxide catalyst onto the surface. Zhong et al. (2011) measured a similar enhancement effect on tungsten-substituted $BiVO_4$, loaded with a cobalt oxide as a cocatalyst, which was attributed to efficient suppression of surface recombination on application of the cocatalyst. Seabold and Choi (2012) attained high photocurrent densities (2-mA/cm^2 short-circuit current density) with stable $BiVO_4$ film in neutral aqueous electrolytes, which was synthesized by an electrodeposition/calcination technique and coated with the iron oxyhydroxide cocatalyst for water oxidation under AM1.5G illumination (Youngblood et al. 2009). Further efforts would be devoted to the suppression of the recombination losses, coupling efficient OER cocatalysts, enhancement in catalytic activity, higher charge collection efficiencies, stability limits of $BiVO_4$ in terms of pH and electrochemical potential in aqueous solutions, and incorporation of a micro- or nanostructured $BiVO_4$ in place of the bulk structure of the same. By empowering the above areas, $BiVO_4$ may emerge as a very promising metal oxide photoanode for water oxidation. Another important vanadate perovskite are silver vanadates $AgVO_3$ that exist in two allotropic forms, namely, α-$AgVO_3$ (E_g = 2.5 eV) and β-$AgVO_3$ (E_g = 2.3 eV), their VB attributed to the 3 d orbital energies in volts (Konta et al. 2003). Both phases are photocatalytically active. However, β-$AgVO_3$ shows better photocatalytic performance than the α-phase. The CB potential of $AgVO_3$ is not sufficient for H_2 evolution, but it is suitable for the degradation of volatile organic compounds (VOCs) and O_2 evolution (Xu et al. 2012a; Sang et al. 2014).

Tantalum, niobium, and vanadium belong to the same group in the periodic table. Perovskite compounds of these elements show decreasing BG and CB potential values. This trend is attributed to the 3d, 4d, and 5d orbital energies in V, Nb, and Ta, respectively. Visible light absorption in $PbBi_2Nb_2O_9$ is due to the transition from O 2p hybridized with Pb_6s/Bi_6s to Nb_4d and is reported to directly cleave water into H_2 and O_2. $InTaO_4$ and $InNbO_4$, with BGs in the range 2.8–2.4 eV, are known to split water. There are two kinds of octahedral InO_6 and NbO_6 (or TaO_6) in a unit cell for both $InNbO_4$ and $InTaO_4$. The difference in unit cell volumes between TaO_6 and NbO_6 leads to a change in the lattice parameters slightly affecting their photocatalytic activity, mainly due to variation in the CB levels formed by Ta_5d TaO_6 and Nb_4d NbO_6.

5.7.3 TITANATE PHOTOCATALYSTS

Titanate perovskites have been studied widely for photocatalytic applications for a long time because of their excellent photostability and corrosion resistance in aqueous solutions. Moreover, their CB potential lies between −0.3 and −0.6 eV, that is, above the water reduction level (Kavan et al. 1996), and the VB level of TiO_2 is far more positive than the oxygen evolution energy level. Therefore, most of the titanate perovskites have BG energy (E_g) value more than 3.0 eV and show superb photocatalytic properties under UV radiation and an external bias (Maeda et al. 2011) but low activity in the visible light region. Some of the titanate perovskites have CB energies more negative than TiO_2, making them more suitable candidates for hydrogen generation without external bias. Doping is the most widely opted tool to alter the optical properties and induce visible light absorption in titanates. The introduction of another metal (or other element) can affect the observed overall photocatalytic behavior due to the (1) modification of the BG; (2) introduction of an intermediate band; (3) introduction of recombination sites or passivation of such sites (on the surface or in the bulk); (4) modification of the carrier mobilities; and (5) modification of the density of carriers in conduction and VBs. Figure 5.17b represents an overview of the band diagram of the $MTiO_3$ systems with respect to water oxidation and reduction potential levels versus NHE; pH = 0 for the few alkali and transition metal

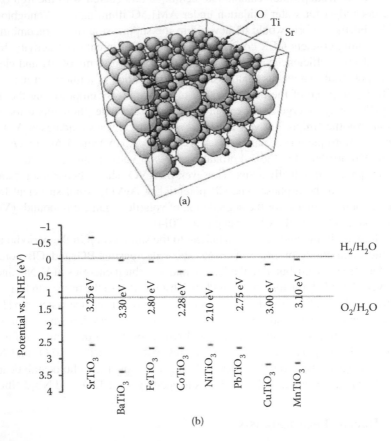

FIGURE 5.17 (a) Crystal structure of $SrTiO_3$ and (b) band edge potentials (vs. NHE; pH = 0) of an $MTiO_3$ system. (From Ye, H. et al., *J. Phys. Chem. C.*, 114, 13322–28, 2010.)

perovskite titanates $MTiO_3$ (M = Sr, Ba, Ca, Mn, Co, Fe, Pb, Cd, Ni) systems (Xu and Schoonen 2000; Ye et al. 2010).

Along with alkali titanates, several transition metal titanates show promise for visible light photocatalysis. Alkali metal titanates such as Ba, Ca, Sr, Pb, and Mn have enough CB potential for hydrogen evolution. However, certain transition metal titanates do not possess the desired CB potential for water reduction, though they have BG values in the visible region (such as Co, Ni, Fe, etc.). These materials could be suitable for the degradation of organics or other photo-oxidation reactions.

Strontium titanate ($SrTiO_3$; STO) is the simplest cubic (Pm_3m, a = 3.9 Å) perovskite with an indirect BG of 3.25 eV (Benthem et al. 2001).

By means of the BG engineering on $SrTiO_3$, the magnitude of the BG decreased, while retaining the position of the conduction band minimum (CBM). Such band engineering was successfully demonstrated by doping of noble metal or transition metals (Konta et al. 2004; Kudo et al. 2007). The rhodium dopants introduce an impurity level located 2.3 eV below the CBM of STO. Visible light can therefore excite electrons from the dopant levels into the CB of STO, and the photocatalytic process is initiated (Lehn et al. 1982; Domen et al. 1986c). Transition/noble-metal doped $SrTiO_3$ photocatalysts work as an H_2-evolving photocatalyst in Z-scheme systems combined with O_2-evolving photocatalysts such as $BiVO_4$, WO_3, and Bi_2MoO_6 for solar water splitting in the presence and absence of electron mediators (Kato et al. 2004; Sasaki et al. 2008). Apparent quantum yields of these three systems ($Pt/SrTiO3:Rh$)–($BiVO_4$), –(Bi_2MoO_6), and –(WO_3) in an Fe^{3+}/Fe^{2+} redox mediator were 0.3%, 0.2%, and 0.2% at 440 nm, respectively. Thus, $SrTiO_3:Rh$ is an attractive material for solar energy conversion and is expected to function as a photoelectrode material for water splitting under visible light irradiation. Doping of the Ti site with Mn, Ru, Rh, and Ir attracts significant interest (Konta et al. 2004). It was found that these dopants induce mid-gap states in the BG allowing visible light absorption (Chen et al. 2012). Mn and Ru doping are found useful for O_2 evolution, while dopants such as Ru, Rh, and Ir are suitable for H_2 evolution. The Rh(1%)-doped $SrTiO_3$ photocatalyst loaded with a Pt cocatalyst (0.1 wt.%) gave 5.2% of the QY at 420 nm for the H_2 evolution reaction (Konta et al. 2004), where 7% Rh-doped $SrTiO_3$ showed 0.18% IPEC efficiency under 420-nm irradiation. Furthermore, when it is loaded with a cocatalyst such as Rh or NiO_x, $SrTiO_3$ shows stoichiometric water splitting under UV radiation and has been studied extensively for developing visible light water-splitting catalysts (Townsend et al. 2012). Irie et al. reported Z-scheme water-splitting systems using only $SrTiO_3$-based photocatalysts (Hara and Irie 2012). Among the materials they prepared, $SrTiO_3:Ga/Bi$ and $SrTiO3:In/V$ were able to reduce and oxidize water to form H_2 and O_2, respectively, in the presence of an IO^{3-}/I^- redox couple, achieving Z-scheme water splitting under UV irradiation (λ < 400 nm). Redox couples I^{3-}/I^-, Fe^{3+}/Fe^{2+}, IO^{3-}/I^-, NO^{3-}/NO^{2-} (Sayama et al. 2006b; Sasaki 2013) serve as redox mediators in various Z-scheme systems that used Rh-doped $SrTiO_3$ as a H_2 evolving photocatalyst. Kudo et al. reported that $[Co(bpy)_3]^{3+/2+}$ and $[Co(phen)_3]^{3+/2+}$ (bpy = 2,2'-bipyridine; phen = 1,10-phenanthroline) redox couples achieved the functionality to mediate electron transfer from $BiVO_4$ to $Ru/SrTiO_3:Rh$, resulting in stable overall water splitting, with a maximum AQY of 2.1% at 420 nm. In a significant demonstration, an electron mediator

[Co(bpy)$_3$]$^{3+/2+}$ was used for Rh-doped SrTiO$_3$ with BiVO$_4$ photocatalyst that showed a solar energy conversion efficiency of 0.06% under daylight. Electron mediators are indispensable and play vital roles in achieving Z-scheme water splitting but can cause several kinds of undesirable backward reactions. Hence, a Z-scheme water-splitting system that does not rely on any redox mediator is highly desirable. This idea of a redox mediator-free system may seem like a photocatalytic material that has a built-in electric field between the interface of p- and n-type semiconductors (Nozik 1976; Kim et al. 2005). The quality of the interface and the band alignment of the two semiconductors are important factors for the successful realization of mediator-free type Z-schemes. In 2009, Kudo et al. reported a unique Z-scheme water-splitting system consisting of Ru/SrTiO$_3$:Rh as a H$_2$ evolution photocatalyst and a wide variety of O$_2$ evolution photocatalysts (such as TiO$_2$:Cr/Sb, BiVO$_4$, AgNbO$_3$, Bi$_2$MoO$_6$, and WO$_3$) (Sasaki et al. 2009). This system was capable of stoichiometrically splitting water into H$_2$ and O$_2$ under visible light without a redox mediator. The activity was strongly dependent on the reaction pH, and the highest activity was obtained at pH 3.5, due to the agglomeration of the photocatalyst powders of Ru/SrTiO$_3$:Rh and BiVO$_4$ (Figure 5.18a). A schematic mechanism of water splitting of an agglomerated Z-scheme photocatalyst is shown in Figure 5.18b. Efforts in Z-scheme photocatalysis

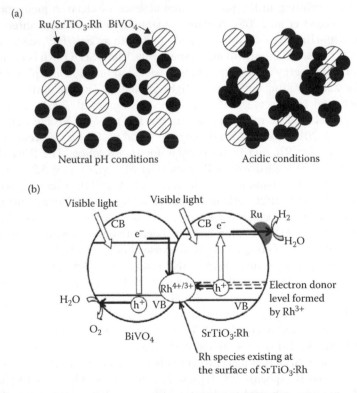

FIGURE 5.18   (a) Schematic microstructure at neutral and 3.5 pH (mediator-free Z-scheme) conditions and (b) band diagram of Z-scheme photocatalysis using Rh-doped SrTiO$_3$ and BiVO$_4$. (From Sasaki, Y et al., *J. Phys. Chem. C*, 113, 17536–42, 2009.)

have also been targeted toward eliminating the need for electron mediators by preparing composite photocatalysts. Composite of 1% Rh-doped $SrTiO_3$ loaded with 0.7% Ru and $BiVO_4$ showed a stoichiometric water-splitting reaction (pH 7) with QY of 1.6% at 420 nm (Jia et al. 2014).

Further, the water-splitting efficiency is dependent on the synthesis method used for Rh-doped $SrTiO_3$. The use of excess Sr in hydrothermal and complex polymerization methods proved useful for improving the apparent yield to 4.2% under 420-nm radiation (Kato et al. 2013). Apart from monodoping, codoping has been employed in $SrTiO_3$ to pursue visible light-driven photocatalysis. Codoping of Sb (1%) and Rh (0.5%) was found useful for visible light photocatalysis and estimated H_2 and O_2 evolution rates for 1-m^2 surface area were 26 and 13 mL/h, respectively (Asai et al. 2014). Further, a composite system was prepared from codoped La–Cr in $SrTiO_3$ and codoped La–Cr Sr_2TiO_4, which showed visible light-driven H_2 evolution (24 µmol/h/g) (Jia et al. 2013). Further, a solid solution $((AgNbO_3)_{1-x} (SrTiO_3)_x;$ with $x = 0.25$) of $AgNbO_3$ and $SrTiO_3$ was discovered to be a visible light active photocatalyst (Wang et al. 2009). $(AgNbO_3)_{0.75}(SrTiO_3)_{0.25}$ showed promising performance for O_2 evolution and isopropanol (IPA) degradation under visible light. DFT-based band structure calculations provide useful insights into the electronic structure and its correlation with photocatalytic activity to understand and design $SrTiO_3$-based photocatalysts. It is shown that Rh doping in $SrTiO_3$ produces bandlike states above the valence band maximum (VBM), which are responsible for visible light absorption (Chen et al. 2012). Further, theoretical work on doped $SrTiO_3$ compounds shows that certain dopants such as La strongly lower the effective mass of electrons and holes (near the VB region), increasing the mobility of the photoexcited carrier (Wunderlich et al. 2009). Along with ground-state band structure calculations, development of low dimensionality, study of electron and hole masses, and a defect study, photoexcited electron transport could be carried out to understand this system in detail, which will be useful to further give new dimension in developing highly efficient (>20%) nontoxic, noncorrosive photocatalysts at low cost.

Another popular titanate $BaTiO_3$ is a wide-gap n-type semiconductor having a BG of 3.0 eV. Doping of Rh species into the lattice of $BaTiO_3$ is used to form a new absorption band in the visible light region. Recently, Maeda prepared a rhodium-doped barium titanate ($BaTiO_3$:Rh) powder by the PC method and examined its photocatalytic activity for H_2 evolution from water after modification with Pt nanoparticulate (Figure 5.19). He found that this system exhibited higher activity than that made by a conventional solid-state reaction method, upon visible light ($\lambda > 420$ nm) irradiation in water containing an electron donor such as methanol and iodide (Maeda 2014).

Being a hydrogen-evolving catalyst, this material has also been used for a Z-scheme with Pt/WO_3 for overall water splitting (Maeda 2014). Overall visible light-driven Z-scheme water splitting was also accomplished using Pt/$BaTiO_3$:Rh as a building block for H_2 evolution in combination with PtO_x-loaded WO_3 as an O_2 evolution photocatalyst in the presence of an IO_3^-/I^- shuttle redox mediator. PEC analysis indicated that a porous $BaTiO_3$:Rh electrode exhibited cathodic photoresponse due to water reduction in a neutral aqueous Na_2SO_4 solution under visible light, at QY of 0.5%, under 420 nm (Maeda 2014). Calcium titanate is

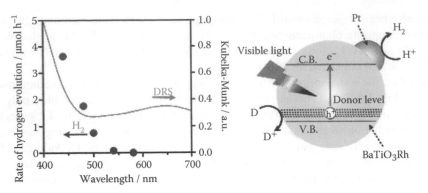

FIGURE 5.19 Representing $BaTiO_3$:Rh for visible light water splitting and reduction in band gap by introduction of an additional donor level of Rh in a $BaTiO_3$ lattice. (From Maeda, K., *ACS Appl. Mater. Interfaces*, 6(3), 2167–73, 2014.)

also counted as one of the common perovskite minerals with a BG of 3.6 eV. Metal doping- or codoping or loading of cocatalyst on $CaTiO_3$ surface is used to induce visible light absorption in this widely studied photocatalyst for visible light-driven photocatalytic water decomposition and hydrogen evolution (Zhang et al. 2010). Cu-doped and NiO_x-cocatalyst loaded $CaTiO_3$ not only exhibit pho-tocatlytic activity in visible light but also enhance the rate of hydrogen evolution. Studies of such doped systems, where dopants enhance the photocatalytic activity of the host materials, are important for the designing the efficient photo-catalysts. Detrimental effects of doping on the properties such as electron–hole recombination, electron/hole effective mass, and reduced crystallinity, should be studied and reported in detail. These studies are useful to gain insights on the photocatalytic activities of the doped systems. Codoping of Ag and La at $CaTiO_3$ has been done to narrow the BG and induce visible light absorption (Zhang et al. 2012a). DFT studies also indicate that like $SrTiO_3$, TiO_2-terminated $CaTiO_3$ surfaces possess the capability of visible light absorption (Fu et al. 2013). $CoTiO_3$ has a BG in the visible region (E_g = 2.28 eV). Recently, this compound has been investigated for photocatalytic O_2 evolution (64 µmol/g/h) reaction without any cocatalyst under visible light (Qu 2014). Similarly, $NiTiO_3$ and $FeTiO_3$ of 2.16 and 2.8 eV have also been reported with light absorption spectra show peaks in visible region corresponding to the crystal field splitting, respectively (Kim et al. 2009; Rawal et al. 2013). Abe et al. used eosin-Y-fixed TiO_2 photocatalyst, using a silane coupling reagent, R_3MO_7 and $R_2Ti_2O_7$ (R = Y, Gd, La; M = Nb, Ta) under visible light irradiation for hydrogen generation by water splitting. Tai et al. prepared mixed oxide $K_2La_2Ti_3O_{10}$ by a complex polymerization process that fol-lowed Au nanoparticle loading, by impregnation (Au-i) and deposition–precipita-tion (Au-d) methods (Tai et al. 2004). Au-i/$K_2La_2Ti_3O_{10}$ has been found more active for water splitting than Ni/$K_2La_2Ti_3O_{10}$. Moreover, Au-i/$K_2La_2Ti_3O_{10}$ shows superior water-splitting properties in the UV and visible region compared with its counterpart Au-d/$K_2La_2Ti_3O_{10}$. This is because of the better crystallinity and surface plasmonic resonance effect of gold nanoparticles than that prepared through Au deposition precipitation. Ishikawa et al. (2002) used stable mixed oxide $Sm_2Ti_2S_2O_5$ for water splitting under visible light irradiation ($\lambda \leq 650$ nm).

5.7.4 NIOBATE PHOTOCATALYSTS

Alkali niobates $KNbO_3$ (E_g = 3.14 eV) and $NaNbO_3$ (E_g = 3.08 eV) are wide BG materials, which have photocatalytic activity in the UV responsive region. But after suitable modifications of the band structure they became active in visible light (Liu et al. 2007). Nitrogen-doped $KNbO_3$ was studied for water splitting, under visible light (Wang et al. 2013). Codoping (La and Bi) was also used to induce visible light responses in $NaNbO_3$ (Liu et al. 2011a). Furthermore, recent work on ferroelectric niobate perovskites of $KNbO_3$–$BaNiNbO_3$ shows that the solid solution of these compounds could absorb six times more light and shows 50 times more photocurrent than others (Grinberg et al. 2013). Photocatalytic properties of this compound confirm it as an attractive candidate for visible light-driven photocatalysis. Replacing an A site alkali metal by silver further reduces the BG (2.7 eV) of this perovskite. Studies have shown that the photocatalytic activity of $AgNbO_3$ strongly depends on the shape of the particles. It was noticed that the polyhedron-shaped particles are favorable for O_2 evolution reactions (Li et al. 2009). p-type intrinsic semiconductor $CuNbO_3$ crystallizes in the monoclinic structure, has a BG of 2.0 eV, and has shown 5% efficiency for photon to electron conversion when used as a photocathode. Being a stable material under irradiation, more investigations should be carried out on this material (Joshi et al. 2011).

Layered perovskite materials such as $K_4Nb_6O_{17}$ and $A_4Ta_xNb_{6-x}O_{17}$ (A = K, Rb) were found by Domen et al. (Sayama et al. 1990) to be much more efficient (QYs of ~5%) photocatalysts for overall water splitting. The greatly improved QY of these layered perovskites has been attributed to the role played by the existing interlayer spaces as reaction sites. Owing to the important role of the interlayer spaces, most of the recent reports have focused on the layered structures (Sayama et al. 1996; Ikeda et al. 1998b). At such reaction sites, the electron–hole recombination process could be retarded by the physical separation of electron and hole pairs generated by photoabsorption, which mainly contributes in yielding higher QYs. Recently, we also found that highly donor-doped (110) layered perovskite materials exhibited higher water-splitting activities than the previously known (100) layered perovskite materials under UV light irradiation (Kim et al. 1999, 2002; Hwang et al. 2000, 2004). An n-type semiconductor, strontium niobium oxynitride ($SrNbO_2N$; E_g = 2.0 eV), coated on fluorine-doped tin oxide (FTO) glass was examined as a photoelectrode for water splitting under visible light in a neutral aqueous solution (Na_2SO_4, pH ~6) and without an externally applied potential. Under visible light ($\lambda > 420$ nm) with an applied potential of +1.0–1.55 V versus a reversible hydrogen electrode (RHE), nearly stoichiometric H_2 and O_2 evolution was achieved using a $SrNbO_2N$/FTO electrode modified with colloidal iridium oxide (IrO_2) as a water oxidation promoter (Figure 5.20a). This is a first example of PEC water splitting involving an n-type semiconductor with a BG smaller than 2.0 eV that does not require an externally applied potential (Maeda et al. 2011).

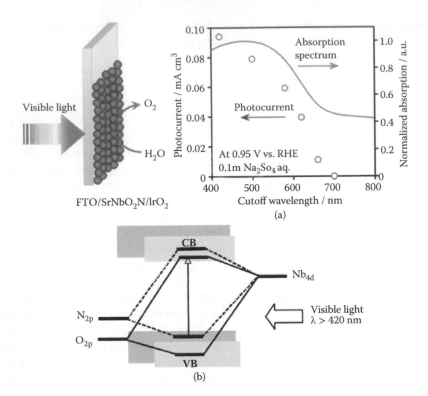

FIGURE 5.20 (a) $SrNbO_2N$ as a water-splitting photoanode with its photocurrent profile on visible light exposure and (b) band model of N-doped $Sr_2Nb_2O_7$ that shows mixing of N 2p with O 2p states near the top of the VB leading to band gap narrowing and visible light activity.

Lee et al. (2009) and Ji et al. (2005) reported N-doped Sr2Nb2O7 for water splitting under visible light irradiation. Ammonia nitridation of the layered perovskite Sr2Nb2O7 precursor, at 800°C, resulted in unlayered cubic oxynitride structure Sr2Nb2O7-x Nx (x = 1.47–2.828). Density of states (DOS) studies suggest that N substitution doping gives rise to BG narrowing by mixing of N 2p with O 2p states near the top of the VB (Figure 5.20b). This BG narrowing enables the material to shift the optical absorption into the visible light range. It has been demonstrated that materials of layered-type structure have higher catalytic activity for water splitting than unlayered materials of the same composition.

5.7.5 Tungstate Photocatalysts

WO_3 is a high-performing oxygen-evolving photoanode capable of generating significant photocurrents at an applied bias when illuminated with light. It can absorb light of ~475 nm ($E_g = 2.75$ eV) wavelength, which is just into the blue part of the solar spectrum, and shows low theoretical STH efficiency (7%). WO_3 possesses a relatively high Hall e⁻ mobility (10 cm²·[V⁻¹·s⁻¹]) (Berak et al. 1970) and long minority carrier diffusion length (0.3–4.2 µm) (Wang et al. 2000; Reyes-Gil et al. 2013) compared with other oxides such as TiO_2 and α-Fe_2O_3. To induce visible light absorption in wide BG semiconductors WO_3, various strategies are utilized such as metal and/or nonmetal doping, but in most cases these absorption events are due to defects, which result in low absorptivity coefficients and high charge carrier recombination. Rather than use a doping approach, fabrication of the ternary tungstate materials (AW_xO_y) that show increased visible light absorption as a result of charge transfer events. This electronic transition is a charge transfer and therefore should result in collectable carriers. $PbWO_4$ particles have been shown to decompose H_2O photocatalytically under UV illumination when properly loaded with a cocatalyst (Kadowaki et al. 2007). Bi_2WO_6 has also received attention as a photocatalyst for H_2 formation and organic dye degradation, and more recently as a photoanode for the oxidation of water (Amano et al. 2008; Zhang and Bahnemann 2013). Finally, when evaluating the wide BG tungstates, the high photocatalytic activity is likely due to the high energy CB (<0-V RHE). These materials are energetically capable of forming the super oxide radical under illumination, which subsequently degrades the water or organic species in solution. The first-row transition metal tungstates AWO_4 (A = Mn, Fe, Co, Ni, Cu, Zn) crystallize into the monoclinic wolframite structure. $CuWO_4$ is an exception since Jahn–Teller distortion causes a transition from monoclinic to triclinic crystal symmetry. These tungstates (except for $ZnWO_4$) absorb in the visible part of the solar spectrum as seen in Figure 5.21a (Cui et al. 2015). $CuWO_4$ is 7% efficient at 400 nm for the reaction with an overall AQY of 0.38% under simulated solar irradiation. $CuWO_4$–WO_3 generates 17 $\mu A/cm^2$ of constant photocurrent over extended periods of illumination and the amount of ferricyanide reduced was quantified using UV–Vis spectroscopy. $Rh:SrTiO_3$ are p-type oxides, which have been studied for PEC hydrogen production. In Z-scheme water splitting, p-type oxide photocatalyst $Rh:SrTiO_3$ is an active powdered catalyst for H_2 production and I^- oxidation and $CuWO_4$ is used for water oxidation. Recently, Cui et al. established an IR-driven photocatalytic water-splitting system based on hybrid conductor material WO_2–Na_xWO_3 (nanowire bundles, $x > 0.25$) for the first time. They suggested a novel ladder type carrier transfer mechanism for this system (Figure 5.21b) (Cui et al. 2015).

Jaing et al. (2010) hydrothermally prepared 0.3 wt.% Pt-loaded ZrW_2O_8 (BG = 4.0 eV) and examined the photocatalytic water splitting under UV light irradiation in the presence of hole scavenger CH_3OH and electron scavenger (ES) $AgNO_3$ that release 23.4-$\mu mol/h$ H_2 and 9.8-$\mu mol/h$ O_2, respectively. This compound showed an average rate of 5.2-$\mu mol/h$ H_2 evolution in the pure water. Li et al. (2010) and Ikeda et al. (2004) utilized pyrochlore-structured-tungsten-based photocatalyst $AMWO_3$ (A = Rb, Cs; M = Ta, Nb) with defect, showed over all water splitting under UV light exposure. Carbon-modified p-type tungsten oxide was used as a photoelectrode for overall water splitting in 0.5-M H_2SO_4 acidic medium with a photoresponse of 2.08 mA/cm^2 under 80-mW/cm^2 Xe light exposure with a highest photoconversion efficiency of 2.16% (Shaban and Khan 2007). Tungsten carbide nanoparticles (5 nm) supported on Na-doped SrTiO(3) was used as an effective cocatalyst for photocatalytic overall water splitting to give H_2 and O_2 in a stoichiometric ratio from H_2O decomposition when supported on a Na-doped $SrTiO_3$ photocatalyst. Moreover, tungsten carbide (on a small scale) has shown a promising and durable catalytic property which can be a good substitute for platinum (Garcia-Esparza et al. 2013). Double-perovskite $Sr_2FeNb_{1-x}W_xO_6$ for $0.0 \leq x < 0.10$ (BG = 2.17 eV), were further investigated for visible light photocatalytic water splitting that gave twice the higher QY for photosplitting in a H_2O–CH_3OH mixture compared with that for $x \sim 0$ under

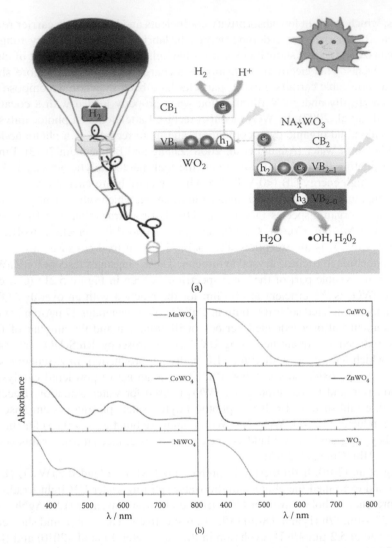

FIGURE 5.21 (a) WO_2–$Na_x WO_3$ hybrid conductor material for IR-driven photocatalytic water splitting. (b) Absorption spectra of AWO_4 and WO_3 powders. (From Cui, G. et al., *Nano Lett.*, 15, 11, 7199–203, 2015.)

visible light ($\lambda \geq 420$ nm) (Borse et al. 2014). Ordered double-perovskite compounds such as Ba_2CoWO_6, Ba_2NiWO_6, Sr_2CoWO_6, and Sr_2NiWO_6 are stable and used as catalysts for O_2 evolution using sacrificial agents (Iwakura et al. 2011).

5.7.6 OTHER OXIDE PHOTOCATALYSTS

5.7.6.1 Graphene Oxide

Graphene-based nanocomposites or super carbons (Castro Neto et al. 2009) can also be proved as a good alternative for water splitting by using renewable energy, that is, solar energy. They can be fabricated using graphite as an initial material, as shown in Figure 5.22a. They can absorb over a wide range of wavelengths, that is, from IR through visible to UV region and can act as electron acceptors and transporters. Graphene is a large aromatic macromolecule of a flat monolayer of sp^2-bonded carbon atoms tightly packed into a 2D honeycomb lattice, which has no BG and has a unique band structure, in which the lowest unoccupied molecular orbital (LUMO) level (π^*) and highest unoccupied molecular orbital (HOMO) level (π) meet up at the Dirac point, as shown in Figure 5.22b. This allows graphene to display quite high conductivity and electron mobility. Furthermore, doping of heteroatom or electrostatic field tuning makes graphene an *n*- or *p*-type semiconductor with a small BG that accompanied by detuning off the Fermi level from the Dirac point near to LUMO and HOMO, respectively (Novoselov et al. 2004). Moreover, it has high thermal conductivity (~5000 Wm/K) (Stankovich et al. 2006), excellent mobility of charge carriers (200,000 $cm^2 \cdot (V^{-1} \cdot s^{-1})$))

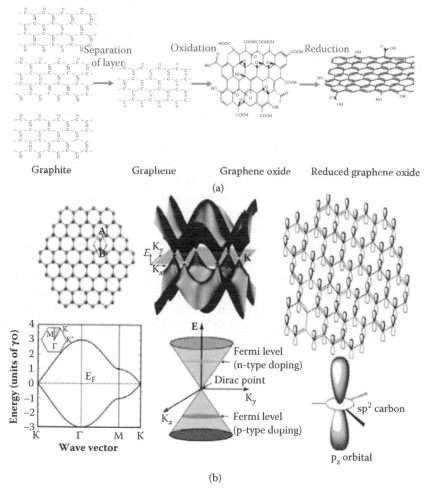

FIGURE 5.22 (a) The making of graphene-based compounds from graphite. (b) Electronic structure of graphene. (From Novoselov, K.S. et al., *Science*, 306, 666–69, 2014.)

(Castro Neto et al. 2009), a large specific surface area (2630 m^2/g) (Kotchey et al. 2011), easily functionalized compared with carbon nanotube, biodegradability (Bianco 2013), nontoxicity and good mechanical stability (Nair et al. 2008), transparency (Yeh et al. 2013), large work function (4.42 eV) (Huang et al. 2012), which enables it to accept photogenerated electrons from CB of most of semiconductors and LUMO of dyes that results in suppression in charge recombination and enhance H$_2$ generation capacity. Accepted electrons can migrate across its 2D plane to a reactive site for hydrogen generation. Therefore, the roles of graphene as an electron acceptor and transporter need to be extensively investigated to enhance PEC/photocatalytic hydrogen generation. Moreover, the lateral size, layer thickness, homogeneity, and purity of the graphene are the qualities which direct the material for high-end uses in electronics and other devices. Furthermore, it might be suitable as a good electrode material in electronic devices due to its high thermal and chemical stability, low resistivity (10–30 per square), light weight, flexibility, and transparency (>90% at 550 nm) (Bae et al. 2010). Looking at our earlier discussion, it has the potential to replace indium tin oxide (ITO) as an electrode material as it is thermally and chemically stable, highly conductive, flexible, and transparent (Tung et al. 2011). Graphene-based systems are classified into three categories: binary, ternary, and dye sensitized.

5.7.6.1.1 Graphene-Based Binary Systems

A few of the notable graphene-based composite systems that contain TiO$_2$ (Bell et al. 2011; Xie et al. 2013), ZnO (Guo et al. 2013), ZnS (Yeh et al. 2013), 6H-SiC (Jayasekera et al. 2012), BiOCl (Tian et al. 2013), Zn$_x$ Cd$_{1-x}$ S, (Zhang et al. 2012b), and CdSe and CdTe (Geng et al. 2010; Lu et al. 2011; Li et al. 2011); TiO$_2$/graphene photocatalyst was first reported for photocatalytic H$_2$ generation, as demonstrated in Figure 5.23a and b. When graphene is combined with the semiconductor, the excited electron of the semiconductor will be transferred from the semiconductor to the graphene matrices, which is evident from the emission decay kinetics of the semiconductor NC decay being much faster and the photoluminescence (PL) lifetime of nanocomposites decreasing with increasing graphene content. This analysis of excited state deactivation lifetimes of graphenes as a function of degree of oxidation and charging in RGO support the above phenomena. After irradiation, the epoxy, the carbonyl, and the hydroxyl groups are gradually removed from GO, resulting in an increase of sp^2 π-conjugated domains and defect carbons with holes for the formed RGO. The RGO conductivity increases due to the restoration of sp^2 π-conjugated domains. Therefore, we can say RGO has a great potency for efficient hydrogen production. Moreover, their modified forms, like GO and RGO, have good chemical potential in comparison to graphene, which also attracts the attention of researchers. Large numbers of the RGO composites were used to generate hydrogen through breaking of water. Some of them are RGO/CdS/0.5%Pt (H$_2$ with a rate of 1.12 mmol/h and efficiency ~22.5% at wavelength of 420 nm (4.84 factors greater than pure CdS/Pt) (Li et al. 2011) RGO/gC$_3$N$_4$ (produces 3.07 times more hydrogen than pure RGO/gC$_3$N$_4$) (Xiang et al. 2011a), RGO/3C–SiC (produces, H$_2$ with rate of 95-μL/h 1.3 times larger than the pure 3C–SiC nanoparticles) (Yang et al. 2013), TiSi$_2$/RGO composite with cocatalyst RuO$_2$ (Mou et al. 2013), N-doped TiO$_2$ with RGO (13.6 times higher than that of P25 photocatalyst) (Pei et al. 2013), RGO/Sr$_2$Ta$_2$O$_{7-x}$ N$_x$ (i.e., N doped) (Mukherji 2011), RGO wrapped anatase TiO$_2$ (Lee et al. 2012), and so on.

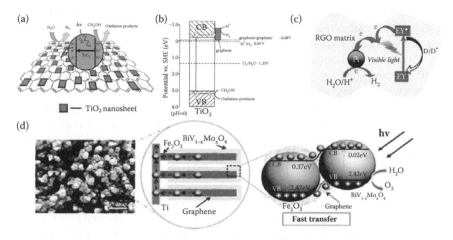

FIGURE 5.23 (a) Schematic illustration for charge transfer and separation in graphene-modified TiO_2 nanosheets system under ultraviolet (UV) light irradiation; (b) proposed mechanism for photocatalytic H_2 production under UV light irradiation. Practical example of the charge transfer are (c) dye (Eosin Y)-doped reduced graphene oxide for highly efficient visible-light-driven water reduction and (d) α-Fe_2O_3 nanorod/ graphene oxide/ $BiV_{1-x}Mo_xO_4$, the core/shell heterojunction array for efficient photoelectrochemical visible light-driven – water splitting. (From Bell, N. J. et al., *J. Phys. Chem.*, C115, 6004–09, 2011.) (c) (From Min, S. and Lu, G., *J. Phys. Chem. C*, 115 (28), 13938–45, 2011.) (d) (From Hou, Y. et al., *Nano Lett.*, 12(12), 6464–73, 2012.)

5.7.6.1.2 Graphene-Based Ternary Systems

In ternary graphene systems, graphene acts as an electron mediator for water splitting in an ingenious Z-scheme that accelerates the transfer of electrons from two different kind of semiconductor to the active sites of the graphene through the 2D plane. In a typical photoreduced graphene oxide (PRGO) is used as a mediator among Z-scheme semiconductors $BiVO_4$ and Ru–$SrTiO_3$:Rh. Under visible light illumination, photogenerated electrons from the CB of $BiVO_4$ transfer to Ru–$SrTiO_3$:Rh via graphene as an electron conductor, as depicted in Figure 5.23d. The electrons in Ru/SrTiO3:Rh reduce water to H_2 on the Ru cocatalyst, while the holes in $BiVO_4$ oxidize water to O_2, accomplishing a complete water splitting. Hou et al. (2012b) synthesized a novel heterojunction array of α-Fe_2O_3/RGO/$BiV_{1-x}Mo_xO_4$ core–shell nanorod, with α-Fe_2O_3 as a core, RGO as an interlayer and $BiV_{1-x}Mo_xO_4$ as a shell. This assembly shows 1.97 mA/cm^2 at 1.0 V versus Ag/AgCl and a high photoconversion efficiency of ~0.53% at −0.04 V versus Ag/AgCl under the irradiation of a Xe lamp (Hou et al. 2012b). Similarly, a CdS@TaON core–shell associated with GO was also used to produced hydrogen at the rate of 633 μmol/h with a high quantum efficiency of 31% (Hou et al. 2012a).

5.7.6.1.3 Graphene-Based Dye-Sensitized Systems

In this system, the dye (Eosin Y [EY], Rose Bengal [RB], Fluorescein sodium, and Rhodamine B, etc.) molecules act as a photosensitizer and are first adsorbed on the surface of RGO sheets with dispersed Pt cocatalyst, which has excellent electron-accepting and electron-transporting properties; RGO can facilitate the electron transfer from excited dye molecules to the Pt cocatalyst, which promotes photocatalytic activity for hydrogen evolution (Figure 5.23c) (Min and Lu 2011). Similarly, an efficient and stable photocatalytic H_2 evolution system was also constructed on Pt-deposited graphene sheets cosensitized by two dyes, EY and Rose Bengal (Yeh et al. 2013). Further efforts are likely to be developed in organic dye-sensitized graphene systems for highly solar photocatalytic H_2 production performance. Zhang et al. (2014a) reported the role of Ru(dcbpy)$_3$-RGO/Pt and 5,10,15,20-tetrakis (4-(hydroxyl)phenyl) porphyrin (TPPH) noncovalently functionalized RGO nanocomposite, that is, TPPH-RGO/Pt, in water splitting Eosin Y-sensitized and cocatalyst Pt-loaded RGO photocatalyst that was used for hydrogen evolution that was raised up to 12.9% under visible light irradiation ($\lambda \geq 420$ nm), and 23.4% under monochromatic light irradiation at 520 nm. A highly efficient and stable catalyst was constructed on Pt-deposited graphene sheets cosensitized by EY and RB under two-beam monochromic light irradiation (520 and 550 nm) that can give a photocatalytic H_2 evolution of high QY up to 37.3% (Min and Lu 2011, 2012).

5.7.6.2 Complex Perovskite Materials

Double-perovskite compounds of general formula $A_2B_2O_6$ have similar crystal structures to simple perovskites. They have the basic framework of corner connected BO_6 octahedra and A cations enclosed within. However, the connectivity of the octahedra may differ from structure to structure. They accommodate different cations at the A or/and B sites, taking a general formula $AA'BB'O_6$. Accommodation of different cations at the A and B sites alters the photophysical properties of the compound significantly. Among the binary oxides, only a few compounds are known to have BG values in the visible region (narrow gap), for example, Fe_2O_3, WO_6, CuO, and Bi_2O_3. However, these materials individually suffer from insufficient CB potential for hydrogen evolution. They also suffer from poor stability and low mobility of photoexcited charges but most of the binary complex compounds offer a possibility of combining the elements from "narrow gap" and "wide gap" binary compounds to exploit the properties of both types of oxides, and thus are potentially useful as visible light photocatalysts. A few visible light active selective double perovskites are discussed in this section. Cubic crystal structured Sr_2FeNbO_6 has a BG of 2.06 eV. In its pristine form, it is a well-known photocatalyst. While 7% Ti-doping at an Fe site has shown two times enhanced hydrogen generation in methanol solution, 28.5 and 650 μmol/h of hydrogen were reported in the presence of sacrificial agents and 0.2% Pt as cocatalyst, respectively (Borse et al. 2012). W-doped $SrFeNbO_6$ has also been studied and demonstrated enhancement to hydrogen evolution under visible radiation (Borse et al. 2014). Hu et al. (2012) studied the rare-earth and bismuth-based double perovskites of formula Ba_2RBiO_6 (R = La, Ce, Pr, Nd, Sm, Eu, Gd, Dy), for visible light photocatalysis in MB degradation. Authors found rare-earth cation dependent photocatalytic performance, where compounds such as Ba_2EuBiO_6, Ba_2SmBiO_6, and Ba_2CeBiO_6 showed high photocatalytic activity. Complex perovskites such as $CaCu_3Ti_4O_{12}$ have also been studied for their photocatalytic performance. $CaCu_3Ti_4O_{12}$ was found to possess an indirect BG of 1.27 eV and Pt-loaded

$CaCu_3Ti_4O_{12}$ shows degradation of MO under radiation greater than 420 nm (Clark et al. 2011). Photophysical properties of certain double-perovskite compounds have been reported; however, more efforts are needed to investigate photocatalytic properties of these materials. Compounds such as Ba_2CoWO_6, Ba_2NiWO_6, Sr_2CoWO_6, and Sr_2NiWO_6 are stable perovskite compounds for O_2 evolution using sacrificial agents (Iwakura et al. 2011). Certain tantalum-based compounds have been studied for degradation of organic compounds. N-doped $K_2Ta_2O_6$ is known to absorb visible light up to 600 nm and degrade formaldehyde under visible light (Zhu et al. 2008). Our group demonstrated that Bi doping in $Na_2Ta_2O_6$ causes visible light absorption and degradation of MB (Kanhere et al. 2013). Although some work has been done with double perovskites as photocatalysts, understanding of their fundamental properties is limited. More work is needed to discover their advantages as visible light photocatalysts and develop novel material systems.

5.7.6.3 Mixed Oxides

A mixture of oxides with nitrides or sulfides has been developed to engineer a suitable band structure of the oxide photocatalysts for visible light absorption. Oxynitride and oxysulfide photocatalysts offer distinctive advantages over their doped counterparts. Earlier reports on oxynitrides have revealed that replacing oxygen by nitrogen at lattice sites, in a stoichiometric manner, narrows the BG of the oxide, by pushing the VBM into the BG (Hara et al. 2004). Such a modification does not produce native point defects, which would otherwise be introduced in the case of N doping. Stoichiometric incorporation of nitrogen also avoids the localized states induced by N doping and reduces the possible electron–hole recombination. A similar composition containing mixtures of sulfur and oxygen, that is, oxysulfides, has been developed for photocatalysts. The mixed oxysulfide perovskite $Sm_2Ti_2S_2O_5$ ($E_g = 2.0$ eV) is known for water oxidation and reduction reaction for oxygen and hydrogen evolution, respectively, in the presence of sacrificial agents under low photon energy wavelengths of 650 nm. The band structure of this phase reveals that the presence of sulfur narrows the BG and enables visible light absorption (Ishikawa et al. 2002). Oxynitride compounds such as $CaNbO_2N$, $SrNbO_2N$, $BaNbO_2N$, and $LaNbON_2$ belong to the perovskite-type crystal structures (Siritanaratkul et al. 2011). Photocatalytic hydrogen evolution has been reported under visible light from methanol solution. $SrNbO2N$ ($E_g = 1.8$ eV) has been investigated in detail, where the photoelectrode of $SrNbO_2N$ on a transparent conducting surface shows water oxidation reaction under no external bias (Maeda et al. 2011). Tantalum counterparts of these compounds were developed and utilized in Z-scheme photocatalysis. Compounds such as $CaTaO_2N$ and $BaTaO_2N$ were loaded with Pt cocatalyst and coupled with Pt/WO_3 for Z-scheme water splitting (Higashi et al. 2009). A solid solution of $BaTaO_2N$ and $BaZrO_3$ was formulated for hydrogen and oxygen evolution, which showed improved performance compared with individual photocatalysts under visible light (Maeda and Domen 2014). $LaTiO_x N_y$ is another perovskite-type compound which shows high photocurrent density under visible light (Paven-Thivet et al. 2009). Apart from the double perovskites belonging to the general formula $AA'BB'O_6$, there are a several other compounds that show crystal structures close to the perovskite-type structure; however, such compounds are not included in the current review. Theoretically, double perovskites offer a wider scope to design photocatalysts by selecting suitable cations and AA′ and BB′ sites in the lattice. Work on design and development of double perovskites is currently limited and synthesis and characterization of new materials in this category are needed.

5.8 OXIDE PHOTOCATALYSTS: CHALLENGES AND PERSPECTIVES

Although the oxide photocatalyst is a very common and famous class of photocatalyst, used for water splitting for a long time, due to their wide BG they have some limitations, on which we need to focus our efforts. First, the study of oxide photocatalysis requires more devoted and goal-oriented efforts in the direction of overall efficient water splitting, as this field is quite complicated and the reported studies are quite scattered. A more systematic practical approach aligned with a theoretical scheme is needed to move toward this direction and to rationalize the major experimental observations. Second, new oxide photocatalyst materials with better performance at lower cost are required. In the long run, the cost of the material will be a driving factor even if the photocatalysis achieves high efficiency and stability under sunlight irradiation. Oxide materials, which have an apt BG with good light-harvesting capacities, are still at the top of the "wanted" list in the field of functional materials. Third, new modified reactor designs are urgently needed in place of conventionally designs (typically slurry reactors and immobilized film reactors) (Wunderlich et al. 2009).

Finally, it will be valuable to extend the wavelength response region to near IR using appropriate supportive material to the main oxide photocatalyst assembly such as plasmonic photocatalysts, cocatalysts, or stand-alone photocatalysts. Most oxide materials are sensitive to the UV region (4% of sunlight) of the light and a handful of oxide photocatalysts can be worked under visible light (45% of sunlight). However, solar light contains a large proportion of energy in the near IR region, which is totally unexplored in this respect. Some studies have already demonstrated IR response (up to 1300 nm) using noble metal nanoparticles but the efficiency is low (Herrmann 2005).

5.9 SUMMARY

To overcome the major challenges in achieving more than 20% quantum efficiency for water splitting with low cost and stable materials, one should work at three levels (theoretical and practical alignment, synthesis technology of photocatalyst, reactor design), simultaneously. Oxides are well known for their stability and abundancy. Therefore, authors discussed pros and cons of the water-splitting phenomena using metal oxides as the main light-harvesting photocatalyst material. This includes designing of visible light responsive probes (doping, dye, QD, Z-scheme oxides, plasmonic material, graphene, etc.) for metal oxide photocatalysts or improving morphology (at nanolevel) or making new materials with metal ion incorporation. Some reaction mechanisms of photocatalytic reactions have also been discussed. At the end of this chapter, challenges and perspectives in this field were discussed.

REFERENCES

Adleman, J. R., D. A. Boyd, D. G. Goodwin, and D. Psaltis 2009. Heterogenous catalysis mediated by plasmon heating. *Nano Lett.* 9: 4417–23.

Al Mazroai, L. S., M. Bowker, P. Davies, et al. 2007. The photocatalytic reforming of methanol. *Catal. Today* 122(1–2): 46–50.

Alvi, N. ul H., P. Eduardo, D. S. Rodriguez, et al. 2015. InN/InGaN quantum dot photoelectrode: Efficient hydrogen generation by water splitting at zero voltage. *Nano Energy* 13: 291–97.

Amano, F., K. Nogami, R. Abe, and B. Ohtani. 2008. Preparation and characterization of bismuth tungstate polycrystalline flake-ball particles for photocatalytic reactions. *J. Phys. Chem. C* 112: 9320–26.

Anpo, M. and M. Takeuchi. 2003. The design and development of highly reactive titanium oxide photocatalysts operating under visible light irradiation. *J. Catal.* 216: 505–16.

Anpo, M., Y. Ichihashi, M. Takeuchi, and H. Yamashita. 1998. Design of unique titanium oxide photocatalysts by an advanced metal ion-implantation method and photocatalytic reactions under visible light irradiation. *Res. Chem. Intermed.* 24: 143–49.

Asai, R., H. Nemoto, Q. Jia, K. Saito, A. Iwase, and A. Kudo. 2014. A visible light responsive rhodium and antimony-codoped $SrTiO_3$ powdered photocatalyst loaded with an IrO_2 cocatalyst for solar water splitting. *Chem. Commun.* 50: 2543–46.

Atwater, H. A. 2007. The promises of plasmonics. *Sci. Am.* 296(4): 56–63.

Awazu, K., M. Fujimaki, C. Rockstuhl, et al. 2008. A plasmonic photocatalyst consisting of silver nanoparticles embedded in titanium dioxide. *J. Am. Chem. Soc.* 130:1676–80.

Bae, S., H. Kim, Y. Lee, et al. 2010. Roll-to-roll production of 30-inch graphene films for transparent electrodes. *Nat. Nanotechnol.* 5: 574–78.

Barnes, W. L., A. Dereux, and T. W. Ebbesen. 2003. Surface plasmon subwavelength optics. *Nature* 424: 824–30.

Beard, M. C., K. P. Knutsen, P. Yu, et al. 2007. MEG in colloidal silicon nanocrystals. *Nano Lett.* 7(8): 2506–12.

Bell, N. J., Y. H. Ng, A. Du, H. Coster, S. C. Smith, and R. Amal. 2011. Understanding the enhancement in photoelectrochemical properties of photocatalytically prepared TiO_2-reduced graphene oxide composite. *J. Phys. Chem. C* 115: 6004–09.

Benthem, K. V., C. Elsässer, and R. H. French. 2001. Bulk electronic structure of $SrTiO_3$: Experiment and theory. *J. Appl. Phys.* 90: 6156–64.

Berak, J. M. and M. J. Sienko. 1970. Effect of oxygen-deficiency of electrical transport properties of tungsten trioxide crystals. *J. Solid State Chem.* 2: 109–33.

Berger, L. I. 2008. *CRC Handbook of Chemistry and Physics* (ed. D. R. Lide). Boca Raton, FL: CRC Press/Taylor and Francis.

Berglund, S. P., D. W. Flaherty, N. T. Hahn, A. J. Bard, and C. B. Mullins. 2011. Photoelectrochemical oxidation of water using nanostructured $BiVO_4$ films. 2011. *J. Phys. Chem. C* 115: 3794–802.

Bianco, A. 2013. Graphene: Safe or toxic? The two faces of the medal. *Angew. Chem. Int. Engl.* 52: 4986–97.

Böhm, M. L., T. C. Jellicoe, M. Tabachnyk, et al. 2015. Lead telluride quantum dot solar cells displaying external quantum efficiencies exceeding 120%. *Nano Lett.* 15(12): 7987–93.

Borse, P. H., K. T. Lim, J. H. Yoon, et al. 2014. Investigation of the physico-chemical properties of $Sr_2FeNb_{1-x}W_xO_6$ ($0.0 \leq x \leq 0.1$) for visible-light photocatalytic water-splitting applications. *J. Korean Phys. Soc.* 64: 295–300.

Butler, M. A. and D. S. Ginley. 1978. Prediction of flatband potentials at semiconductor-electrolyte interfaces from atomic electronegativities. *J. Electrochem. Soc.* 125(2): 228–32.

Califano, M., A. Zunger, and A. Franceschetti. 2004a. Direct carrier multiplication due to inverse Auger scattering in CdSe quantum dots. *Appl. Phys. Lett.* 84(13): 2409.

Califano, M., A. Zunger, and A. Franceschetti. 2004b. Efficient inverse Auger recombination at threshold in CdSe nanocrystals. *Nano Lett.* 4(3): 525.

Castro Neto, A. H., F. Guinea, N. M. R. Peres, K. S. Novoselov, and A. K. Geim. 2009. The electronic properties of graphene. *Rev. Mod. Phys.* 81: 109–62.

Chakrapani, V., K. Tvrdy, and P. V. Kamat. 2010. Modulation of electron injection in CdSe–TiO$_2$ system through medium alkalinity. *J. Am. Chem. Soc.* 132: 1228–29.

Chen, H.-C., C.-W. Huang, J. C. S. Wu, and S.-T. Lin. 2012. Theoretical investigation of the metal-doped SrTiO$_3$ photocatalysts for water splitting. *J. Phys. Chem. C* 116: 7897–903.

Chen, S., Y. Qi, T. Hisatomi, et al. 2015. Efficient visible-light-driven Z-scheme overall water splitting using a MgTa$_2$O$_{6-x}$ N y/TaON heterostructure photocatalyst for H$_2$ evolution. *Angew. Chem. Int. Ed.* 54(29): 8498–501.

Chen X., H. Zhu, J. Zhao, Z. Zheng, and X. Gao. 2008. Visible-light-driven oxidation of organic contaminants in air with gold nanoparticle catalysts on oxide supports. *Angew. Chem. Int. Edn Engl.* 47: 5353–56.

Choi, W. Y., A. Termin, and M. R. Hoffmann. 1994. The role of metal ion dopants in quantum-sized TiO$_2$: Correlation between photoreactivity and charge carrier recombination dynamics. *J Phys Chem* 84: 13669–79.

Christopher P, H. Xin, and S. Linic. 2011. Visible-light-enhanced catalytic oxidation reactions on plasmonic silver nanostructures. *Nature Chem.* 3: 467–72.

Clark, J. H., M. S. Dyer, R. G. Palgrave, et al. 2011. Visible light photo-oxidation of model pollutants using CaCu$_3$Ti$_4$O$_{12}$: An experimental and theoretical study of optical properties, electronic structure, and selectivity. *J. Am. Chem. Soc.* 133: 1016–32.

Cowan, A. J., J. W. Tang, W. H. Leng, J. R. Durrant, and D. R. Klug. 2010. Water splitting by nanocrystalline TiO$_2$ in a complete photoelectrochemical cell exhibits efficiencies limited by charge recombination. *J. Phys. Chem. C* 114: 4208–14.

Cox, P. A., 2010. *Transition Metal Oxides: An Introduction to Their Electronic Structure and Properties.* Oxford, UK: Clarendon Press.

Cui, G., W. Wang, M. Ma, et al. 2015. IR-driven photocatalytic water splitting with WO$_2$–Na x WO$_3$ hybrid conductor material. *Nano Lett.* 15(11): 7199–203.

De Krol, R. V., Y. Liang, and J. Schoonman. 2008. Solar hydrogen production with nanostructured metal oxides. *J. Mater. Chem.* 18: 2311–20.

Dholam, R. N. Patel, M. Adami, and A. Miotello. 2009. Hydrogen production by photocatalytic water-splitting using Cr- or Fe-doped TiO$_2$ composite thin films photocatalyst. *Int. J. Hydrogen Energy* 34: 5337–46.

Domen, K., A. Kudo, A. Shinozaki, A. Tanaka, K. Maruya, and T. Onishi. 1986a. Photodecomposition of water and hydrogen evolution from aqueous methanol solution over novel niobate photocatalysts. *J. Chem. Soc. Chem. Commun.* (4): 356–57.

Domen, K., A. Kudo, and T. Onishi. 1986b. Mechanism of photocatalytic decomposition of water into H$_2$ and O$_2$ over NiO-SrTiO$_3$. *J. Catal.* 102: 92–8.

Domen, K., A. Kudo, T. Onishi, N. Kosugi, and H. Kuroda. 1986c. Photocatalytic decomposition of water into hydrogen and oxygen over nickel(II) oxide-strontium titanate (SrTiO$_3$) powder. 1. Structure of the catalysts. *J. Phys. Chem.* 90: 292–95.

Domen, K., S. Naito, M. Soma, T. Onishi, and K. Tamaru. 1980. Photocatalytic decomposition of water vapour on NiO–SrTiO$_3$ catalyst. *J. Chem. Soc. Chem. Commun.* (12): 543–44.

Domen, K., S. Naito, T. Onishi, and K. Tamaru. 1982. Photocatalytic decomposition of liquid water on a NiO-SrTiO$_3$ catalyst. *Chem. Phys. Lett.* 92: 433–34.

Dvoranova, D., V. Brezova, M. Mazur, and M. Malati. 2002. Investigations of metal-doped titanium dioxide photocatalysts. *Appl. Catal. B Environ.* 37: 91–105.

Ellingson, R. J., M. C. Beard, J. C. Johnson, et al. 2005. Highly efficient multiple exciton generation in colloidal PbSe and PbS quantum dots. *Nano Lett.* 5: 865.

Eng, H. W., P. W. Barnes, B. M. Auer, and P. M. Woodword. 2003. Investigation of the electronic structure of d^0 transition metal oxides belonging to the perovskite family. *J. Solid State Chem.* 175: 94–109.

Esposito, D. V., R. V. Forest, Y. C. Chang, et al. 2012. Photoelectrochemical reforming of glucose for hydrogen production using a WO_3-based tandem cell device. *Energy Environ. Sci.* 5: 9091–99.

Feng, N. D., A. M. Zheng, Q. A. Wang, et al. 2011. Boron environments in B-doped and (B, N)-codoped TiO_2 photocatalysts: A combined solid-state NMR and theoretical calculation study. *J. Phys. Chem. C* 115: 2709–19.

Franking, R., L. S. Li, M. A. Lukowski, et al. 2013. Facile post-growth doping of nanostructured hematite photoanodes for enhanced photoelectrochemical water oxidation. *Energy Environ. Sci.* 6: 500–12.

Frydrych, J., L. Machala, J. Tucek, et al. 2012. Facile fabrication of tin-doped hematite photoelectrodes: Effect of doping on magnetic properties and performance for light-induced water splitting. *J. Mater. Chem.* 22: 23232–39.

Fu, Q., J. L. Li, T. He, and G. W. Yang. 2013. Band-engineered $CaTiO_3$ nanowires for visible light photocatalysis. *J. Appl. Phys.* 113: 104303.

Fujishima, A. and K. Honda. 1972. Electrochemical photolysis of water at a semiconductor electrode. *Nature* 238: 37–8.

Garcia-Esparza, A. T., D. Cha, Y. Ou, J. Kubota, K. Domen, and K. Takanabe. 2013. Tungsten carbide nanoparticles as efficient cocatalysts for photocatalytic overall water splitting. *Chem. Sus. Chem.* 6(1): 168–81.

Geng, X., L. Niu, Z. Xing, et al. 2010. Aqueous-processable noncovalent chemically converted graphene–quantum dot composites for flexible and transparent optoelectronic films. *Adv. Mater.* 22: 638–42.

Grinberg, I., D. V. West, M. Torres, et al. 2013. Perovskite oxides for visible-light-absorbing ferroelectric and photovoltaic materials. *Nature* 503: 509–12.

Hameed, A., M. A. Gondal, and Z. H. Yamani. 2004. Effect of transition metal doping on photocatalytic activity of WO_3 for water splitting under laser illumination: Role of 3D-orbitals. *Catalysis Commun.* 5(11): 715–19.

Han, P., X. Wang, Y. H. Zhao, and C. Tang. 2009. Electronic structure and optical properties of non-metals (N, F, P, Cl, S)—Doped cubic $NaTaO_3$ by density functional theory. *Adv. Mater. Res.* 79–82: 1245–48.

Hara, M. T. T., J. N. Kondo, and K. Domen. 2004. Photocatalytic reduction of water by TaON under visible light irradiation. *Catal. Today* 90: 313–17.

Hara, S. and H. Irie. 2012. Band structure controls of $SrTiO_3$ towards two-step overall water splitting. *Appl. Catal. B* 115–116: 330–35.

Hara, S., M. Yoshimizu, S. Tanigawa, L. Ni, B. Ohtani, and H. Irie. 2012. Hydrogen and oxygen evolution photocatalysts synthesized from strontium titanate by controlled doping and their performance in two-step overall water splitting under visible light. *J. Phys. Chem. C* 116: 17458–63.

Hashimoto, K., H. Irie, and A. Fujishima. 2005. TiO_2 photocatalysis: A historical overview and future prospects. *Jpn. J. Appl. Phys.* 44: 8269.

Hensel, J., G. M. Wang, Y. Li, and J. Z. Zhang. 2010. Synergistic effect of CdSe quantum dot sensitization and nitrogen doping of TiO_2 nanostructures for photoelectrochemical solar hydrogen generation. *Nano Lett.* 10: 478–83.

Herrmann, J.-M. 2005. Heterogeneous photocatalysis: State of the art and present applications. *Top. Catal.* 34: 49–65.

Higashi, M., R. Abe, T. Takata, and K. Domen. 2009. Photocatalytic overall water splitting under visible light using $ATaO_2N$ (A) Ca, Sr, Ba) and WO_3 in a IO^{3-}/I^- shuttle redox mediated system. *Chem. Mater.* 21: 1543–49.

Hill J. C. and K. S. Choi. 2012. Effect of electrolytes on the selectivity and stability of n-type WO_3 photoelectrodes for use in solar water oxidation. *J. Phys. Chem. C* 116: 7612–20.

Hoang, S., S. P. Berglund, N. T. Hahn, A. J. Bard, and C. B. Mullins. 2012a. Enhancing visible light photo-oxidation of water with TiO_2 nanowire arrays via cotreatment with H_2 and NH_3: Synergistic effects between Ti^{3+} and N. *J. Am. Chem. Soc.* 134: 3659–62.

Hoang, S., S. W. Guo, N. T. Hahn, A. J. Bard, and C. B. Mullins. 2012b. Visible light driven photoelectrochemical water oxidation on nitrogen-modified TiO_2 nanowires. *Nano Lett.* 12: 26–32.

Hoffmann, M. R., S. T. Martin, W. Y. Choi, and D. W. Bahnemann. 1995. Environmental applications of semiconductor photocatalysis. *Chem. Rev.* 95: 69–96.

Holmes, M. A., T. K. Townsend, and F. E. Osterloh. 2012. Quantum confinement controlled photocatalytic water splitting by suspended CdSe nanocrystals. *Chem. Commun.* 48: 371–73.

Hong, S. J., S. Lee, J. S. Jang, and J. S. Lee. 2011. Heterojunction $BiVO_4/WO_3$ electrodes for enhanced photoactivity of water oxidation. *Energy Environ. Sci.* 4: 1781–87.

Hou, J., Z. Wang, W. Kan, S. Jiao, H. Zhu, and R. V. Kumar. 2012a. Efficient visible-light-driven photocatalytic hydrogen production using CdS@TaON core–shell composites coupled with graphene oxide Nanosheets. *J. Mater. Chem.* 22: 7291–99.

Hou, X., S. Jiang, Y. Li, J. Xiao, and Y. Li. 2015. Ni-doped InN/GaZnON composite, catalyst for overall water splitting under visible light irradiation. *Int. J. Hydrogen Energy* 40(45): 15448–53.

Hou, Y., F. Zuo, A. Dagg, and P. Feng. 2012b. Visible light-driven α-Fe_2O_3 nanorod/graphene/BiV_1 core/shell heterojunction array for efficient photoelectrochemical water splitting. *Nano Lett.* 12(12): 6464–73.

Hu, R., C. Li, X. Wang, et al. 2012. Photocatalytic activities of $LaFeO_3$ and La_2FeTiO_6 in p-chlorophenol degradation under visible light. *Catal. Commun.* 29: 35–39.

Hu, Y. S., A. K. Shwarsctein, G. D. Stucky, and E. W. McFarland. 2009. Improved photoelectrochemical performance of Ti-doped α-Fe_2O_3 thin films by surface modification with fluoride. *Chem. Commun.* (19): 2652–54.

Huang, X., Z. Zeng, Z. Fan, J. Liu, and H. Zhang. 2012. Graphene-based electrodes. *Adv. Mater.* 24: 5979–6004.

Huda, M. N., M. M. Al-Jassim, and J. A. Turner. 2011. Mott insulators: An early selection criterion for materials for photoelectrochemical H_2 production. *J. Renew. Sustain. Energy* 3(5): 053101.

Hwang, D. W., H. G. Kim, J. Kim, and J. S. Lee. 2000. Photocatalytic water splitting over highly donor-doped (110) layered perovskites. *J. Catal.* 193: 40–8.

Hwang, D. W., H. G. Kim, J. S. Jang, S. W. Bae, S. M. Ji, and J. S. Lee. 2004. Photocatalytic decomposition of water–methanol solution over metal-doped layered perovskites under visible light irradiation. *Catal. Today* 93: 845–50.

Hwang, Y. J., A. Boukai, and P. D. Yang. 2009. High density n-Si/n-TiO_2 core/shell nanowire arrays with enhanced photoactivity. *Nano Lett.* 9: 410–15.

Ikeda, S., M. Hara, J. N. Kondo, K. Domen, H. Takahashi, T. Okubo, and M. Kakihana. 1998b. Density, viscosity, refractive index, and speed of sound in binary mixtures of methyl acetate + ethylene glycol or + poly(ethylene glycol) in the temperature interval (298.15–308.15) K. *J. Mater. Res.* 13: 852–55.

Ikeda, S., T. Itani, K. Nango, and M. Mutsumura. 2004. Overall water splitting on tungsten based photocatalysts with defect pyrochlore structure. *Catal. Lett.* 98(4): 229–32.

Inoue, Y., T. Kubokawa, and K. Sato. 1990. Photocatalytic activity of sodium hexatitanate, $Na_2Ti_6O_{13}$, with a tunnel structure for decomposition of water. *J. Chem. Soc. Chem. Commun.*: 1298–99.

Ishikawa, A., T. Takata, J. N. Kondo, M. Hara, H. Kobayashi, and K. Domen. 2002. Oxysulfide $Sm_2Ti_2S_2O_5$ as a stable photocatalyst for water oxidation and reduction under visible light irradiation ($\lambda \le 650$ nm). *J. Am. Chem. Soc.* 124: 13547–53.

Iwakura, H., H. Einaga, and Y. Teraoka. 2011. Photocatalytic properties of ordered double perovskite oxides. *J. Novel Carbon Resour. Sci.* 3: 1–5.

Jang J. S., H. G. Kim, V. R. Reddy, S. W. Bae, S. M. Ji, and J. S. Lee. 2005. Photocatalytic water splitting over iron oxide nanoparticles intercalated in $HTiNb(Ta)O_5$ layered compounds. *J. Catal.* 231: 213–22.

Jayasekera, T., K. W. Kim, and M. B. Nardelli. 2012. Electronic and structural properties of turbostratic epitaxial graphene on the 6H-SiC (000-1) surface. *Mater. Sci. Forum* 717–720: 595–600.

Ji, S. M., P. H. Borse, H. G. Kim, et al. 2005. Photocatalytic hydrogen production from water-methanol mixtures using N-doped $Sr_2Nb_2O_7$ under visible light irradiation: Effects of catalyst structure. *Phys. Chem. Chem. Phys.* 7(6): 1315–21.

Jia, Q., A. Iwase, and A. Kudo. 2014. $BiVO_4$-Ru/$SrTiO_3$:Rh composite Z-scheme photocatalyst for solar water splitting. *Chem. Sci.* 5: 1513–19.

Jia, Y., S. Shen, and D. Wang, et al. 2013. Composite Sr_2TiO_4 /$SrTiO_3$ (La,Cr) heterojunction based photocatalyst for hydrogen production under visible light irradiation. *J. Mater. Chem. A* 1: 7905–12.

Jiang, L., H. Liu, J. Yuan, and W. Shangguan. 2010. Hydrothermal preparation and photocatalytic water splitting properties of ZrW_2O_8. *J. Wuhan Univ. Technol. Mater. Sci. Ed.* 25(6): 919–23.

Joshi, U. A., A. M. Palasyuk, and P. A. Maggard. 2011. Photoelectrochemical investigation and electronic structure of a p-type $CuNbO_3$ photocathode. *J. Phys. Chem. C* 115: 13534–39.

Kadowaki, H., N. Saito, H. Nishiyama, H. Kobayahsi, Y. Shimodaira, and Y. Inoue. 2007. Overall splitting of water by RuO_2-loaded $PbWO_4$ photocatalyst with $d^{10}s^2$-d^0 configuration. *J. Phys. Chem. C* 111: 439–44.

Kamat, P. V. 2008. Quantum dot solar cells. Semiconductor nanocrystals as light harvesters. *J. Phys. Chem. C* 112: 18737.

Kanhere, P., J. Nisar, Y. Tang, et al. 2012a. Electronic structure, optical properties, and photocatalytic activities of $LaFeO_3$–$NaTaO_3$ solid solution. *J. Phys. Chem. C* 116: 22767–73.

Kanhere, P., J. Zheng, and Z. Chen. 2012b. Visible light driven photocatalytic hydrogen evolution and photophysical properties of Bi^{3+} doped $NaTaO_3$. *Int. J. Hydrogen Energy* 37: 4889–96.

Kanhere, P., P. Shenai, S. Chakraborty, R. Ahuja, J. Zheng, and Z. Chen. 2014. Mono- and co-doped $NaTaO_3$ for visible light photocatalysis. *Phys. Chem. Chem. Phys.* 16: 16085–94.

Kanhere, P., Y. Tang, J. Zheng, and Z. Chen. 2013. Synthesis, photophysical properties, and photocatalytic applications of Bi doped $NaTaO_3$ and Bi doped $Na_2Ta_2O_6$ nanoparticles. *J. Phys. Chem. Solids* 74: 1708–13.

Kanhere, P. D., J. Zheng, and Z. Chen. 2011. Site specific optical and photocatalytic properties of Bi-doped $NaTaO_3$. *J. Phys. Chem. C* 115: 11846–53.

Kato, H. and A. Kudo. 2001. Water splitting into H_2 and O_2 on alkali tantalate photocatalysts $ATaO_3$ (A = Li, Na, and K). *J. Phys. Chem. B* 105: 4285–92.

Kato, H., K. Asakura, and A. Kudo. 2003. Highly efficient water splitting into H_2 and O_2 over lanthanum-doped $NaTaO_3$ photocatalysts with high crystallinity and surface nanostructure. *J. Am. Chem. Soc.* 125: 3082–89.

Kato, H., M. Hori, R. Konta, Y. Shimodaira, and A. Kudo. 2004. Construction of Z-scheme type heterogeneous photocatalysis systems for water splitting into H_2 and O_2 under visible light irradiation. *Chem. Lett.* 33: 1348–49.

Kato, H., Y. Sasaki, N. Shirakura, and A. Kudo. 2013. Synthesis of highly active rhodium-doped $SrTiO_3$ powders in Z-scheme systems for visible-light-driven photocatalytic overall water splitting. *J. Mater. Chem. A* 1: 12327–33.

Kavan, L., M. Grätzel, S. Gilbert, C. Klemenz, and H. Scheel. 1996. Electrochemical and photoelectrochemical investigation of single-crystal anatase. *J. Am. Chem. Soc.* 118: 6716–23.

Kay, A., I. Cesar, and M. Gratzel. 2006. New benchmark for water photooxidation by nanostructured α-Fe_2O_3 films. *J. Am. Chem. Soc.* 128: 15714–21.

Kelly, K. L., E. Coronado, L. L. Zhao, and G. C. Schatz. 2003. The optical properties of metal nanoparticles: The influence of size, shape and dielectric environment. *J. Phys. Chem. B* 107: 668–77.

Kim, H. G., D. W. Hwang, J. Kim, and J. S. Lee. 1999. Highly donor-doped (110) layered perovskite materials as novel photocatalysts for overall water splitting. *Chem. Commun.* 1077–78.

Kim, H. G., P. H. Borse, W. Choi, and J. S. Lee. 2005. Photocatalytic nanodiodes for visible-light photocatalysis. *Angew. Chem. Int. Ed.* 44: 4585–89.

Kim, J., D. W. Hwang, H. G. Kim, S. W. Bae, and J. S. Lee. 2002. Nickel-loaded $La_2Ti_2O_7$ as a bifunctional photocatalyst. *Chem. Commun.* (21): 2488–89.

Kim, K., M.-J. Kim, S.-I. Kim, and J.-H. Jang. 2013. Towards visible light hydrogen generation: Quantum dot-sensitization via efficient light harvesting of hybrid-TiO_2. *Sci. Rep.* 3: 3330.

Kim, Y., J. B. Gao, S. Y. Han, et al. 2009. Heterojunction of $FeTiO_3$ nanodisc and TiO_2 nanoparticle for a novel visible light photocatalyst. *J. Phys. Chem. C* 113: 19179–84.

Kobayashi, R., S. Tanigawa, T. Takashima, B. Ohtani, and H. Irie, 2014. Silver-inserted heterojunction photocatalysts for Z-scheme overall pure-water splitting under visible-light irradiation. *J. Phys. Chem. C* 118(39): 22450–456.

Konta, R., H. Kato, H. Kobayashi, and A. Kudo. 2003. Photophysical properties and photocatalytic activities under visible light irradiation of silver vanadates. *Phy. Chem. Chem. Phys.* 5: 3061–65.

Konta, R., T. Ishii, H. Kato, and A. Kudo. 2004. Photocatalytic activities of noble metal ion doped $SrTiO_3$ under visible light irradiation. *J. Phys. Chem. B* 108: 8992–95.

Kotchey, G. P., B. L. Allen, H. Vedala, et al. 2011. The enzymatic oxidation of graphene oxide. *ACS Nano* 5: 2098–108.

Krasnovskii, A. A. and G. P. Brin. 1962. Inorganic models of hills reaction. *Dokl. Akad. Nauk. SSSR* 147: 656–65.

Kronawitter, C. X., L. Vayssieres, S. H. Shen, et al. 2011. A perspective on solar-driven water splitting with all-oxidehetero-nanostructures. *Energy Environ. Sci.* 4: 3889–99.

Kudo, A., A. Tanaka, K. Domen, K. Maruya, K. Aika, and T. Onishi. 1988. Photocatalytic decomposition of water over NiO-$K_4Nb_6O_{17}$ catalyst. *J. Catal.* 111: 67–76.

Kudo, A., R. Niishiro, A. Iwase, and H. Kato. 2007. Effects of doping of metal cations on morphology, activity, and visible light response of photocatalysts. *Chem. Phys.* 339: 104–10.

Kudo, A. and Y. Miseki. 2009. Heterogeneous photocatalyst materials for water splitting. *Chem. Soc. Rev.* 38: 253–78.

Lee, J. S., K. H. You, and C. B. Park. 2012. Highly photoactive, low bandgap TiO_2 nanoparticles wrapped by graphene. *Adv. Mater.* 24: 1084–88.

Lehn, J. M., J. P. Sauvage, R. Zissel, and L. Hilaire. 1982. Water photolysis by UV irradiation of rhodium loaded strontium titanate catalysts. Relation between catalytic activity and nature of the deposit from combined photolysis and ESCA studies. *Isr. J. Chem.* 22: 168–72.

Li, G., S. Yan, Z. Wang, et al. 2009. Synthesis and visible light photocatalytic property of polyhedron-shaped $AgNbO_3$. *J. Chem. Soc. Dalton Trans.* 40: 8519–24.

Li, M., J. Zhang, W. Dang, S. K. Cushing, D. Guo, N. Wu, and P. Yin. 2013. Photocatalytic hydrogen generation enhanced by band gap narrowing and improved charge carrier mobility in $AgTaO_3$ by compensated co-doping. *Phys. Chem. Chem. Phys.* 15: 16220–26.

Li, Q., B. Guo, J. Yu, et al. 2011. Highly efficient visible-light-driven photocatalytic hydrogen production of CdS-cluster-decorated graphene nanosheets. *J. Am. Chem. Soc.* 133: 10878–84.

Li, Z., X. Cui, H. Hao, M. Lu, and Y. Lin. 2015. Enhanced photoelectrochemical water splitting from Si quantum dots/TiO$_2$ nanotube arrays composite electrodes. *Mater. Res. Bull.* 66: 9–15.

Lin, Y. G., Y. K. Hsu, Y. C. Chen, L. C. Chen, S. Y. Chen, and K. H. Chen. 2012. Visible-light-driven photocatalytic carbon-doped porous ZnO nanoarchitectures for solarwater-splitting. *Nanoscale* 4: 6515–19.

Ling, Y. C., G. M. Wang, D. A. Wheeler, J. Z. Zhang, and Y. Li. 2011. Sn-doped hematite nanostructures for photoelectrochemical water splitting. *Nano Lett.* 11: 2119–25.

Linic, S., P. Christopher, and D. B. Ingram. 2011. Plasmonic-metal nanostructures for efficient conversion of solar to chemical energy. *Nat. Mater.* 10: 911–21.

Litter, M. I. 1999. Heterogeneous photocatalysis transition metal ions in photocatalytic systems. *Appl. Catal. B Environ.* 23: 89–114.

Liu, G., S. Ji, L. Yin, G. Xu, G. Fei, and C. Ye. 2011. Visible-light-driven photocatalysts: (La/Bi + N)-codoped NaNbO$_3$ by first principles. *J. Appl. Phys.* 109: 063103.

Liu, J. W., G. Chen, Z. H. Li, and Z. G. Zhang. 2007. Hydrothermal synthesis and photocatalytic properties of ATaO$_3$ and ANbO$_3$ (A = Na and K). *Int. J. Hydrogen Energy* 32: 2269–72.

Liu, Z., W. Hou, P. Pavaskar, M. Aykol, and S. B. Cronin. 2011. Plasmon resonant enhancement of photocatalytic water splitting under visible illumination. *Nano Lett.* 11(3): 1111–16.

Lu, N., X. Quan, J. Y. Li, S. Chen, H. T. Yu, and G. H. Chen. 2007. Fabrication of boron-doped TiO$_2$ nanotube array electrode and investigation of its photoelectrochemical capability. *J. Phys. Chem. C* 111: 11836–42.

Lu, X. H., G. M. Wang, S. L. Xie, 2012. Efficient photo-catalytic hydrogen evolution over hydrogenated ZnO nanorod arrays. *Chem.Commun.* 48: 7717–19.

Lu, Z., C. X. Guo, H. B. Yang, et. al. 2011. One-step aqueous synthesis of graphene -CdTe quantum dot-composed nanosheet and its enhanced photoresponses. *J Colloid and Iter. Science.* 353(2): 588–592.

Luo, J., J- H. Im, M. T. Mayer etal. 2014. Water photolysis at 12.3% efficiency via perovskite photovoltaics and Earth-abundant catalysts. *Science* 345(6204):1593–96.

Luo, W., Z. Wang, L. Wan, Z. Li, T. Yu, and Z. Zou. 2010. Synthesis, growth mechanism and photoelectrochemical properties of BiVO$_4$ microcrystal electrodes. *J. Phys. D Appl. Phys.* 43: 405402.

Maeda, K. 2011. Photocatalytic water splitting using semiconductor particles: History and recent developments. *J. Photchem. Photobiol. C* 12: 237–68.

Maeda, K. 2013. Z-scheme water splitting using two different semiconductor photocatalysts. *ACS Catal.* 3: 1486–503.

Maeda, K. 2014. Rhodium-doped barium titanate perovskite as a stable p-type semiconductor photocatalyst for hydrogen evolution under visible light. *ACS Appl. Mater. Interfaces* 6: 2167–73.

Maeda, K. and K. Domen. 2014. Preparation of BaZrO$_3$–BaTaO$_2$N solid solutions and the photocatalytic activities for water reduction and oxidation under visible light. *J. Catal.* 310: 67–74.

Maeda, K., M. Higashi, B. Siritanaratkul, R. Abe, and K. Domen. 2011. SrNbO$_2$N as a water-splitting photoanode with a wide visible-light absorption band. *J. Am. Chem. Soc.* 133: 12334–37.

Maeda, K., M. Higashi, D. L. Lu, R. Abe, and K. Domen. 2010. Efficient nonsacrificial water splitting through two-step photoexcitation by visible light using a modified oxynitride as a hydrogen evolution photocatalyst. *J. Am. Chem. Soc.* 132: 5858–68.

Min, S. and G. Lu. 2011. Dye-sensitized reduced graphene oxide photocatalysts for highly efficient visible-light-driven water reduction. *J. Phys. Chem. C* 115(28): 13938–45.

Min, S. and G. Lu. 2012. Dye-cosensitized graphene/Pt photocatalyst for high efficient visible light hydrogen evolution. *Int. J. Hydrogen Energy* 37(14): 10564–74.

Mou, Z., S. Yin, M. Zhu, et al. 2013. $RuO_2/TiSi_2$/graphene composite for enhanced photocatalytic hydrogen generation under visible light irradiation. *Phys. Chem. Chem. Phys.* 15: 2793–99.

Mubeen S., G. Hernandez-Sosa, D. Moses, J. Lee, and M. Moskovits. 2011. Plasmonic photosensitization of a wide band gap semiconductor: converting plasmons to chargecarriers. *Nano Lett.* 11: 5548–52.

Mukherji, A., B. Seger, G. Q. M. Lu, and L. Wang. 2011. Nitrogen doped $Sr_2Ta_2O_7$ coupled with graphene sheets as photocatalysts for increased photocatalytic hydrogen production. *ACS Nano* 5: 3483–92.

Murphy, A. B., P. R. F. Barnes, L. K. Randeniya, et al. 2006a. Efficiency of solar water splitting using semiconductor electrodes. *Int. J. Hydrogen Energy* 31: 1999–2017.

Murphy, J. E., M. C. Beard, A. G. Norman, et al. 2006b. PbTe colloidal nanocrystals: Synthesis, characterization, and multiple exciton generation. *J. Am. Chem. Soc.* 128: 3241.

Nair, R. R., P. Blake., A. N. Grigorenko, et al. 2008. Fine structure constant defines visual transparency of graphene. *Science* 320: 1308.

Nakato Y. and H. Tsubomura. 1985. Structures and functions of thin metal layers on semiconductor electrodes. *J. Photochem.* 29(1–3): 257–66.

Nakato, Y., T. Ohnishi, and H. Tsubomura. 1975. Photo-electrochemical behaviors of semiconductor electrodes coated with thin metal films. *Chem. Lett.* 4: 883–92.

Nann, T., S. K. Ibrahim, P.-M. Woi, S. Xu, C. J. Ziegler, and J. Pickett. 2010. Water splitting by visible light: A nanophotocathode for hydrogen production. *Angew. Chem. Int. Ed.* 49: 1574–77.

Ng, Y. H., A. Iwase, A. Kudo, and R. Amal. 2010. Reducing graphene oxide on a visible-light $BiVO_4$ photocatalyst for an enhanced photoelectrochemical water splitting. *J. Phys. Chem. Lett.* 1: 2607–12.

Ni, L., M. Tanabe, and H. Irie 2013. A visible-light-induced overall water-splitting photocatalyst: Conduction-band-controlled silver tantalate. *Chem. Commun.* 49: 10094–96.

Niitsoo, O., S. K. Sarkar, C. Pejoux, et al. 2006. Chemical bath deposited CdS/CdSe sensitized porous TiO_2 solar cells. *J. Photo Chem. PhotoBio. A* 181: 306.

Novoselov, K. S., A. K. Geim, S. V. Morozov, et al. 2004. Electric field effect in atomically thin carbon films. *Science* 306: 666–69.

Nozik, A. J. 2001. Spectroscopy and hot electron relaxation dynamics in semiconductor quantum wells and quantum dots. *Annual Rev. Phys. Chem.* 52: 193.

Osterloh, F. E. 2013. Inorganic nanostructures for photoelectrochemical and photocatalytic water splitting. *Chem. Soc. Rev.* 42: 2294–320.

Osterloh, F. E. 2016. Nanoscale effects in water splitting photocatalysis. *Top. Curr. Chem.* 371: 105–42.

Paola, A. D., G. Marci, L. Palmisano, et al. 2002. Preparation of polycrystalline TiO_2 photocatalysts impregnated with various transition metal ions: Characterization and photocatalytic activity for degradation of 4-nitrophenol. *J. Phys. Chem. B* 106: 637–45.

Park, Y., W. Kim, H. Park, T. Tachikawa, T. Majima, and W. Choi. 2009. Carbon-doped TiO_2 photocatalyst synthesized without using an external carbon precursor and the visible light activity. *Appl. Catal. B Environ.* 91: 355–61.

Paven-Thivet, C. L., A. Ishikawa, A. Ziani, et al. 2009. Photoelectrochemical properties of crystalline perovskite lanthanum titanium oxynitride films under visible light. *J. Phys. Chem. C* 113: 6156–62.

Pei, F., Y. Liu, S. Xu, J. Lü, C. Wang, and S. Cao. 2013. Nanocomposite of graphene oxide with nitrogen-doped TiO_2 exhibiting enhanced photocatalytic efficiency for hydrogen evolution. *Int. J. Hydrogen Energy* 38: 2670–78.

Peter, L. M., K. G. U. Wijayantha, D. J. Riley, and J. P. Waggett. 2003. Band-edge tuning in self-assembled layers of Bi_2S_3 nanoparticles used to photosensitize nanocrystalline TiO_2. *J. Phys. Chem. B* 107: 8378–81.

Pilli, S. K., T. E. Furtak, L. D. Brown, T. G. Deutsch, J. A. Turner, and A. M. Herring. 2011. Cobalt-phosphate (Co-Pi) catalyst modified Mo-doped $BiVO_4$ photoelectrodes for solar water oxidation. *Energy Environ. Sci.* 4: 5028–34.

Primo A, A. Corma, and H. Garćia. 2011. Titania supported gold nanoparticles as photocatalyst. *Phys. Chem. Chem. Phys.* 13: 886–910.

Qu, Y., W. Zhou, and H. Fu 2014. Porous cobalt titanate nanorod: A new candidate for visible light-driven photocatalytic water oxidation. *Chem. Cat. Chem.* 6: 265–70.

Rawal, S. B., S. Bera, D. Lee, D.-J. Jang, and W. I. Lee. 2013. Design of visible-light photocatalysts by coupling of narrow bandgap semiconductors and TiO_2: Effect of their relative energy band positions on the photocatalytic efficiency. *Catal. Sci. Technol.* 3: 1822–30.

Reber J. F. and K. Meier. 1984. Photochemical production of hydrogen with zinc sulfide suspensions. *J. Phys. Chem.* 88(24): 5903–13.

Reyes-Gil, K. R., C. Wiggenhorn, B. S. Brunschwig, and N. S. Lewis. 2013. Comparison between the quantum yields of compact and porous WO3 photoanodes. *J. Phys. Chem. C* 117: 14947–57258.

Robel, I., M. Kuno, and P. V. Kamat. 2007. Size-dependent electron injection from excited CdSe quantum dots into TiO_2 nanoparticles. *J. Am. Chem. Soc.* 129: 4136–37.

Rockafellow, E. M., L. K. Stewart, and W. S. Jenks. 2009. Is sulfur-doped TiO_2 an effective visible light photocatalyst for remediation? *Appl. Catal. B Environ.* 91: 554–62.

Salvador, P. 2001. Semiconductors' photoelectrochemistry: A kinetic and thermodynamic analysis in the light of equilibrium and nonequilibrium models. *J. Phys. Chem. B* 105(26): 6128–41.

Sang, Y., L. Kuai, C. Chen, Z. Fang, and B. Geng. 2014. Fabrication of a visible-light-driven plasmonic photocatalyst of $AgVO_3$@AgBr@Ag nanobelt heterostructures. *ACS Appl. Mater. Interfaces* 6: 5061–68.

Sant P. A. and P. V. Kamat. 2002. Interparticle electron transfer between size-quantized CdS and TiO_2 semiconductor nanoclusters. *Phys. Chem. Chem. Phys.* 4: 198–203.

Sasaki, Y., A. Iwase, H. Kato, and A. Kudo 2008. The effect of co-catalyst for Z-scheme photocatalysis systems with an Fe^{3+}/Fe^{2+} electron mediator on overall water splitting under visible light irradiation. *J. Catal.* 259: 133–37.

Sasaki, Y., H. Kato, and A. Kudo. 2013. $[Co(bpy)_3]^{3+/2+}$ and $[Co(phen)_3]^{3+/2+}$ electron mediators for overall water splitting under sunlight irradiation using Z-scheme photocatalyst system. *J. Am. Chem. Soc.* 135: 5441–49.

Sasaki, Y., H. Nemoto, K. Saito, and A. Kudo. 2009. Solar water splitting using powdered photocatalysts driven by Z-schematic interparticle electron transfer without an electron mediator. *J. Phys. Chem. C* 113: 17536–42.

Sasikala, R., A. R. Shirole, V. Sudarsan, et al. 2010. Enhanced photocatalytic activity of indium and nitrogen co-doped TiO_2–Pd nanocomposites for hydrogen generation. *Appl. Catal. A Gen.* 377(1–2): 47–54.

Sato, J., H. Kobayashi, K. Ikarashi, N. Saito, H. Nishiyama, and Y. Inoue. 2004. Photocatalytic activity for water decomposition of RuO_2-dispersed Zn_2GeO_4 with d^{10} configuration. *J. Phys. Chem. B* 108: 4369–75.

Sato, J., N. Saito, H. Nishiyama, and Y. Inoue. 2001. New photocatalyst group for water decomposition of RuO_2-loaded p-block metal (In, Sn, and Sb) oxides with d^{10} configuration. *J. Phys. Chem. B* 105: 6061–63.

Sayama, K., A. Nomura, T. Arai, et al. 2006a. Photoelectrochemical decomposition of water into H_2 and O_2 on porous $BiVO_4$ thin-film electrodes under visible light and significant effect of Ag ion treatment. *J. Phys. Chem. B* 110: 11352–60.

Sayama, K., A. Tanaka, and K. Domen. 1996. Photocatalytic water splitting on nickel intercalated $A_4TaxNb_{6-x}O_{17}$ (A = K, Rb). *J Catal. Today* 28: 175–82.

Sayama, K., A. Tanaka, K. Domen, K. Maruta, and T. Onishi. 1990. Photocatalytic decomposition of water over a Ni-Loaded $Rb_4Nb_6O_{17}$ catalyst. *J. Catal.* 124: 541–47.

Sayama, K., K. Masuka, R. Abe, Y. Abe, and H. Arakawa. 2001. Stiochiometric water splitting into H_2 and O_2 using a mixture of two different photocatalysts and an IO^{3-}/I^- shuttle redox mediator under visible light irradiation. *Chem. Commun.* 2416–17.

Sayama, K., R. Abe, H. Arakawa, and H. Sugihara. 2006b. Decomposition of water into H_2 and O_2 by a two-step photoexcitation reaction over a $Pt-TiO_2$ photocatalyst in $NaNO_2$ and Na_2CO_3 aqueous solution. *Catal. Commun.* 7: 96–9.

Scaife, D. E. 1980. Oxide semiconductors in photoelectrochemical conversion of solar energy. *Sol. Energy* 25: 41–54.

Schaller, R. D., M. Sykora, J. M. Pietryga, and V. I. Klimov. 2006. Seven excitons at a cost of one: Redefining the limits for conversion efficiency of photons into charge carriers. *Nano Lett.* 6: 424.

Schaller, R. D. and V. I. Klimov. 2004. High efficiency carrier multiplication in PbSe nanocrystals: Implications for solar energy conversion. *Phys. Rev. Lett.* 92(18): 186601.

Seabold, J. A. and K.-S. Choi. 2012. Efficient and stable photo-oxidation of water by a bismuth vanadate photoanode coupled with an iron oxyhydroxide oxygen evolution catalyst. *J. Am. Chem. Soc.* 134: 2186–92.

Serpone, N. 2006. Is the band gap of pristine TiO_2 narrowed by anion- and cation-doping of titanium dioxide in second-generation photocatalysts? *J. Phys. Chem. B* 110: 24287–93.

Shaban, Y. A. and S. U. M. Khan. 2007. Visible light active carbon modified (CM) p-type WO_3 thin film electrode for photosplitting of water. *ECS Trans.* 6(2): 93–100.

Sheikh, A., A. Yengantiwar, M. Deo, S. Kelkar, and S. C. Ogale. 2013. Near-field plasmonic functionalization of light harvesting oxide–oxide heterojunctions for efficient solar photoelectrochemical water splitting: The $AuNP/ZnFe_2O_4/ZnO$ system. *Small* 9(12): 2091–96.

Shen, S., C. X. Kronawitter, J. Jiang, P. Guo, L. Guo, and S. S. Mao. 2013. A ZnO/ZnO:Cr isostructural nanojunction electrode for photoelectrochemical water splitting. *Nano Energy* 2: 958–65.

Shimizu, K., Y. Tsuji, M. Kawakami, et al. 2002. Photocatalytic water splitting over spontaneously hydrated layered tantalate $A_2SrTa_2O_7 \cdot nH_2O$ (A=H, K, Rb). *Chem. Lett.* 31: 1158–59.

Singh, A. P., S. Kumari, R. Shrivastav, S. Dass, and V. R. Satsangi. 2008. Iron doped nanostructured TiO_2 for photoelectrochemical generation of hydrogen. *Int. J. Hydrogen Energy* 33: 5363–68.

Singh, P., R. Singh, R. Gale, K. Rajeshwar, and J. Dubow. 1980. Surface charge and specific ion adsorption effects in photoelectrochemical devices. *J. Appl. Phys.* 51: 6286–91.

Siritanaratkul, B., K. Maeda, T. Hisatomi, and K. Domen. 2011. Synthesis and photocatalytic activity of perovskite niobium oxynitrides with wide visible-light absorption bands. *Chem. Sus. Chem.* 4: 74–8.

Sivula, K., F. Le Formal, and M. Gratzel. 2011. Solar water splitting: Progress using hematite (α-Fe_2O_3) photoelectrodes. *Chem. Sus. Chem.* 4: 432–49.

Smalley, R. E. 2003. Top ten problems of humanity for next 50 years. In: *Energy and NanoTechnology Conference.* Rice University, Houston, TX.

Stankovich, S., D. A. Dikin, G. H. B. Dommett, et al. 2006. Graphene based composite materials. *Nature* 442: 282–286.

Su, F., J. Lu, Y. Tian, X. Ma, and J. Gong. 2013. Branched TiO_2 nanoarrays sensitized with CdS quantum dots for highly efficient photoelectrochemical water splitting. *Phys. Chem. Chem. Phys.* 15: 12026–32.

Sun S., W. Wang, L. Zhang, M. Shang, and L. Wang. 2009. Ag@C core/shell nanocomposite as a highly efficient plasmonic photocatalyst. *Catal. Commun.* 11: 290–93.

Suzuki, T. M., A. Iwase, H. Tanaka, S. Sato, A. Kudo, and T. Morikawa. 2015. Z-scheme water splitting under visible light irradiation over powdered metal-complex/semiconductor hybrid photocatalysts mediated by reduced graphene oxide. *J. Mater. Chem. A* 3: 13283–90.

Tabata, M., Maeda, K., Higashi, M., et al. 2010. Modified Ta_3N_5 powder as a photocatalyst for O_2 evolution in a two-step water splitting system with an iodate/iodide shuttle redox mediator under visible light. *Langmuir* 26(12): 9161–65.

Tai, Y. W., J. S. Chen, C. C. Yang, and B. J. Wan. 2004. Preparation of nano-gold on $K_2La_2Ti_3O_{10}$ for producing hydrogen from photo-catalytic water splitting. *Catal. Today* 97: 95–101.

Takata, T., Y. Furumi, K. Shinohara, et al. 1997. Photocatalytic decomposition of water on spontaneously hydrated layered perovskites. *Chem. Mater.* 9: 1063–64.

Tang, J. W., J. R. Durrant, and D. R. Klug. 2008. Mechanism of photocatalytic water splitting in TiO_2. Reaction of water with photoholes, importance of charge carrier dynamics, and evidence for four-hole chemistry. *J. Am. Chem. Soc.* 130: 13885–91.

Thimsen, E., F. Le Formal, M Grätzel, and S. C. Warren. 2011. Influence of plasmonic Au nanoparticles on the photoactivity of Fe_2O_3 electrodes for water splitting. *Nano Lett.* 11: 35–43.

Thomann, I., B. A. Pinaud, Z. Chen, B. M. Clemens, T. F. Jaramillo, and M. L. Brongersma. 2011. Plasmon enhanced solar-to-fuel energy conversion. *Nano Lett.* 11: 3440–46.

Tian, L, J. Liu, C. Gong, L. Ye, and L. Zan. 2013. Fabrication of reduced graphene oxide–BiOCl hybrid material via a novel benzyl alcohol route and its enhanced photocatalytic activity. *J. Nanopart. Res.* 15: 1917.

Tielrooij, K. J., J. C. W. Song, S. A. Jensen, et al. 2013. Photoexcitation cascade and multiple hot-carrier generation in graphene. *Nat. Phys.* 9: 248–52.

Torimoto, T., H. Horibe, T. Kameyama, et al. 2011. Plasmon enhanced photocatalytic activity of cadmium sulfide nanoparticle immobilized on silica-coated gold particles. *J. Phys. Chem. Lett.* 2: 2057–62.

Townsend, T. K., N. D. Browning, and F. E. Osterloh. 2012. Overall photocatalytic water splitting with NiO_x-$SrTiO_3$—A revised mechanism. *Energy Environ. Sci.* 5: 9543–50.

Trevisan, R., P. Rodena, V. G.-Pedro, et al. 2013. Harnessing infrared photons for photoelectrochemical hydrogen generation. A PbS quantum dot based "quasi-artificial leaf". *J. Phys. Chem. Lett.* 4(1): 141–46.

Tung, V. C., J. H. Huang, I. Tevis, et al. 2011. Surfactant-free water-processable photoconductive all-carbon composite. *J. Am. Chem. Soc.* 133: 4940–47.

Van Benthem, K., C. Elsässer, and R. H. French. 2001. Bulk electronic structure of $SrTiO_3$: Experiment and theory. *J. Appl. Phys.* 90: 6156–64.

Walsh, A., Y. Yan, M. N. Huda, M. M. Al-jassim, and S.-H. Wei. 2009. Band edge electronic structure of $BiVO_4$: Elucidating the role of the Bi s and V d orbitals. *Chem. Mater.* 21: 547–51.

Wang D., T. Kako, and J. Ye. 2009. New series of solid-solution semiconductors $(AgNbO_3)_{1-x}(SrTiO_3)_x$ with modulated band structure and enhanced visible-light photocatalytic activity. *J. Phys. Chem. C* 113: 3785–92.

Wang R. H., J. H. Z. Xin, Y. Yang, H. F. Liu, L. M. Xu, and J. H. Hu. 2004. The characteristics and photocatalytic activities of silver doped ZnO nanocrystallites. *Appl. Surf. Sci.* 227: 312–17.

Wang, B., P. Kanhere, Z. Chen, J. Nisar, B. Pathak, and R. Ahuja. 2013a. Anion-doped $NaTaO_3$ for visible light photocatalysis. *J. Phys. Chem. C* 117: 22518–24.

Wang, G. M., Y. C. Ling, H. Y. Wang, et al. 2012a. Hydrogen-treated WO_3 nanoflakes show enhanced photostability. *Energy Environ. Sci.* 5: 6180–87.

Wang, G. M. Y. C. Ling, X. H. Lu, et al. 2013b. Computational and photoelectrochemical study of hydrogenated bismuthvanadate. *J. Phys. Chem. C* 117: 10957–64.

Wang, G. M., H. Y. Wang, Y. C. Ling, et al. 2011a. Hydrogen-treated TiO_2 nanowire arrays for photoelectrochemical water splitting. *Nano Lett.* 11: 3026–33.

Wang, G. M., Y. C. Ling, D. A. Wheeler, et al. 2011b. Facile synthesis of highly photoactive α-Fe_2O_3-based films for water oxidation. *Nano Lett.* 11: 3503–09.

Wang, G. M., X. Y. Yang, F. Qian, J. Z. Zhang, and Y. Li. 2010a. Double-sided CdS and CdSe quantum dot co-sensitized ZnO nanowire arrays for photoelectrochemical hydrogen generation. *Nano Lett.* 10: 1088–92.

Wang, G. M., Y. C. Ling, X. H. Lu, et al. 2013c. A mechanistic study into the catalytic effect of $Ni(OH)_2$ on hematite for photoelectrochemical water oxidation. *Nanoscale* 5: 4129–33.

Wang, H., T. Lindgren, J. He, A. Hagfeldt, and S.-E. Lindquist. 2000. Photoelectrochemistry of nanostructured WO_3 thin film electrodes for water oxidation: Mechanism of electron transport. *J. Phys. Chem. B* 104: 5686–96.

Wang, H., W. Chen, J. Zhang, C. Huang, and L. Mao. 2015. Nickel nanoparticles modified CdS–A potential photocatalyst for hydrogen production through water splitting under visible light irradiation. *Int. J. Hydrogen Energy* 40(1): 340–345.

Wang, H., Y. S. Bai, H. Zhang, Z. H. Zhang, J. H. Li, and L. Guo. 2010b. CdS quantumdots-sensitized TiO_2 nanorod array on transparent conductive glass photo-electrodes. *J. Phys. Chem. C* 114: 16451–55.

Wang, H. Y., G. M. Wang, Y. C. Ling, et al. 2012b. Photoelectrochemical study of oxygen deficient TiO_2 nanowire arrays with CdS quantum dot sensitization. *Nanoscale* 4: 1463–66.

Wang, L. W., M. Califano, A. Zunger, and A. Franceschetti. 2003. Pseudopotential theory of auger processes in CdSe quantum dots. *Phys. Rev. Lett.* 91(5): 0564047.

Wang, M., X. Shang, X. Yu, et al. 2014. Graphene-CdS quantum dots-polyoxometalate composite films for efficient photoelectrochemical water splitting and pollutant degradation. *Phys. Chem. Chem. Phys.* 16(47): 26016–23.

Wang, R., Y. Zhu, Y. Qiu, et al. 2013d. Synthesis of nitrogen-doped $KNbO_3$ nanocubes with high photocatalytic activity for water splitting and degradation of organic pollutants under visible light. *Chem. Eng. J.* 226: 123–30.

Wang, Z., J. Liu, and W. Chen. 2012c. Plasmonic Ag/AgBr nanohybrid: Synergistic effect of SPR with photographic sensitivity for enhanced photocatalytic activity and stability. *Dalton Trans.* 41: 4866–70.

Wang, C.-I, Z. Yang, A. P. Periasamy, and H.-T. Chang. 2013e. High-efficiency photochemical water splitting of CdZnS/CdZnSe nanostructures. *J. Mater.* 2013: 703985 (7 p.).

Wang, S., M. Khafizov, X. Tu, M. Zheng, and T. D. Krauss. 2010c. Multiple exciton generation in single-walled carbon nanotubes. *Nano Lett.* 10(7): 2381–86.

Wang, P., B. Huang, Y. Dai, and M.-H Whangbo. 2012d. Plasmonic photocatalysts: harvesting visible light with noble metal nanoparticles. *Phys. Chem. Chem. Phys.* 14: 9813–25.

Wilke, K. and H. D. Breuer. 1999. The influence of transition metal doping on the physical and photocatalytic properties of titania. *J. Photochem. Photobiol. A Chem.* 121(1): 49–53.

Wu Z.-C., Y. Zhang, T.-X. Tao, L. Zhang, and H. Fong 2010. Silver nanoparticles on amidoxime fibers for photo-catalytic degradation of organic dyes in waste water. *Appl. Surf. Sci.* 257: 1092–97.

Wu, J., S. Uchida, Y. Fujishito, S. Yin, and T. Sato. 1999. Synthesis and photocatalytic properties of $HNbWO_6/TiO_2$ and $HNbWO_6/Fe_2O_3$ nanocomposite. *J. Photochem. Photobiol. A Chem.* 128: 129–33.

Wunderlich, W., H. Ohta, and K. Koumoto. 2009. Enhanced effective mass in doped $SrTiO_3$ and related perovskites. *Physica B* 404: 2202–212.

Würfel, P. 2005. *Physics of solar cells*, p 244. Wiley-VCH, Weinheim.

Xiang, Q., J. Yu, and M. Jaroniec, 2011a. Preparation and enhanced visible-light photo-catalytic H_2-production activity of graphene/C_3N_4 composites. *J. Phys. Chem. C* 115: 7355–63.

Xiang, Q. J., J. G. Yu, and M. Jaroniec. 2011b. Nitrogen and sulfur co-doped TiO_2 nanosheets with exposed {001} facets: Synthesis, characterization and visible-light photocatalytic activity. *Phys. Chem. Chem. Phys.* 13: 4853–61.

Xie, G., K. Zhang, B. Guo, Q. Liu, L. Fang, and J. R. Gong. 2013. Graphene-based mate-rials for hydrogen generation from light-driven water splitting. *Adv. Mater.* 25(28): 3820–39.

Xu, A. W., Y. Gao, and H. Q. Liu. 2002. The preparation characterization and their photocata-lytic activities of rare earth doped TiO_2 nanoparticles. *J. Catal.* 207: 151–57.

Xu, J., C. Hu, Y. Xi, B. Wan, C. Zhang, and Y. Zhang. 2012a. Synthesis and visible light pho-tocatalytic activity of β-$AgVO_3$ nanowires. *Solid State Sci.* 14: 535–39.

Xu, J. C., Y. L. Shi, J. E. Huang, B. Wang, and H. L. Li. 2004. Doping metal ions only onto the catalyst surface. *J. Mol. Catal. A Chem.* 219: 351–55.

Xu, P., M. Da, H. Y. Wu, D. Y. Zhao, and G. F. Zheng. 2012b. Controlled Sn-doping in TiO_2 nanowire photoanodes with enhanced photoelectrochemical conversion. *Nano Lett.* 12: 1503–08.

Xu, Y. and M. A. Schoonen. 2000. The absolute energy positions of conduction and valence bands of selected semiconducting minerals. *Am. Miner.* 85: 543–56.

Yamakata, A. I. T. A., H. Kato, A. Kudo, and H. Onishi. 2003. Photodynamics of $NaTaO_3$ catalysts for efficient water splitting. *J. Phys. Chem. B* 107: 14383–87.

Yamashita H., M. Harada, J. Misaka, M. Takeuchi, K. Ikeue, and M. Anpo. 2002. Degradation of propanol diluted in water under visible light irradiation using metal ion-implanted titanium dioxide photocatalysts. *J. Photochem. Photobiol. A* 148: 257–61.

Yang, J., X. Zeng, L. Chen, and W. Yuan. 2013. Photocatalytic water splitting to hydrogen production of reduced graphene oxide/SiC under visible light. *Appl. Phys. Lett.* 102: 083101.

Yang, M., X. Huang, S. Yan, Z. Li, T. Yu, and Z. Zou 2010. Improved hydrogen evolution activities under visible light irradiation over $NaTaO_3$ codoped with lanthanum and chro-mium. *Mater. Chem. Phys.* 121: 506–10.

Yang, H. B., J. Miao, S.-F. Hung, F. Huo, H. M. Chen, and B. Liu. 2014. Stable quantum dot photoelectrolysis cell for unassisted visible light solar water splitting. *ACS Nano* 8(10): 10403–413.

Ye, H., H. S. Park, and A. J. Bard. 2011. Screening of electrocatalysts for photoelectrochemi-cal water oxidation on W-doped $BiVO_4$ photocatalysts by scanning electrochemical microscopy. *J. Phys. Chem. C* 115: 12464–70.

Ye, H., J. Lee, J. S. Jang, and A. J. Bard. 2010. Rapid screening of $BiVO_4$-based photocata-lysts by scanning electrochemical microscopy (SECM) and studies of their photoelec-trochemical properties. *J. Phys. Chem. C* 114: 13322–28.

Yeh, T. F., S. J. Chen, C. S. Yeh, and H. Teng. 2013. Tuning the electronic structure of graphite oxide through ammonia treatment for photocatalytic generation of H_2 and O_2 from water splitting. *J. Phys. Chem. C* 117: 6516–24.

Yeh, T.-F., C.-Y. Teng, S.-J. Chen, and H. Teng. 2014. Nitrogen-doped graphene oxide quan-tum dots as photocatalysts for overall water-splitting under visible light illumination. *Adv. Mater.* 26(20): 3297–303.

Yi, Z. G. and J. H. Ye. 2009. Band gap tuning of $Na_{1-x}La_xTa_{1-x}Co_xO_3$ for H_2 generation from water under visible light irradiation. *J. App. Phys.* 106: 074910.

Yi, Z. G. and J. H. Ye. 2007. Band gap tuning of $Na_{1-x}La_xTa_{1-x}Co_xO_3$ solid solutions for visible light photocatalysis. *Appl. Phys. Lett.* 91: 254108.

Yoffe, A. D. 2001. Semiconductor quantum dots and related systems: Electronic, optical, lumi-nescence and related properties of low dimensional systems. *Adv. Phys.* 50: 1–208.

Youngblood, W. J., S-H. A. Lee, K. Maeda, and T. E. Mallouk. 2009. Visible light water splitting using dye-sensitized oxide semiconductors. *Acc. Chem. Res.* 42: 1966–73.

Yu, J., G. Dai, and B. Huang. 2009. Fabrication and characterization of visible-light-driven plasmonic photocatalyst $Ag/AgCl/TiO_2$ nanotube arrays. *J. Phys.Chem. C* 113:16394–401.

Zhang, C., Y. Z. Jia, Y. Jing, Y. Yao, J. Ma, and J. Sun. 2013a. Effect of non-metal elements (B, C, N, F, P, S) mono-doping as anions on electronic structure of $SrTiO_3$. *Comput. Mater. Sci.* 79: 69–74.

Zhang, H., G. Chen, X. He, and J. Xu. 2012a. Electronic structure and photocatalytic properties of Ag–La codoped $CaTiO_3$. *J. Alloys Compd.* 516: 91–95.

Zhang, H., G. Chen, Y. Li, and Y. Teng. 2010. Electronic structure and photocatalytic properties of copper-doped $CaTiO_3$. *Int. J. Hydrogen Energy* 35: 2713–16.

Zhang, J., J. Yu, M. Jaroniec, and J. R. Gong. 2012b. Noble metal-free reduced graphene oxide-Zn x Cd$_{1-}x$ S nanocomposite with enhanced solar photocatalytic H_2-production performance. *Nano Lett.* 12: 4584–89.

Zhang, L. and D. Bahnemann. 2013. Synthesis of nanovoid Bi_2WO_6-2D ordered arrays as photoanodes for photoelectrochemical water splitting. *ChemSusChem.* 6: 283–90.

Zhang, W., Y. Li, S. Peng, and X. Cai, 2014a. Enhancement of photocatalytic H_2 evolution of eosin Y-sensitized reduced graphene oxide through a simple photoreaction. *Beilstein J. Nanotechnol.* 5: 801–11.

Zhang, X., Y. L. Chen, R.-S. Liu, and D. P. Tsai. 2013b. Plasmonic photocatalysis. *Rep. Prog. Phys.* 76: 046401.

Zhang, N., J. Shi, S. S. Mao, and L. Guo. 2014b. Co_3O_4 quantum dots: Reverse micelle synthesis and visible-light-driven photocatalytic overall water splitting. *Chem. Commun.* 50: 2002–04.

Zhao, Z., R. Li, Z. Li, and Z. Zou 2011. Photocatalytic activity of La-N-codoped $NaTaO_3$ for H_2 evolution from water under visible-light irradiation. *J. Phys. D Appl. Phys.* 44: 165401.

Zhdanov, V. P., C. Hagglund, and B. Kasemo. 2005. Relaxation of plasmons in nm-sized metal particles located on or embedded in an amorphous semiconductor. *Surf. Sci.* 599: L372–75.

Zhong, D. K., S. Choi, and D. R. Gamelin. 2011. Near-complete suppression of surface recombination in solar photoelectrolysis by Co-Pi catalyst-modified $W:BiVO_4$. *J. Am. Chem. Soc.* 133: 18370–77.

Zhou, X., J. Shi, and C. Li. 2011. Effect of metal doping on electronic structure and visible light absorption of $SrTiO_3$ and $NaTaO_3$ (Metal = Mn, Fe, and Co). *J. Phys. Chem. C* 115: 8305–11.

Zhu, S., H. Fu, S. Zhang, L. Zhang, and Y. Zhu. 2008. Two-step synthesis of a novel visible-light-driven $K_2Ta_2O_{6-x}N_x$ catalyst for the pollutant decomposition. *J. Photochem. Photobiol. A* 193: 33–41.

6 Fundamental Understanding of the Photocatalytic Mechanisms

6.1 INTRODUCTION

Solar light is the most abundant source of energy (15,000 GW annual) on Earth. The estimated practical solar power that is received by Earth is 40 times greater than the present global energy need (Fujishima and Honda 1972). Artificial photosynthesis (AP) is a key strategy for conversion of solar energy into chemically stored energy, that is, hydrogen. Hydrogen, being a clean source of energy, is predicted to be the "fuel of the future" (Dicks 1996; Armor 1999). Water proved itself as a carbon-free renewable energy source for AP and attracted immense attention from researchers. But splitting of water needs a smart material (semiconductor/photocatalyst) to reduce aqueous protons to H_2 fuel and oxidize water to make O_2. Both redox reactions are multielectron processes that need two and four electrons for reduction and oxidation of water to molecular hydrogen and oxygen, respectively. Understanding how this two electron or four-electron reaction works in detail is important for the development of improved robust catalysts made of Earth-abundant materials. In this chapter, we try to unveil the mechanistic part of water splitting by identifying the intermediates in the aforementioned processes and their changes in physical and chemical behavior that belong to different catalytic sites. Knowledge of the structure and kinetics of surface intermediates will enable the design of improved metal oxide materials for more efficient water cleavage for hydrogen and oxygen generation. Mimicking natural photosynthesis, recent efforts have also provided insights into photocatalytic systems for oxygen generation (Sirjoosingh and Hammes-Schiffer 2011; Iyer et al. 2012). Periodic work has been conducted on several photocatalysts, although, to date, no published work dealing with photoreactions (with proof) is known. Before discussing some of these results, it is worthwhile to give a brief introduction on the surfaces of the photocatalysts. This will be followed by the brief introduction of the chemical and physical processes considered to be involved in photocatalytic reactions over a semiconductor.

6.2 MECHANISM OF PHOTOCATALYTIC CLEAVAGE OF WATER IN ELECTROLYTES (ELECTRON SCAVENGER AND HOLE SCAVENGER)

6.2.1 SCAVENGERS OR SACRIFICIAL ELECTROLYTES

During photocatalytic overall water-splitting a photocatalyst can evolve oxygen and hydrogen. Splitting of water became difficult due to rapid recombination of the photogenerated conduction band (CB) electrons and valence band (VB) holes at the surface of the photocatalyst that interrupted the production of hydrogen and oxygen in distilled water. To suppress this recombination process, the provision for addition of electron donors (sacrificial reagents or hole scavengers) is made to react irreversibly with the photogenerated VB holes, which can enhance the photocatalytic electron/hole separation, resulting in higher quantum efficiency (QE). Since electron donors are consumed in a photocatalytic reaction, continual addition of electron donors/electron scavengers is required to sustain hydrogen/oxygen production. Generally, these sacrificial/scavenger electrolytes are the materials with lower potentials than those of the electrode (cathode/anode) material. They corrode faster than those of the electrode (cathode/anode) material, resulting in protecting the surface at electrodes. Sacrificial electrolytes are technical material of interest due to their capability of controlling electrode reactions. A sacrificial reagent (electron donor) provides an efficient electron/hole separation as it reacts irreversibly with photogenerated holes (Figure 6.1), resulting in higher QEs getting rid of the problem of separation of O_2 and H_2. Nevertheless, the continued addition of electron donors is required to maintain hydrogen production since sacrificial reagents are consumed during the photocatalytic reaction. Acetic acid, lactic acid, acetaldehyde, formaldehyde, ethylenediaminetetraacetic acid (EDTA), Ce^{4+}/Ce^{3+}, S^{2-}/SO_3^{2-}, IO_3^-/I^-, NaI, methanol, and ethanol have been tested and proved to be effective sacrificial reagents to improve hydrogen production (Dickinson et al. 1999; Wu et al. 2004; Meng et al. 2007). The degree of hydrogen production enhancement capability was found to be EDTA > methanol > ethanol > lactic acid (Nada et al. 2005). Some of the commonly used hole scavengers are sulfites (Na_2SO_3), hydrazine ($H_2N–NH_2$), carbohydride ($H_2N–NH–CO–HN–NH_2$), dimethyl hydroxyl amine (($C_2H_5)_2NOH$), and hydroquinone (HO–C_6H_4–OH). If we are taking oxygen evolution, electrode materials, such as WO_3, $BiVO_4$, and so on, are in consideration in the presence of electrolytes such as $AgNO_3$ (solution), Ce^{4+}/Ce^{3+}, S^{2-}/SO_3^{2-}, and IO_3^-/I^-. Ions in solution are intended to prevent the backward electron/hole recombination over a metal cocatalyst (Abe et al. 2003a) and are used as electron acceptors or electron scavengers in the systems. Although chalcogenides such as CdSe QDs are efficient for water splitting, they are highly prone to corrosion in light. Therefore, CdSe QD-sensitized TiO_2 was treated with polysulfide electrolyte of 0.5 M Na_2S, 0.1 M S, and 0.05 M CuSCN in ethanol/water (8:2 by volume) to prevent the possibility of corrosion (Jun et al. 2013). Hole scavenger couple K_2SO_3 and Na_2S, under solar stimulated photoirradiation, generated hydrogen from Pt-loaded $(AgIn)_{0.22}Zn_{1.56}S_2$ with 20% QE at 420 nm, by using the following mechanism (Tsuji et al. 2004). The electron/hole scavengers or sacrificial electrolyte (K_2SO_3

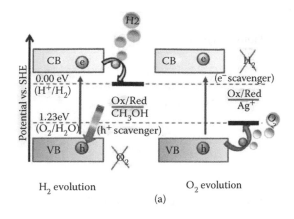

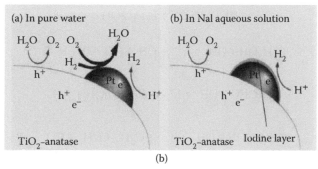

FIGURE 6.1 (a) Basic principles of photocatalytic water splitting in the presence of scavenger/sacrificial electrolyte. (b) A platinum cocatalyst also catalyzes the reverse reaction. This can be prevented by adding an iodine layer on the platinum surface. (From Abe, R. et al., *Chem. Phys. Lett.*, 371, 360–364, 2003a.)

and Na_2S, NaI, Na_3N, C_2H_5OH, and $AgNO_3$) are used to control the water-splitting process with their respective mechanism, as follows:

1. K_2SO_3 and Na_2S

$$Photocatalyst + h\nu \rightarrow e^-(CB) + h^+(VB) \tag{6.1}$$

$$H_2O + 2e^- \rightarrow H_2 + 2OH^- \tag{6.2}$$

$$SO_3^{2-} + H_2O + 2h^+(VB) \rightarrow SO_4^{2-} + 2H^+ \tag{6.3}$$

$$2S^{2-} + 2h^+(VB) \rightarrow S_2^{2-} \tag{6.4}$$

$$S_2^{2-} + SO_3^{2-} \rightarrow S_2O_3^{2-} + S^{2-} \tag{6.5}$$

$$SO_3^{2-} + S^{2-} + 2h^+(VB) \rightarrow S_2O_3^{2-} \tag{6.6}$$

2. NaI:

$$Photocatalyst + h\nu \rightarrow e^-(CB) + h^+(VB) \tag{6.7}$$

$$H_2O + 2e^- \rightarrow H_2 + 2OH^- \tag{6.8}$$

$$2I^- + 2h^+(VB) \rightarrow I_2 \tag{6.9}$$

3. Na$_3$N:

$$2N^{3-} + 6h^+(VB) \rightarrow N_2 \tag{6.10}$$

4. C$_2$H$_5$OH:

$$Photocatalyst + h\nu \rightarrow e^-(CB) + h^+(VB) \tag{6.11}$$

$$H_2O + 2e^- \rightarrow H_2 + 2OH^- \tag{6.12}$$

$$C_2H_5OH + OH^- + h^+ \rightarrow CH_3CHO + H_2 + H_2O \tag{6.13}$$

$$CH_3CHO + OH^- + h^+ \rightarrow CH_3COOH + \frac{1}{2}H_2 \tag{6.14}$$

$$CH_3COOH + OH^- + h^+ \rightarrow C_2H_6 + CH_4 + CO_2 \tag{6.15}$$

5. AgNO$_3$:

$$AgNO_3 + e^- \rightarrow Ag + NO_3^-(CB) \tag{6.16}$$

$$2H_2O + 4h^+ \rightarrow O_2 + 4H^+(VB) \tag{6.17}$$

$$H^+ + NO_3^- \rightarrow HNO_3 \tag{6.18}$$

6.3 PHOTOCORROSION

Oxide semiconductors such as ZnO and TiO_2 are good options for water split-ting with their excellent optical properties but they usually suffer from inefficient water-splitting activity due to their wide band gap (Eg ~3.2 eV) that can capture only the ultraviolet (UV) portion of the sunlight (4% of the whole). Consequently, to enhance solar energy conversion efficiency, we need to extend the solar energy absorption range from the visible to the infrared (IR) region. There are two signif-icant ways to extend the solar energy absorption range to longer wavelengths (vis-ible region [~45%] and near-infrared (NIR) [~50%]): first, by creation of impurity states near the VB edge (O 2p) and/or the CB edge (Ti $3dt_{2g}$) of the oxides by dop-ing them with nonmetal anions such as C, N, and S or with transition metal cations such as Fe^{3+}, V^{4+}, Mn^{2+}, Ni^{2+}, and Cr^{3+}, respectively. Second, dye or II–VI semicon-ductors quantum dots (QDs) such as CdS (Robel et al. 2005; Sheeney-Haj-Khia et al. 2005; Ardalan et al. 2011), CdTe, CdSe (Robel et al. 2006; Leschkies et al. 2007;

Brown and Kamat 2008), ZnTe, PbS (Plass et al. 2002; Gao et al. 2011), PbSe (Beard et al. 2009; Choi et al. 2009; Law et al. 2009; Leschkies et al. 2009), and InP (Zaban et al. 1998) have been used to sensitize metal oxides, like nanocrystalline TiO_2 or SnO_2, by absorbing the most intense regions of the solar spectrum (Kamat et al. 2010). InAs (Yu et al. 2006; Wei et al. 2009) and the alloys thereof can have nearly ideal band gaps with band-edge positions for the production of H_2 and O_2 via water splitting. But the photoanodes of II–VI semiconductors are also renowned for their notorious instability toward photocorrosion when in contact with aqueous electrolytes and light. The photocorrosion process is the deterioration of a material due to chemical reactions/ interactions between the material and its surroundings (light, water, and gases), resulting in oxidation of the material. Meissner et al. (1988) have given the mechanism of the well-known photocorrosion reaction of CdS (Equations 6.19 through 6.25).

$$CdS + 2H_2O \rightarrow H_2 + Cd^{2+} + S + 2OH^- \tag{6.19}$$

Photooxidation of CdS gives sulfate as the main product in the presence of oxygen under anodic polarization with the following stoichiometry:

$$CdS + 4h^+ + 2H_2O + O_2 \rightarrow Cd^{2+} + SO_4^{2-} + 4H^+ \tag{6.20}$$

where h^+ designates a photo hole and photo e^-. The countercathodic reactions occur as follows:

$$O_2 + 4e^- + 2H^+ \rightarrow 2OH^- \tag{6.21}$$

This leads to an overall stoichiometry for the photoprocess at solar light illuminated CdS particles:

$$CdS + 2O_2 \rightarrow Cd^{2+} + SO_4^{2-} \tag{6.22}$$

The cathodic hydrogen evolution at metal-covered (RuO_2 or Pt) CdS suspended in water is as follows:

$$2H_2O + 4e^- \rightarrow H_2 + 2OH^- \tag{6.23}$$

The well-known anodic photocorrosion reaction of CdS in the absence of oxygen is as follows:

$$CdS + 2h^+ \rightarrow Cd^{2+} + S \tag{6.24}$$

With increasing corrosion time the concentration of Cd^{2+} increases at the cathode, which leads to the formation of cadmium metal as follows:

$$Cd^{2+} + 2e^- \rightarrow Cd^0 \tag{6.25}$$

6.3.1 Chemical Passivation for Photocorrosion Protection

To protect CdS photoanodes from photocorrosion, the surface of the film can be modified with a catalyst that promotes the water-splitting reaction, or a hole scavenger is added to the electrolyte. In both cases, the photocorrosion is suppressed through kinetic competition. And photopassivation involves the use of a light coat of a protective material, such as a metal oxide, to create a shell against corrosion. Unlike their dye counterparts, most of the chalcogenide QDs undergo photodegradation when used in aqueous I_2/I^{3-} medium in a photoelectrochemical cell (Figure 6.2i) (Liu et al. 2009). In the quantum dot sensitized solar cell (QDSSC), the optimum redox couple was found to be sulfide/polysulfide (S^{2-}/S_n^{2-}) (Mahapatra and Dubey 1994; Hodes 2008). However, aqueous Na_2S and K_2SO_3 are used to provide the hole scavenger to protect the photooxidation of chalcogenide QDs loaded on the photoanode. But sulfide ions react very strongly with platinum, the catalyst of the counterelectrode (CE), and can poison the surface, thereby reducing the catalytic activity (Paal et al. 1997; Mohtadi et al. 2003). Therefore, alternative CE materials such as

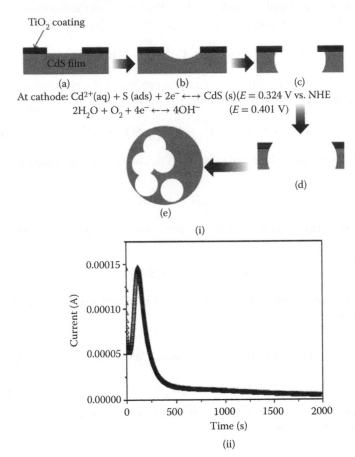

FIGURE 6.2 (i) Schematic view of the corrosion mechanism where (a) the corrosion starts with the presence of a pinhole or crack in the pristine sample (b) at which the corrosion is initiated. (c) The corrosion then expands in all directions until the substrate underneath the CdS is reached, (d) after which the corrosion pit expands in a lateral direction (e) until it merges with one or more neighboring corrosion pits. (ii) Current development with time in 0.1 M KOH upon illumination of a CdS sample coated with 200 atomic layer deposition (ALD) cycles of TiO_2. (From Liu, C. et al., *J. Phys. Chem. C*, 113, 14478–14481, 2009; Tachan, Z. et al., *J. Phys. Chem. C*, 115, 6162–6166, 2011.)

CoS, Cu_2S, NiS, carbon, and PbS have been explored for polysulfide solution (Hodes et al. 1977, 1980; Yu et al. 2002; Yang et al. 2010; Zhang et al. 2010). To overcome the attack of iodine in the electrolyte in I_2/I^{3-}, Zaban et al. have designed a strategy to employ a TiO_2 passivation layer over CdS QDs by using electrophoretic deposition (Tachan et al. 2011). With the amorphous TiO_2 passivation shell, they obtained a CdS QDSSC that is more stable for I_2/I^{3-}, as shown in Figure 6.2ii.

Lichterman et al. (2014) developed a passive, intrinsically safe, and efficient system for solar-driven water splitting by using a 140-nm TiO_2 thin film, loaded with either thin films or islands of Ni oxide to robustly protect Si, GaAs, and other III–V materials from photocorrosion. An n-CdTe/TiO_2/Ni oxide electrode–based oxygen evolution electrocatalyst produced 435 ± 15 mV of photovoltage with a photocurrent density of 21 ± 1 mA/cm² under 100 mW/cm² of simulated air mass 1.5 illumination. The power conversion efficiencies (PCEs) of the solar cell fabricated with the regular array of TiO_2 nanotubes (NTs) anchored with CdSSe/CdS or CdSe/CdS QDs (i.e., [CdSSe/CdS/TiO_2 NTs] or [CdSe/CdS/TiO_2 NTs]) are 3.49% and 2.81% under illumination at 100 mW/cm², respectively. But Park et al. (2013) introduced ZnS onto a CdSSe/CdS surface to protect the photocorrosion of the solar cell (CdSSe/CdS/TiO_2 NTs) or (CdSe/CdS/TiO_2 NTs) from the electrolyte and to suppress recombination of the photocarrier, with the PCE 4.67% for ZnS/CdSSe/CdS/TiO_2 NTs/Ti, under 100 mW/cm² illumination (because ZnS nanocrystals [NCs] provide the rapid generation of electron–hole pairs by photoexcitation along with the highly negative reduction potentials of excited electrons) (Yanagida et al. 1986).

A high current is observed immediately after switching on the light in a photoelectrochemical cell. It is due to the excitation of electron–hole pairs. But since the flux of holes to the surface is larger than the rate at which they can be consumed by chemical reactions, they accumulate near the surface. As a result, recombination increases, which leads to a decrease of the current. This decrease in current resembles the discharge of a capacitor over time, where the charge of the capacitor corresponds to the number of accumulated holes. This can be expressed as the recombination current by using Equation 6.26:

$$I_{recomb.} = I_0 \exp\left(-\frac{\tau}{\tau}\right) \tag{6.26}$$

where I_0 is the initial current (A), t is time (s), and τ is the time constant of the decay (s). The photocorrosion model bears a strong resemblance to the Avrami model (1940) of crystal growth in metals, which predicts the fraction Φ of material that is converted into a new phase (the corrode part of the material) as a function of time t, written in Equation 6.27:

$$\Phi = 1 - \exp(-kt^n) \tag{6.27}$$

where the constant k is the product of a shape factor, known as the effective number of nuclei, and the direction-averaged growth rate. The exponent n is the sum of the dimension of the crystal growth process (1 for needle-like, 2 for plate-like, and 3 for 3D growth, in this case the itching will mostly take place in the lateral direction) and an integer value describing the nucleation rate (1 for constant nucleation rate,

0 for the absence of nucleation). If we assume that O_2 production at the photoanode is negligible, the fraction of material that has been corroded can be described by integrating the corrosion current I_{corr}, shown in Equation 6.28:

$$\phi = \frac{1}{dA\rho Nq} \int \frac{1}{\eta} I corr \, dt \tag{6.28}$$

where d is the thickness of the film (in m), A is the total surface area of the electrode (2.82×10^{-5} m^2), ρ is the molar density (mol/m^3), N is Avogadro's number, q is the elementary charge (C), and η the number of electrons consumed per reaction (–). The values of I_{corr} and integral yield ϕ were obtained from Equations 6.28 and 6.29:

$$I_{corr} = \alpha nkt^{n-1}(\exp(-kt^n)) \tag{6.29}$$

In the preceding equation

$$\alpha = dA\rho q\eta \tag{6.30}$$

The total current can be expressed as the sum of the recombination current, $I_{recomb.}$, the corrosion current, I_{corr}, and a linear equation ($at + b$) that is used to fit the sloping part of the curve that is visible at the second half of the curve:

$$I = \alpha nkt^{n-1}(\exp(-kt^n)) + I_0 \exp\left(-\frac{\tau}{\tau}\right) + at + b \tag{6.31}$$

The fit parameters are α, n, k, τ, a, and b.

The equation was used to fit the experimental data given in Figure 6.2ii, which shows the photocurrent transient of a CdS sample coated with 200 cycles ALD of TiO$_2$ in 0.1 M KOH (Tachan et al. 2011).

Shi et al. studied the photocorrosion of BiTaO$_4$ and BiNbO$_4$ for methylene blue (MB) degradation (Figure 6.3a and b) under visible light irradiation and found that the powder of BiTaO$_4$ was quite stable during the aforementioned photocatalytic process (78% after 10 hours), but BiNbO$_4$ is not stable. TiO$_{2-x}$N$_X$ exhibited a high value of photodegradation of MB of 87% in 10 hours and was taken as the standard. The differences in the photocorrosion rate for BiTaO$_4$ and BiNbO$_4$ could be attributed to the conduction curvature and bandwidth, leading to the differences in the mobility of the photogenerated electrons in CBs, which mainly consisted of Ta 5d and Nb 4d orbitals, respectively.

A heterostructured photoanode based on NIR-active core@shell QDs of the PbS@CdS/ZnS QDs-functionalized TiO$_2$ mesoporous film is applied in PEC cells in S$_2^-$ / SO$_3^{2-}$ solution (Jin et al. 2016). The hybrid heterostructured photoanode can produce a photocurrent density as high as 11.2 mA/cm^2 (hydrogen generation rate around 112 mL/cm^2/d^2 H$_2$) by cyclic voltammetry in a three-electrode configuration under 100 mW/cm^2 AM 1.5G illumination and is a good candidate for solar hydrogen generation. Wang and Wang (2014) investigated a plasmonic metal gold-enhanced photocatalyst consisting of an ultrathin (20 nm) Cu$_x$O photocatalyst (CuO/Cu$_2$O composite) loaded with a plasmonic metallic (Au with work function < 5 eV or Pt with work function = 5.65 eV), on the W substrate; this was used for photocatalytic pure water splitting. As depicted in the energy band diagram (Figure 6.4c) (Didden 2015),

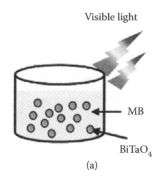

(a)

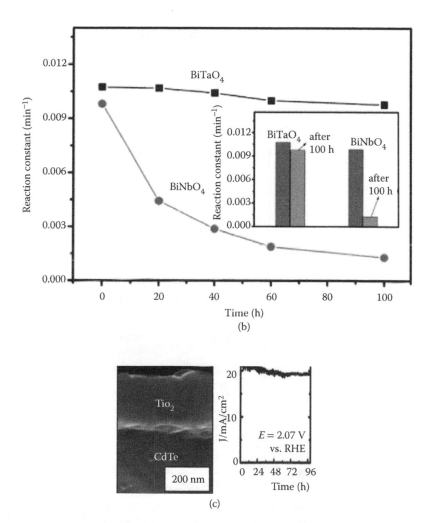

FIGURE 6.3 (a) Visible light photocatalytic degradation of BiTaO$_4$. (b) The photostability of BiTaO$_4$ and BiNbO$_4$ samples. (From Shi, R. et al., *J. Phys. Chem. C*, 114, 14, 6472–6477, 2010; Didden, A. et al., *Int. J. Photoenergy*, 457980, 8 pp., 2015.)

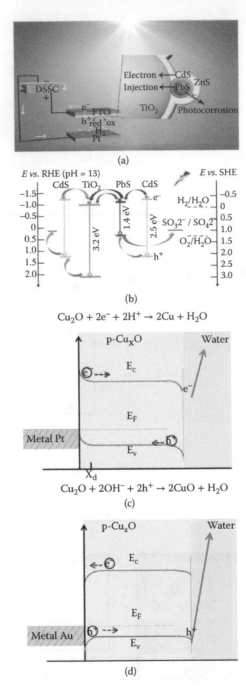

FIGURE 6.4 (a) Electron transfer (ET) and (b) band-energy diagram of the PbS@CdS/ZnS core@shell quantum dots (QDs)–functionalized TiO₂ mesoporous film, through electrophoretic deposition in S_2^-/SO_3^{2-} and Cu_xO photocatalyst (c) on Pt- and (d) on Au-plasmonic metals, respectively. (From Jin, L. et al., *Adv. Sci.*, 1500345, 2016; Wang, C.-M. and Wang, C.-Y., *J. Nanophoton.*, 8, 084095, 2014.)

at the Pt/Cu$_x$O interface, the barrier height at the Pt/Cu$_2$O heterojunction interface is 0.65 eV. There exists a built-in potential for separating the photogenerated electron–hole pairs that consequently suppress the recombination rate. With this situation, the electrons go toward the water and the holes go toward the metal. Thus, H$_2$ and O$_2$ can be generated at the Cu$_x$O/water interface and metal/water interface, respectively. The electrons drifted toward the Cu$_x$O surface. At this time, the electrons accumulate at the Cu$_x$O/water interface. These accumulated electrons contribute to the reduction reaction of Cu$_2$O to Cu on the Pt surface as depicted in Equation 6.32:

$$Cu_2O + 2e^- + 2H^+ \rightarrow 2Cu + H_2O \qquad (6.32)$$

When the film of Cu$_x$O-loaded plasmonic metal Au is deposited on the W surface, it can be seen that the O^{-1}/O^{-2} peak ratio dramatically drops after the photocatalytic water-splitting reaction. Because the work function is <5 eV for the plasmonic metal Au, the electron–hole pairs propagate in the opposite direction (Figure 6.4d). The holes drifted toward Cu$_x$O surface and started accumulating at the Cu$_x$O/water interface. Therefore, Cu$_2$O is oxidized to CuO by photocorrosion, as shown in Equation 6.33 and Figure 6.4b (Wang and Wang 2014):

$$Cu_2O + 2OH^- + 2h^+ \rightarrow 2CuO + H_2O \qquad (6.33)$$

The novel systems, that is, WO$_3$/TiO$_2$ and SnO$_2$/TiO$_2$, create a photocatalytic effect in oxidation of water as shown in Figure 6.5. This system offers the unique advantage of environmental cleaning by superoxide and hydroxyl radicals, generated by UV radiations using Equations 6.34 through 6.39 (Ohko et al. 2001):

$$TiO_2 + h\nu\ (UV) \rightarrow TiO_2^x(e + h) \qquad (6.34)$$

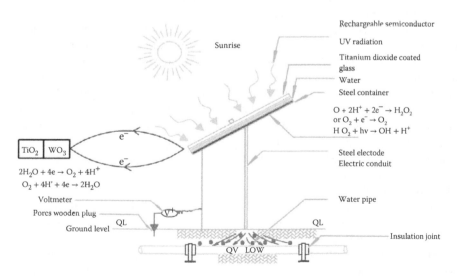

FIGURE 6.5 The photocatalytic degradation of water by using a TiO$_2$ and SnO$_2$ couple. (From Ohko, Y. et al., *J. Electrochem. Soc.*, 148, B24–B28, 2001.)

$$2H_2O + 4h^+ \rightarrow O_2 + 4H^+ \tag{6.35}$$

$$WO_3 + xe^- + xNa \rightarrow Na_x WO_3 \tag{6.36}$$

$$O_2 + e^- \rightarrow {}^*O_2^- \tag{6.37}$$

$$H_2O + h_{vb}^+ \rightarrow {}^*OH + H^+ \tag{6.38}$$

$$2({}^*OH) \rightarrow H_2O_2 \tag{6.39}$$

In the preceding reactions, a TiO_2 and WO_3 couple was used (Figure 6.5) to explain the effect of heterogeneous catalysis. Despite some shortcomings, the system offers potential advantages, such as the use of maintenance-free nonsacrificial anodes that eliminate the need for a periodic replacement of anode, and presents cathodic protection and environmental cleaning effects. These are the true assets of the system.

6.4 MECHANISM OF HETEROGENEOUS ELECTROCATALYSIS

The world economy is greatly affected by the presence of heterogeneous catalyst because most of the chemical manufacturing processes (>90%) need solid/liquid catalyst. Catalysts are also vital in converting hazardous waste into less harmful products such as dye bleaching or example, in the automobile exhaust system. The efficient, controlled, and cost-effective design of catalysts is thus a goal of great importance, one that promises to supersede present trial and error approaches. Catalysis is also one of the scientific disciplines in which even small advances, when based on fundamental research, can have a very significant impact on society. Due to the enormous scale of commercial applications (chemical synthesis, pharmaceutical, automotive, and power generation industries), progress in catalysis can have a positive economic as well as environmental impact. Heterogeneous electrocatalysis is a regime of the core interfacial science and basically involves chemical reactions that take place at the electrochemical interface (solid/polymer, solid/liquid, and gas/solid electrolyte) at the meeting zone of the electrolyte solution and the surface of an electrode in an electrochemical cell. This field holds all promises of the heterogeneous catalysis along with the benefits of the variation in electrode potential and its effect on heterogeneous electron transfer (ET) (non-faradaic electrochemical modification of the catalytic activity). In looking to the current importance of the clean energy that is going hand in hand with scarcity of the traditional fuels, we have great hope in clean hydrogen fuel water. The clean fuel has intimate relationship with clean energy producers such as, solar cells, fuel cells, and heterocatalyst that used in hydrogen generation through water splitting.

Figure 6.6 represents the various fields of heterogeneous electrocatalysis, which is mainly dominated by electricity generation and chemical productions. Many of the promising electrocatalysis-driven technologies are already in use or in an advanced preapplication phase or still on the research and development (R&D) level and are manifested by heterogeneous electrocatalysis. We can name these popular fields as phosphoric acid fuel cells (PAFCs; Ghouse et al. 2000), solid oxide fuel cells (SOFCs; Park

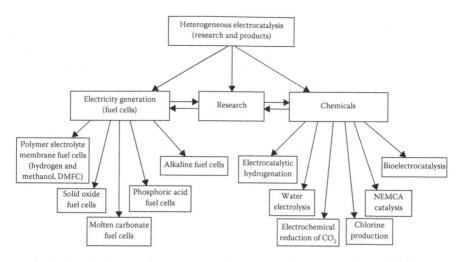

FIGURE 6.6 Practical outcome of heterogeneous electrocatalytic reactions. The significance of research both in electricity generation and in the category of production of chemicals is highlighted. (From Lu, G.-Q. and Wieckowski, A., *Curr. Opin. Colloid Interface Sci.*, 5(1–2), 95–100, 2000.)

et al. 2000), molten carbonate fuel cells (MCFCs; Mathur et al. 1999), water electrolysis on advanced (catalytic) electrode materials (Trasatti 1999), and chlorine chemistry [the use of dimensionally stable anodes (Vijayaraghavan et al. 1999), electrocatalytic hydrogenation (da Silva et al. 1999), electrochemical reduction of CO (Kaneco et al. 1999), and polymer electrolyte membrane (PEM) fuel cells (McNicol et al. 1999; Lee et al. 1999), etc.]. The rate of these reactions is commanded by the electrode surface structure, materials used in the proton exchange membrane, electrode potential (directly or indirectly), and other variables specific to the field of electrochemistry. From the economic and technical perspective, electrode kinetics is of prime importance. The high surface area electrocatalysts are the real gas-phase systems that draw high attention, particularly to develop an understanding of porous electrode structures and the measurements of kinetic parameters for highly dispersed materials. These highly dispersed heterogeneous electrocatalysts exhibit crystallite size effects, which are electronic in nature rather than geometric. Because electrochemical reactions involve charge-transfer reactions through strongly bonded surface species, they are most sensitive to crystallite sizes. In heterogeneous catalysis, the reaction occurs at the electrode and electrolyte interface. For the expected comprehensive gains, both electrocatalysis and gas-phase catalysis may be combined with results of the solid-state studies of surfaces under ideal conditions. Hydrogen molecule evolution on the catalytic surface includes the most spectacular exposed surface reactions in both gas-phase and electrochemical systems (dual-site dissociative chemisorption). Thus, there is a direct proportionality relationship between the faradic current (i) and the electrolysis rate (dN/dt) for electrode reactions (Equation 6.40).

$$\text{Thus, rate of electrolysis (mol/s)} = dN/dt = i/(nF) \text{ (homogeneous)} \quad (6.40)$$

where n is the stoichiometric number of electrons that is consumed during the electrode reaction and F is the Faraday constant. Unlike a homogeneous reaction, which occurs everywhere within the medium at a uniform rate, the heterogeneous reaction occurs only at the electrode–electrolyte interface. Therefore, the reaction rates are usually described in the units of mol/s per unit area:

$$\text{Rate of electrolysis (mol/s/cm}^2) = i/(nFA) = j/(nF) \text{ (heterogeneous)} \quad (6.41)$$

where $j = i/A$ is the current density (A/cm^2). From a thermodynamic point of view, the charge transfer at each electrode is characterized in terms of an equilibrium potential. For a general cathodic charge transfer, the electrode reaction can be schematically represented using Equation 6.42:

$$M + e^- \leftrightarrow M^- \quad (6.42)$$

At equilibrium, the rate of the forward reactions is equal to the rate of the backward reactions so that there is no net chemical reaction observed in absence of the faradaic current. Besides the net zero current at the equilibrium, the balanced faradaic activities are visualized, which can be expressed in terms of the exchange current, i_0, with magnitude equal to either component current: cathodic (representing forward reaction), or anodic (representing backward reaction). Being a heterogeneous reaction, the exchange current is normalized to unit area to provide the exchange current density j_0:

$$j_0 = i_0/A \quad (6.43)$$

In order to have an appreciable cathodic reaction, the actual applied potential at the cathode must be of a higher magnitude (more negative) than that implied by the equilibrium potential. In fact, one drives the reaction by supplying the activation energy electrically; that extra potential is known as overpotential (η). The overpotential has a distinguishing role in deciding the efficiency of the particular electrode–electrolyte system. Besides the electrode reactions, it depends upon mass-transfer effects, chemical reactions preceding or succeeding the electrode charge transfer, and other surface phenomena like adsorption, desorption, and so on.

In contrast to reactions that transpire at the solid/gas interface, heterogeneous reactions allow one to harvest electrons and holes at the catalytic surface where transformation takes place at the electrochemical interface, which may lead to new product formation (in catalytic electrosynthesis) (Lu and Wieckowski 2000). Tafel studied the electrode kinetics of the water-splitting process by using several heterogeneous solid systems that involved hydrogen evolution reactions (HERs) at metallic electrode surfaces through the following primary steps (Trasatti 1992):

$$HA + * \rightarrow e + H* + A \quad \text{(discharge or Volmer step)} \quad (6.44)$$

$$H* + HA \rightarrow e + H_2 + A \quad \text{(atom + ion or Heyrovsky step)} \quad (6.45)$$

$$2H* \rightarrow H_2 \quad \text{(atom + atom or Tafel step)} \quad (6.46)$$

where the asterisk (*) indicates a surface adsorption site and Equation 6.44 is commonly known as the "discharge" or Volmer step, Equation 6.45 is known as the "atom + ion" or Heyrovsky step, and Equation 6.46 is known as the "atom + atom" or Tafel step. Hydrogen evolution occurs through Equation 6.44 followed by either Equation 6.45 or 6.46. These heterogeneous mechanisms are analogous to the mononuclear and binuclear pathways for homogeneous hydrogen evolution, which was identified at 10^{-8} Torr on a highly dispersed Pt-supported graphitic carbon, loaded on clean polycrystalline Pt wire in 96 wt.% H_3PO_4 at 160°C (Trasatti 1992).

Several mechanistic models and hypotheses of the sequence of elementary steps based on atomic-scale experiments have been proposed during the course of five decades for getting a view of oxygen evolution reaction (OER)/HER. The main hindrance for the direct atomic-scale imminent study of OER/HER is the reactive intermediates that largely elude direct spectroscopic observations, classical current–potential–time analyses, and the Tafel slopes. Tafel slopes are mostly used to obtain the mechanistic information and the rate-determining step of the overall half-cell reactions. The mechanistic schemes of heterogeneous electrochemical water splitting based on the experimental results are RuO_2-based electrodes in acidic environments (Lodi et al. 1978; Castelli et al. 1986), Ru–MO_2 mixed oxide films, and Ti/IrO_2 electrodes (Trasatti 2000; Fierro 2007); these are outlined in Equations 6.47 through 6.51 (Damjanovic 1969):

$$* + H_2O \rightarrow * - OH + H^+ + e^- \tag{6.47}$$

$$* - OH \rightarrow * - OH^{\#} \tag{6.48}$$

$$* - OH \rightarrow * - O + H^+ + e^- \tag{6.49}$$

$$* - OH + * - OH \rightarrow * - O + * + H_2O \tag{6.50}$$

$$* - O + * - O \rightarrow O_2 + 2* \tag{6.51}$$

In this mechanistic hypothesis, * denotes a catalytically active surface site and # denotes the radical state of the species. Equations 6.49 and 6.50, running parallel to each other, represent parallel reaction pathways. An additional second chemical step, such as the hypothetical rearrangement of the *–OH species, is shown in Equation 6.48 (Castelli 1986).

Nowadays, the development of heterocatalytic material is in progress toward optimum functionality by practicing rational designs by controlling size, shape, support, composition, and oxidation state effects (Nørskov et al. 2009; Semagina and Kiwi-Minsker 2009; Cuenya 2010; Yuan et al. 2012; Guo et al. 2014). However, it still remains a grand challenge because of the innumerable complexities of the catalytic phenomena (Schlögl and Abd Hamid 2004).

McCrory et al. made a prominent efficiency comparison on the basis of the overpotential for different metal oxide OER electrocatalysts (CoO_x, CoPi, $CoFeO_x$, NiO_x, $NiCeO_x$, $NiCoO_x$, $NiCuO_x$, $NiFeO_x$, and $NiLaO_x$) with respect to IrO_x in basic and acidic mediums to achieve a fixed current, that is, 10 mA/cm² (McCrory et al. 2013). The activity of the electrocatalysts was evaluated based on criteria such as stability

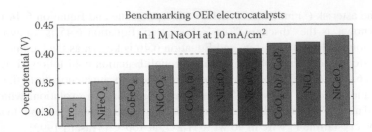

FIGURE 6.7 Efficiency as a function of overpotential for different metal oxide oxygen evolution reaction (OER) electrocatalysts (CoO_x, CoPi, $CoFeO_x$, NiO_x, $NiCeO_x$, $NiCoO_x$, $NiCuO_x$, $NiFeO_x$, and $NiLaO_x$) with respect to the reference IrO_x, to achieve a fixed current, that is, 10 mA/cm² in aqueous 1 M NaOH. (From McCrory, C.C.L., *J. Am. Chem. Soc.*, 135, 16977–16987, 2013.)

and Faradaic efficiency for electrodeposited oxygen-evolving electrocatalysts, under conditions relevant to an integrated solar water-splitting device in alkaline and acidic solution. The device has been investigated and every nonnoble metal system attained 10 mA/cm² current densities at operating overpotential between 0.35 and 0.43 V as presented in Figure 6.7.

6.5 MECHANISM OF HOMOGENEOUS MOLECULAR CATALYSIS

Homogeneous catalysts function in the same phase as the reactants, but the mechanistic principles summoned for heterogeneous catalysis are usually applicable for them. Normally homogeneous catalysts are dissolved in a solvent with the substrates. Homogeneous molecular catalysts are well known for their high activity and tunability, but their solubility and limited stability restrict their use in practical applications. Homogeneous catalysts may be classified into two categories: organometallic catalysts and acid base catalysts, and they can facilitate many chemical reactions. A true catalytic system for efficient water splitting is operated with four consecutive proton-coupled electron transfer (PCET) steps to generate oxygen and two-electron reduction of water to hydrogen, but overall water splitting with high turnover number (TON) and turnover frequency (TOF) is yet to be achieved. Electrochemical splitting of water involves two concurrent catalytic half-cell reactions: (1) water reduction catalysis (WRC) and (2) water oxidation catalysis (WOC). Various homogeneous complexes have been investigated for both half reactions in a homogeneous environment at photo- or electrochemical conditions. Metal complexes synthesized for homogeneous WOC/WRC are amenable to be spectroscopically/crystallographically characterized up to the atomic resolution. Moreover, ease of synthetic variations to design-specific work oriented metal complex structures to investigate the relation between structure and reactivity. These experimental advantages lead to a strong foundation for better understanding of fundamental mechanistic aspects of WOC/WRC and their respective modes of redox-potential leveling. Compared with the increasing number of studies on heterogeneous water-splitting systems with integrated WOC and WRC catalysis, a visible light–driven homogeneous WOC/WRC catalyst for overall water splitting has yet to be described well. The bioinspired homogeneous catalysts are exceptionally promising as due to their manifold structural tenability with self-repair options, robustness, stability, and economically viability, they can render AP setups close to the high performance of natural photosystems.

Most of the investigations are directed to the WOC part of water splitting and negligible work is done for WRC molecular catalysts. And the most challenging task is a search for the "missing chemical links" between WRCs and WOCs, namely common photosensitizers and electron shuttles. Cutting-edge *in situ* studies of the aforementioned reaction pathways will provide the mechanistic pathways to bring life to the concept of homogeneous water splitting via the systematic development of recyclable and robust ET agents (Patzke et al. 2014). Recent studies facilitate the major breakthroughs in this area and disclosed that the dimeric or higher-order assemblies of multiple metal centers are not essential for WOC/WRC, but mononuclear systems can work fine. The easier structural tailoring can allow more rational synthesis of the simpler mononuclear systems/model systems with respect to the redox-potential leveling, structure–function analyses, and mechanistic studies. To meet the efficiency of the natural photosynthesis process, artificial devices might be designed by controlled ligand decomposition. Immobilization of molecular complexes on surfaces may increase drastically catalytic performance and lifetime due to suppression of bimolecular decomposition. Stabilized matrices or surfaces of molecular complexes in collaboration with self-assembly and self-repairing quality make the homogeneous catalyst relevant and important in the development of efficient catalysts for water splitting. Several mechanistic schemes are being proposed for these types of reaction processes. Clearly, a four-electron process oxygen evolution is more complex compared with a two-electron hydrogen evolution process and both involve several surface-adsorbed intermediate stages. Although both WRC and WOC processes contribute to the voltage and efficiency losses, the electrocatalysis of water splitting in principle refers to both. Homogeneous catalysis of water splitting has involved systems which liberate either hydrogen or oxygen (or, in rare cases, both). Some of the examples are discussed in this chapter.

6.5.1 Tetramanganese–Oxo Cluster Complex for O_2 Generation

The most accepted example of homogeneous catalysis is photosynthetic dioxygen evolution in photosystem II; the tetramanganese–oxo cluster complex is referred to as the active site and responsible for efficient catalytic water splitting and rapid evolution of oxygen (Wydrzynski and Satoh 2005). Structural arrangement and functioning of this homogenous tetramanganese–oxo catalyst are considered the heart of these reactions. This is a tiny metal oxide cluster composed of four manganese ions, one calcium ion, and five oxygen atoms (Mn_4O_5Ca/Sr), which can take five oxidation states (S_0–S_4) by the redox of the Mn ions and stores four redox equivalents, as shown in Figure 6.8a. X-ray absorption spectroscopy of the Mn_4O_5Ca/Sr cluster of the WOC of photosystem II (PSII) provides information about the oxidation states and cluster geometry in samples that are active in O_2 evolution. Advances in the Kok cycle from S_0 to S_1 and from S_1 to S_2 show clear evidence of Mn oxidation, whereas the advance from S_2 to S_3 does not show any change. Significant changes in the

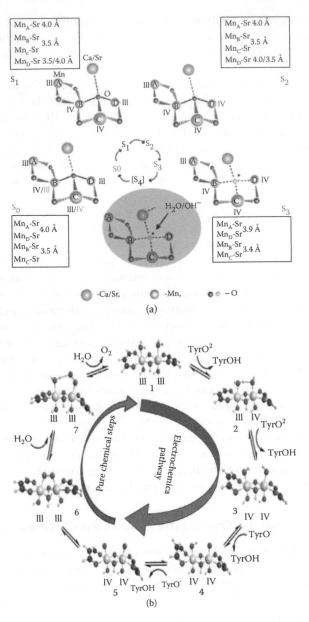

FIGURE 6.8 (a) Cluster arrangements of four Mn atoms bridged by O atoms, exhibiting three di-μ-oxo bridged Mn–Mn pairs in Mn$_4$O$_5$Ca/Sr. (b) Optimized structures of Mn$_4$O$_5$Ca/Sr for the water-splitting mechanism. 1, 2, 3, 4, and 5, numbers indicate the electrochemical pathway and the 6 and 7 numbered structures are generated from purely chemical steps. While small white spheres represent hydrogen atoms and large gray spheres represent manganese atoms. Carbon and oxygen atoms are represented by small dark black and gray spheres that directly connected to Mo atoms. [(a) From Sauer, K. et al., *Photosynth. Res.*, 85, 73–86, 2005. (b) From Pushkar, Y. et al., *Proc. Natl. Acad. Sci. U S A*, 105, 1879–1884, 2008. (c) From Busch, M. et al., *Phys. Chem. Chem. Phys.*, 13, 15069–15076, 2011.]

structure of the complex have been detected especially in the Mn–Mn and Ca(Sr)–Mn bond distances, on the S_2-to-S_3 and S_3-to-S_0 transitions. These results implicate the involvement of at least one common bridging oxygen atom between the Mn–Mn and Mn–Ca(Sr) atoms in the O–O bond formation (Sauer et al. 2005). These conclusions, originally based on x-ray absorption near edge structure (XANES) measurements, have been fortified by the results of x-ray emission (XES) studies. The application of extended x-ray absorption fine structure (EXAFS) analysis has established the importance of di-μ-oxo bridged pairs of Mn atoms in the cluster. The catalytic mechanism of the water oxidation in the WOC is still not fully understood.

But Busch et al. used complete-active-space self-consistent-field (CASSCF) theory to investigate a water-splitting reaction mechanism on a model manganese dimer electrocatalyst as demonstrated by Figure 6.8b. His group examined the occupied orbital status of the manganese and oxygen atoms involved in water splitting and obtained information regarding the electronic details and contributions. Apart from the resting state of the catalyst, a singly occupied ring p orbital on the ligands can be seen throughout the catalytic cycle. We also hypothesize that this ring p population may be a reason for the out of plane orientation of the respective ligands. In structure 3 with two Mn(IV) atoms, an out of plane bend of both ligands in opposite directions was observed. This became more pronounced upon the formation of Mn(IV) oxidation states on two manganese atoms in structure 5. However, with the formation of the oxygen–oxygen bridge between the Mn(III) atoms on structure 6, these ligands tend to bend in the same direction forming a butterfly-like structure, which later distorts with the detachment of the oxo bridge from one manganese atom on structure 7. Radical properties of a Mn–O moiety have also been observed in this direct coupling pathway. The p orbital contribution to the active space is observed with the formation of the Mn(IV)–O group and is much more obvious in structure 5 with two Mn(IV)–O. For structure 5 with two Mn(IV) states, we have observed the prominent contribution of p orbitals of both oxo groups to the singly occupied active space. This then yields the oxo bridge in structure 6. The p* orbitals were seen to be occupied in the oxygen–oxygen bond in structures 6 and 7 (Busch et al. 2011).

6.5.2 RUTHENIUM COMPLEXES FOR O_2 GENERATION

Meyer et al. first reported an oxo-bridged polypyridyl dinuclear ruthenium complex *cis*, *cis*-(bpy)$_2$(H$_2$O) RuORu(OH$_2$) (bpy)$_2$]$^{4+}$ (bpy: 2,2′-bipyridine), so called "blue dimer" as a homogeneous WOC molecular catalyst for water splitting (Gersten et al. 1982). The "blue dimer" evolves oxygen gas in the presence of an excess sacrificial oxidant such as (NH$_4$)$_2$[CeIV(NO$_3$)$_6$] [ceric ammonium nitrate (CAN)] or Co(III) with a maximum yield and TON of 87% and 13.2%, respectively (Collin and Sauvage 1986). The low oxidation state Ru^{III}–O–Ru^{III} form of the ruthenium blue dimer undergoes oxidative activation by PCET in which stepwise loss of electrons and protons occurs. PCET is essential because it allows for the buildup of multiple oxidative equivalents at a single site or cluster without building up positive charge (Huynh and Meyer 2007; Liu et al. 2008a). Figure 6.9a demonstrates the comparative photocurrent density versus applied potential and O_2 evolution rate versus time for

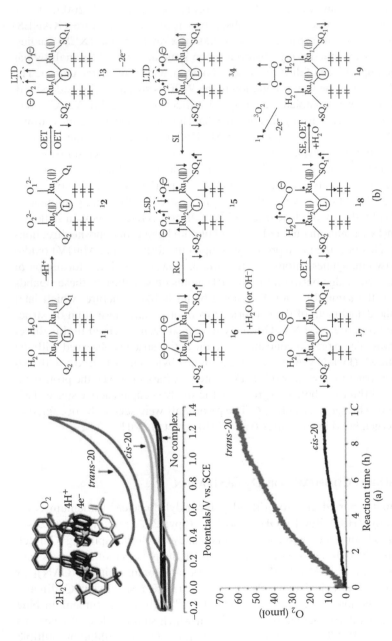

FIGURE 6.9 (a) Catalytic dioxygen evolution by the complexes at pH 4 and dioxygen evolutions by (*trans*-[Ru(trpy)(pad)(OH₂)]⁺ = *trans*-20⁺ and *cis*-[Ru(trpy)(pad)(OH₂)]⁺ = *cis*-20⁺) in the electrochemical method at pH 4 and potential 1.40 V (vs. saturated calomel electrode [SCE]). (b) Radical coupling mechanism for O–O bond formation in water splitting by [[Ru^{II}₂(btpyan)(Bu₂Q₂)(OH)₂]²⁺ (btpyan: 1,8-bis(2,2′:6′,2″-terpyrid-4′-yl) anthracene; Bu₂Q: 3,6-di-*tert*-butyl-1,2-benzoquinone)], proposed by Tanaka et al. L: btpyan; Q (Sq): 3,6-di-tert-butyl-1,2-benzo(semi)quinone. (From Tanaka, K. et al., *Proc. Natl. Acad. Sci. U S A*, 109, 15600–15605, 2012.)

the mononuclear complexes, that is, *trans*-[Ru(trpy)(pad)(OH$_2$)]$^+$ = *trans*-20$^+$ and *cis*-[Ru(trpy)(pad)(OH$_2$)]$^+$ = *cis*-20$^+$, in the electrochemical method at pH 4 and potential 1.40 V (vs. saturated calomel electrode [SCE]).

Noteworthy catalytic activity of [Ru$_2$(btpyan)(Bu$_2$Sq)$_2$(OH)$_2$]$^{2+}$ compels researchers performing theoretical studies to expose the reaction mechanism of the water oxidation and the most crucial step, that is, the O–O bond formation (Tanaka et al. 2012). The most conceivable mechanism of the O–O bond formation is based on the radical coupling (RC) between the two oxyl radicals, stimulated by spin inversion (SI) on the Ru(III) centers (Figure 6.9b) (Tanaka et al. 2012). Acid–base equilibrium of the aqua ligands coupled with an ET to the quinone ligands gives the tetraradical complex [Ru^{II}$_2$(btpyan)(Bu$_2$Sq)$_2$(O$^-$)$_2$]. O–O bond formation may be suppressed in the tetraradical species because of a local triplet diradical (LTD) state of the two oxyl radicals (Hund's rule). The transfer of two electrons from the two closed-shell Ru(II) centers to open-shell Ru(III) produces [[Ru^{II}$_2$(btpyan)(Bu$_2$Sq)$_2$(O$^-$)$_2$]$^{2+}$ with six unpaired electrons. The computational study exhibits that the unpaired electron of the oxyl radical, present in aforementioned species, couples ferromagnetically with the adjacent Ru(III) center. The heavy atom effect associated with the Ru(III) facilitates the SI, which encourages simultaneous inversion of the magnetically coupled spin on the neighboring O$^-$. As a result, the two oxyl radicals produce a local singlet diradical (LSD) pair that readily couples to form an O–O bond (RC) to give a peroxide species (O$_2$$^{2-}$). Activity of the catalytic system is governed by the pathway of water activation to form an O–O bond. The bulky btpyan and dioxolene ligands cause a strong steric hindrance, which enforces the substitution of the one oxygen atom of the peroxide ligand by water, as a rate-limiting step in the catalytic cycle. After this nucleophilic substitution, one electron is transferred from the terminal oxygen to the Ru(III) center that offered a superoxide anion (O^{2-}) to the species. Before another ET and H$_2$O attack, a spin exchange (SE) between the Ru(III) center and the semiquinone ligand should occur, and thus a release of triplet dioxygen will go hand in hand with the recovery of the initial catalyst [Ru$_2$(btpyan)(Bu$_2$Sq)$_2$(OH)$_2$]$^{2+}$. The mononuclear counterpart does not exhibit catalytic activity for water oxidation because of the absence of intermolecular RC due to strong electrostatic repulsion between the anionic oxyl radicals.

6.5.3 Manganese Porphyrin Dimer Complexes for O$_2$ Generation

The paradigm of the "blue dimer" was followed by the unique class of the homogeneous Mn-based water oxidation catalysts, initially discovered by Naruta et al. in 1994 (Naruta et al. 1994; Shimazaki et al. 2013). Electrolysis of a pincer-shaped Mn porphyrin dimer 2 in CH$_3$CN containing 5% H$_2$O at the potential range of +1.2 to +2.0 V (vs. Ag/Ag+) results in the evolution of O$_2$ gas with a current efficiency of 5%–17% depending on the applied potential and a TON of 9.2 at a maximum potential.

6.5.4 Dinuclear Co^{III}–Pyridylmethylamine Complex for O$_2$ Generation

A bis-hydroxo-bridged dinuclear Co^{III}-pyridylmethylamine complex (1) was synthesized by Ishizuka et al. that can act as a homogeneous catalyst for visible light–driven water oxidation with persulfate (S$_2$O$_8^{2-}$) as an oxidant and [Ru^{II}(bpy)$_3$]$^{2+}$ (bpy = 2,2′-bipyridine) as a photosensitizer, as shown in Figure 6.10a. Complex (1) offers a high quantum yield (44%) with a large TON (TON = 742) for O$_2$ formation without forming Co-oxide NPs (nanoparticles). DFT calculations suggested that complex (1) experiences PCET oxidation as the rate-determining step in the water oxidation process that ultimately forms a putative dinuclear bis-μ-oxyl Co^{III} complex. Prior to O$_2$ evolution, formation of an intramolecular O–O bond in the two-electron-oxidized bis-μ-oxyl intermediate was confirmed in aforementioned catalytic water oxidation by complex (1).

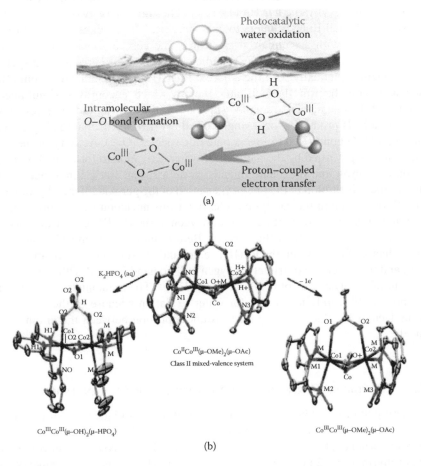

(a)

(b)

FIGURE 6.10 (a) Dinuclear Co^{III}/Co^{III} pyridylmethylamine complex for homogeneous photocatalytic water oxidation with a through proton-coupled electron transfer (PCET) and (b) conversion of dinuclear and trinuclear Co(III)Co(III) complexes from [LCo^{II}(μ-carboxylato) bis(μ-methoxo)Co^{III}L](ClO$_4$)$_2$ (L = N-methyl-N,N-bis(2-pyridyl methyl) amine) mixed-valence complex for water splitting. (From Ishizuka, T. et al., *Inorg. Chem.*, 55, 1154–1164, 2016; Luo, J. et al., *Inorg. Chem.*, 50, 6152–6157, 2011.)

Another example of cobalt complex is the dinuclear Co(II)Co(III) mixed-valence complex, [LCo^{II}(μ-carboxylato)bis(μ-methoxo)Co^{III}L](ClO$_4$)$_2$, containing μ-methoxo and μ-carboxylato bridging ligands with the tridentate ligand N-methyl-N,N-bis(2-pyridylmethyl) amine (L), which has been employed for water oxidation. Oxidation of this Co(II)Co(III) complex leads to formation of the corresponding Co(III) Co(III) complex, that is, dinuclear and trinuclear μ-hydroxo Co(III) complexes in the presence of phosphate anions and absence of methanol, respectively—suggesting the presence of an additional bridging ligand to stabilize the Co^{III}bis(μ-hydroxo) Co^{III} fragment. Moreover, the ability of the mixed-valence Co(II)Co(III) complex versus the dinuclear and trinuclear μ-hydroxo Co(III)/Co(III) complexes has been investigated for the electrocatalytic oxidation of the water (Figure 6.10b) (Luo et al. 2011). Inadequate water oxidation catalytic ability for these Co(III)/Co(III) systems compared with a multinuclear Co cluster and/or presence of O-rich ligands results in efficient molecular Co-based water oxidation catalysts for the generation of oxygen.

6.5.5 HOMOGENOUS METAL COMPLEX FOR HYDROGEN GENERATION THROUGH WATER SPLITTING

Synthetic complexes of nickel (DuBois and DuBois 2009; Helm et al. 2011), cobalt (Connolly and Espenson 1986; Jacobsen et al. 2008; Jacques et al. 2009; Stubbert et al. 2011), iron (Liu and Darensbourg 2007; Gloaguen and Rauchfuss 2009; Kaur-Ghumaan et al. 2010), and molybdenum (Chao and Espenson 1978; Appel et al. 2005; Karunadasa et al. 2010) have been developed recently as electrocatalysts for the production of hydrogen. It was observed that the Co–diglyoxime complexes have shown hydrogen generation capacity from protic solutions at relatively modest over-potentials. The field has attracted much researcher interest and various experimental (Schrauzer and Holland 1971; Baffert et al. 2007; Lazarides et al. 2009; Dempsey et al. 2010; Solis and Hammes-Schiffer 2011; Bhattacharjee 2012) and theoretical (Muckerman and Fujita 2011; Solis and Hammes-Schiffer 2011a) investigations have been made to trace the mechanism of hydrogen formation.

Besides all of the aforementioned works, Marinescu et al. reported a new homogenous cobalt(I) complex that reacts with tosylic acid to evolve hydrogen with a minimal driving force of 30 meV/Co, because protonation of Co^I produced a transient Co^{III}–H complex that decayed by second-order kinetics with an inverse dependence on acid concentration. Their kinetic study suggested that Co^{III}–H produces hydrogen by two competing pathways: a slower homolytic route involving two Co^{III}–H species and a dominant heterolytic channel in which a highly reactive Co^{II}–H transient is generated by Co^I reduction of Co^{III}–H. Both involved protonation of a Co^I complex to form Co^{III}H. H$_2$ evolution can occur via protonation of Co^{III}–H or upon bimolecular combination of two Co^{III}–H species or Co^{III}–H can be reduced by Co^I to a mixture of Co^{II}–H and Co^{II}, as illustrated by Figure 6.11a.

Another example of homogeneous catalyst is tris-[1-(4methoxyphenyl)-2-phenyl-1,2-ethylenodithiolenic-S,S'] tungsten (Figure 6.11b) reported by researchers at the University of Athens (Katakis et al. 1994). This tris-dithiolene-W catalyst possesses a unique and highly desirable characteristic of formation of both hydrogen and oxygen from water without consumption of any (sacrificial) donor (Heyduk and Nocera

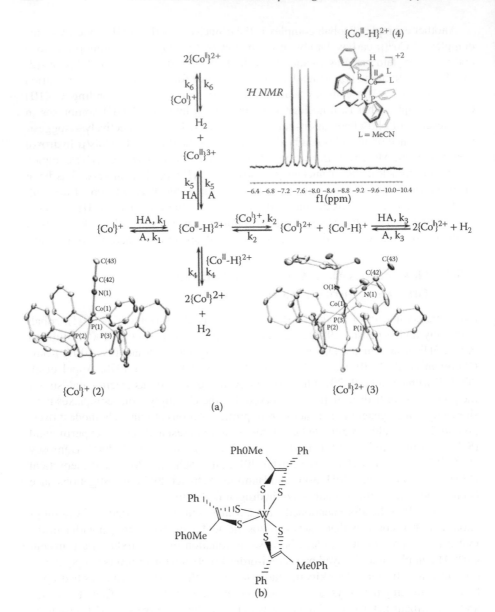

FIGURE 6.11 (a) Reaction pathways for the evolution of H_2 from the reaction of a Co^I complex with acid (HA). Models based on the x-ray crystal structures of Co^I (2) and Co^{II} (3) triphos complexes appear in the lower portion of the figure. Thermal ellipsoids are drawn at 50% probability; hydrogen atoms, out-of-sphere anions, and solvent molecules are omitted for clarity. The ^{1}H-NMR (nuclear magnetic resonance) spectrum and structural model of the Co^{III}–H (4) intermediate formed in the reaction of two with $TsOH \cdot H_2O$ appear in the upper right. (b) Photocatalyst catalyst (PCC) tris-[1-(4methoxyphenyl)-2-phenyl-1,2-ethylenodithiolenic-*S,S'*-tungsten. (From Marinescu, S.C. et al., *Proc. Natl. Acad. Sci. U S A*, 109, 15127–15131, 2012; Autrey, T. et al., *Fuel Chem.*, 47, 2, 752–754, 2002.)

2001). Water is reduced to hydrogen and oxidized to oxygen in a sequential pathway. The energy storage efficiencies of these trisdithiolenes complexes have been reported as 7%–11%, near to the economic break point on the path of solar production of hydrogen. After studying the stability and efficiency factors of the various derivatives and isomers of aforementioned *tris*-dithiolenes complex at various angles, the mechanism given in Equations 6.52 through 6.56 leads to hydrogen and oxygen formation:

$$H_2O - M - MV^{2+} + h\nu \rightarrow (H_2O - M - MV^{2+})* \tag{6.52}$$

$$(H_2O - M - MV^{2+})* \rightarrow (H_2O - M^+ - MV^{+\cdot}) \tag{6.53}$$

$$(H_2O - M^+ - MV^{+\cdot}) \rightarrow (HO - MH - MV^{+\cdot}) \tag{6.54}$$

$$(HO - MH - MV^{+\cdot}) \rightarrow H_2 + OM + MV^{2+} \tag{6.55}$$

$$OM + OM \rightarrow O_2 + 2\,M \tag{6.56}$$

where M = tris-[1-(4methoxyphenyl)-2-phenyl-1,2-ethylenodithiolenic-S,S'] tungsten, MV^{2+} = reversible electron acceptor methylviologen, and $h\nu$ = sunlight radiation.

6.6 BRIDGING THE GAP BETWEEN HETEROGENEOUS ELECTROCATALYSIS AND HOMOGENEOUS MOLECULAR CATALYSIS

Basically, electrochemistry is heterogeneous by nature as an ET between two phases usually occurs between an electrolyte (liquid) and an electron conductor (solid). Thus, most electrocatalytic processes such as hydrogen oxidation/evolution occuring at a Pt electrode or oxygen reduction reactions are typical heterogeneous processes. On the other hand, in molecular catalysts such as organics, organometallic, acid–base, enzyme-mediated systems and salt as the catalysts, electrons shuttled between an electrode and the catalyst (the enzyme) are the typical homogeneous catalysts for photoelectrochemical cell/electrochemical cell/fuel cells/mediator. Actually, the sequence of reactions that involve the same phase of the catalytic system as the reactant are known as a homogeneous process. But the products can belong to different phases than the catalyst and the reactant—gases, liquids, or solids. Therefore, there is no existence of the term homogeneous catalysis, when we consider the reaction as a whole including all three components of the reaction, that is, reactant (water), catalyst (solid/liquid), and product (gaseous/solid/liquid). Or we can say that electrocatalysis/photoelectrocatalysis can be combined with the advantages of both homogeneous and heterogeneous catalyses. If the quantification of the extent of modification of the electrode surface in terms of increase in current density or increased rate of H_2/O_2 evolved can be done with performed experiments/designed experiments, then these values will be dictated to the leading contributors: (1) homogeneous, (2) synergy/mixed, and (3) heterogeneous.

Thus, we will study the synergy/mixed catalyst in this section in place of pure homogeneous and heterogeneous. Table 6.1 delivers the comparative behavioral information of homogeneous and heterogeneous catalysts.

TABLE 6.1
Comparative Behavior of Homogeneous and Heterogeneous Catalysts

Serial No.	Characteristic Behavior	Homogeneous	Heterogeneous
1.	Phase	Single phase: same as reactant	At least two phases: different from reactant
2.	Solvent	Required	Not required
3.	Active center	All metal atoms: No phase boundary and good contact with reactant. Might provide more reactive and attacking species	Only a few surface atoms: phase boundary is always there
4.	Selectivity	High: because no phase boundary between reactant and catalyst	Poor: due to the presence of the phase boundary between reactant and catalyst
5.	Concentration	Low	High
6.	Diffusion problem	Absent	Present: mass transfer controlled reactions
7.	Reaction temperature and stability	Mild: 50°C–200°C; decomposed at high temperature <100°C	Severe: >250°C; highly stable at reaction conditions
8.	Applicability	Limited	Wide
9.	Recyclable	Difficult to recycle: separation of the product and catalyst difficult because temperature required for distillation can destroy the catalyst	Easy to recycle: separation of the product are easy
10.	Effectiveness	High: high TON and TOF	Poor: not very high TON and TOF
11.	Synthesis	Easily tunable: can be designed as complex structures for specific jobs	Poor degree of synthetic control
12.	Characterization	Amenable to complete spectroscopic characterization	Challenging to be studied
13.	Cost	Ligand cost dominates the catalyst cost	Depend upon the metal used
14.	Examples	Metal organometallic complex, organic molecules, salt and acid–base, as molecular catalyst	Usually metal, metal oxides, sulfides, nitrides or their modified versions

Notes: TOF, turnover frequency; TON, turnover number.

Heterogeneous catalysts for water splitting were known much earlier than their homogeneous analogs due to their thermodynamic stability, recyclability, high activity, in some cases, longevity, easy large-scale production and use of solid-state materials for commercial devices, and so on (Mills 1989). Moreover, they can be modified to afford high surface area bulk materials with nanoscale features that further enhance the activity (Nakagawa et al. 2009; Gorlin and Jaramillo 2010). Heterogeneous catalysis always involves the mass transfer and diffusion process among the reactants, products, and solid catalysts. Moreover, heterogeneous catalysts with their active sites are more complex and relevant mechanisms are not well understood for most cases. In usual fashion, homogeneous catalytic reactions have well-defined active sites that provide relatively high activity and selectivity. However, most of the homogeneous catalytic processes meet with difficulty in large-scale applications including recycling of the catalyst, stability, and handling in industrial processes. Furthermore, homogeneous catalysis research is struggling with a long-standing and confusing ambiguity as to whether a given catalyst is homogeneous or simply a precursor to a catalytically active heterogeneous material. Unfortunately, the field of homogeneous and heterogeneous catalyses had been developing independently, possibly because the scientists of these fields have been working in separated disciplines: homogeneous catalysis researchers mostly belong to organic chemistry, while heterogeneous catalysis researchers major in chemical engineering or inorganic or physical chemistry. Although strong scientific temperament and expertise are necessary for the aforementioned type of catalysis, devoted efforts of researchers are also required to bridge the gap between homogeneous and heterogeneous catalyses. To take advantage of both homogeneous and heterogeneous catalyses, we need to develop heterogeneous catalysis as solid supports on which the immobilization of molecular catalysts can take place via chemical bonding or physical adsorption. By means of this approach, the immobilized molecular catalysts generally exhibit lower performance than their homogeneous complements, which is the result of leaching of catalysts because of unstable linkage between the catalyst and support that may occasionally cause the low efficiency of the device. A better understanding of the relationship between homogeneous and heterogeneous catalyses is required for appropriate design and preparation of advanced catalysts with the merits of both types of catalysis. However, an activation model of enzyme catalysis falls somewhere at the broader line between the homogeneous and heterogeneous catalyses. Therefore, it is required to get acquainted with the working of enzyme catalysis by relating enzyme catalysis with traditional chemical catalysis. Strategies are being developed for bridging the gap between highly active but less studied homogeneous catalysis and highly studied but less active heterogeneous catalysis.

- Acid–base cooperative catalysis by designing solid surfaces with organo-functional groups
- Catalytic reactions in or by ionic liquids
- Heterogeneous catalysis with organic–inorganic hybrid materials
- Homogeneous asymmetric catalysis using immobilized chiral catalysts
- Asymmetric catalysis with organometallics
- Catalysis in and on water
- Green chemistry strategy: fluorous catalysis
- Emulsion catalysis: interface between homogeneous and heterogeneous catalyses
- Identification of binding and reactive sites in metal cluster catalysts
- Catalysis in porous material–based nanoreactors
- Asymmetric phase-transfer catalysis in organic synthesis
- Catalysis in supercritical fluids
- Hydroformylation of olefins in aqueous–organic biphasic catalytic systems
- Enzyme catalysis in reverse micelles

On the basis of the phases of reactants and catalysts and their method of preparation, we can broadly divide these interface catalysis processes into the following biphasic classes.

6.6.1 SOLID–LIQUID

This class of catalysts deals with molecular catalysts (in liquid form), immobilized onto solid materials, including bifunctional acid–base cooperative catalysis by organofunctionalized solid surfaces, functional organic–inorganic hybridized mesoporous materials, immobilized chiral catalyst on soluble polymers (linear polymers, dendrimers, and helical polymer), highly efficient homo- and heterogeneous chiral organometallic catalysts, and porous material–based nanoreactors exhibiting advantages from both homogeneous and heterogeneous catalyses.

6.6.2 SOLID–GAS

This describes metal cluster catalysis using solid–gas biphasic catalytic reactions.

6.6.3 LIQUID–LIQUID SYSTEM

Liquid–liquid biphasic systems involve ion-supported liquid catalysts, ionic liquid–supported catalytic reactions at room temperature, fluorous catalysts, supercritical fluids as media for catalytic reactions, metal-catalyzed reactions carried out in water, emulsion catalysis using aqueous–oil/organic biphasic reactions, various types of chiral phase-transfer catalysts, and reverse micelles catalysis with macroscopic homogeneous but microscopic heterogeneous biocatalysis/enzyme catalysis.

Out of the three main classes, some important interface catalysts are discussed in detail in the following.

6.6.4 FLUOROUS CATALYSTS

To pursue a more economical chemical process research design, more catalytic systems combine the high selectivity and mild reaction conditions of homogeneous catalysis with the easy catalyst and product separation of heterogeneous catalysis. Unique properties of the perflourocarbon fluids attracted researchers to use them as solvents in biphasic systems. This novel concept is based on the limited miscibility of partially or fully fluorinated compounds with nonfluorinated compounds, utilized for stoichiometric and catalytic chemical transformations. A fluorous biphasic system (FBS) contains one fluorous phase (fluorocarbon: mostly perfluorinated alkanes, ethers, and tertiary amines) containing a dissolved reactant or catalyst and another phase with common organic or nonorganic solvent that possesses limited or no solubility in the fluorous phase. The process required an FBS compatible reagent or catalyst that contains enough fluorous moieties, which are preferentially soluble only in the fluorous phase. The most effective fluorous moieties are linear or branched perfluoroalkyl chains with high carbon number; they may also contain heteroatoms. The chemical transformation may occur either in the fluorous phase or at the interface of the two phases. The ability to separate a catalyst or a reagent from the products completely at mild conditions (Figure 6.12a) could lead to commercial application of these homogeneous catalysts

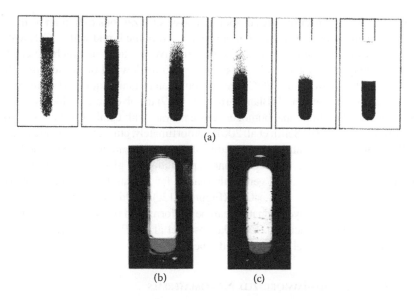

(a)

(b)   (c)

FIGURE 6.12 (a) Hand warm phase separation of *n*-hexane/toluene/perflouromethyl cyclohexane mixture by cooling the mixture from 36°C to 25°C. (b) An ionic liquid/supercritical CO_2 (scCO_2) biphasic solvent system at 210 bar (the ionic liquid is *N*,*N*-butylmethylimidazolium hexafluorophosphate) and (c) poly(ethylene glycol) under scCO_2 at 172 bar and 40°C. In both biphasic systems, the dye Nile Red can be seen visually to be entirely in the lower (poly(ethylene glycol) [PEG] or ionic liquid) phase. (From Horváth, I. and Rába, J., *Science*, 266, 72–75, 1994; Heldebrant, D.J. and P.G. Jessop, *J. Am. Chem. Soc.*, 125, 5600–5601, 2013.)

or reagents by using the environmentally benign processes. Fluorous metallic compounds, fluorous palladacycle, fluorous Pincer ligand–based catalysts, fluorous immobilized NPs, fluorous palladium–*N*-heterocyclic carbene (NHC) complexes, fluorous phosphine–based palladium, Grubbs' fluorous silver, fluorous Wilkinson catalyst, and many more are used to perform reactions like asymmetric aldol condensation, Morita–Baylis–Hillman reactions, asymmetric Michael addition reactions, catalytic oxidation, catalytic acetalization, catalytic condensation, catalytic asymmetric fluorination, and so on (Flannigan 2002).

6.6.5 Liquid Poly(ethylene Glycol) and Supercritical Carbon Dioxide: A Benign Biphasic Solvent System

Although homogeneous catalysts are difficult to recycle, biphasic catalysis is an important immobilization technology for making homogeneous catalysts recyclable, which requires liquids of different properties (polarity), that is, one polar solvent (dissolves and retains the catalyst) and another nonpolar solvent (dissolves the products). As a nonpolar solvent, nonvolatile organic carbon (VOC) solvents are preferred over volatile or halogenated solvents. The coupling of nonpolar VOC solvents with liquid water and supercritical CO_2 (scCO_2) is important for biphasic homogeneous catalysis since the H_2O/scCO_2 system is the most effective for the water-soluble polar catalytic phase (Bhanage et al. 1999a, 1999b). The catalyst-bearing phase in a biphasic solvent system is preferably nonvolatile because this eliminates evaporative losses (hazardous to human health and environment) and allows the extraction of products from the liquid with scCO$_2$ without concomitant extraction of the solvent. For the effective reactions in which the catalyst, substrate, or reagent is water-insoluble or sensitive to the low pH of the aqueous phase (Bonilla et al. 2000). Recently, VOC-free combination of ionic liquids (ILs) and scCO$_2$ (Figure 6.12b) were discovered (Blanchard et al. 1999) to solve these problems, and this allowed the use of hydrophobic homogeneous catalysis with catalyst recycling (Brown et al. 2001; Liu et al. 2001; Sellin et al. 2001). Unfortunately, currently available interleukins (ILKs) are expensive and the knowledge of their environmental impact and human health is limited. Therefore, non-VOC, that is, poly(ethylene glycol) (PEG) is treated as an inexpensive, nonvolatile, and gentle alternative for the catalyst-bearing nonvolatile phase in biphasic catalysis with scCO$_2$ (Figure 6.12c) (Heldebrant and Jessop 2003). Homogeneous-catalyzed hydrogenation can be performed in the molten PEG, followed by extraction of the product by scCO$_2$ (Jacobson et al. 1999). The catalyst-containing PEG phase left in the vessel can be reused repeatedly.

6.6.6 IONIC LIQUID–IMMOBILIZED NANOMATERIALS

IL could be employed in nanomaterial synthesis, organometallic synthesis, and catalysis, and electrochemistry immobilized superfine NPs are macroscopically homogeneous and microscopically heterogeneous. To relate homogeneous and heterogeneous catalyses, nanometal particles act as a reservoir for the catalytically active metal complex. On the contrary, the porous material supported ILs and ILs confined in nanoscale are heterogeneous macroscopically and homogeneous microscopically. IL-incorporated catalyses would be developed as a functional material that can be hybridized between hetero- and homogeneous phases. Use of ILs in preparation of nanostructured materials with control of size, size distribution, and shape, is a very effective technique because of their unique properties such as extremely low volatility, wide temperature range in liquid state (room temperature to 400°C), good dissolving ability, high thermal stability, excellent microwave absorbing ability, high ionic conductivity, wide electrochemical window, nonflammability, and so on (Antonietti et al. 2004; Suh et al. 2004). The ionic liquids were successfully used as solvents in synthesis of metallic catalytic NPs, solar cell electrolytes, especially those originating from the diethyl and dibutyl-alkylsulfonium iodides, cocatalyst. The highest overall conversion efficiency of almost 4% was achieved by a dye-sensitized, NC solar cell using (Bu$_2$MeS)I:I$_2$ (100:1) as electrolyte (air mass 1.5 spectrum at 100 W/m^2) with the fastest electron transports as well as the highest charge accumulation, quite compatible with the standard efficiencies provided by organic solvent-containing cells. No general improvements could be observed with the addition of amphiphilic coadsorbents to the dyes or NPs.

6.6.7 PHASE-BOUNDARY CATALYST

Phase-boundary catalysts (PBCs), in contrast to conventional emulsion catalytic systems, are applicable to the reactions at the interface of an aqueous phase and organic phase. With a PBC, the catalyst acts at the interface between the aqueous and organic phases. The reaction medium of phase boundary catalysis systems for the catalytic reaction of immiscible aqueous and organic phases consists of three phases: an organic liquid phase, containing most of the substrate; an aqueous liquid phase containing most of the substrate in aqueous phase; and the solid catalyst.

A conventional emulsion catalytic system requires continuous stirring and mass transfer from the organic to the aqueous phase and vice versa. Conversely, in PBC, stirring and addition of a cosolvent is not required to drive liquid–liquid phase transfer because the mass transfer is not the rate-determining step in this catalytic system, as shown in Figure 6.13a and b (Ikeda et al. 2001; Nur et al. 2001). The active site located on the external surface of the solid catalyst plays an important role in controlling effectiveness of the observed PBC system. Although the use of interfacial catalysts has been reported in previous studies, they either used liquid surfactants to stabilize emulsions (Lu et al. 2006; Lu and Li 2007) or did not generate emulsions (Nur et al. 2001). Carbon nanotubes (NTs) fused to oxide NPs are able to stabilize emulsions and, when properly functionalized, catalyze reactions at the oil–water interface (i.e., liquid/solid/liquid interfacial catalysis) (Shen and Resasco 2009; Crossley et al. 2010). The amphiphilic nanohybrids stabilize a high interfacial area, which results in higher conversions, and it does not require the use of a liquid surfactant that could not be separated from the reaction mixture. In contrast to homogeneous catalysts previously used, the recoverable solid stabilizes a water-in-oil emulsion and catalyzes reactions. This combination of heterogeneous and phase-transfer catalysis has several advantages (Ungureanu et al. 2008): (1) increased interfacial area and (2) enhanced mass transfer of molecules and their application for the oxidative desulfurization, Lewis acid–catalyzed organic reactions, organocatalytic reactions, and other catalytic reactions.

6.6.8 EXAMPLES

Cape et al. (2009) investigated the water-oxidation catalysts by bridging homogeneous molecular systems for water splitting due to their straightforward synthesis, ease of characterization, and tunable properties. The catalytic mechanism is also more easily investigated in solution. Furthermore, through modification of the ligands, homogeneous catalysts can be incorporated into more complex molecular structures including molecular photosensitizer scaffolds designed to enable solar water splitting (Wasielewski 2009).

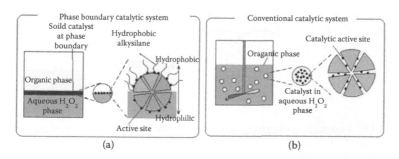

FIGURE 6.13 Schematic representation of the working of the (a) phase-boundary catalysis and (b) conventional catalytic system. (From Nur, H. et al., *J. Catal.*, 204, 402–408, 2001; Ikeda, S. et al., *Langmuir*, 17, 7976–7979, 2001.)

Recently, Blakemore et al. (2011) used electrochemical quartz crystal nanobalance (EQCN) studies to distinguish the homogeneous or heterogeneous nature of oxygen-evolving electrocatalysts by using $[Cp*Ir(H_2O)_3]^{2+}$ complex, by measuring the electrode mass in real time with piezoelectric gravimetry. They reported that the tris-aqua complex (1) (Figure 6.14a and b) shows a prominent catalytic wave in voltammetry and is shown by EQCN to be a precursor for a heterogeneous catalyst in the form of an amorphous blue layer (BL) of iridium oxide material, which is a robust heterogeneous water-oxidation catalyst. Conversely, complex (2) catalyzes water oxidation without evidence of deposition, giving dioxygen as observed by

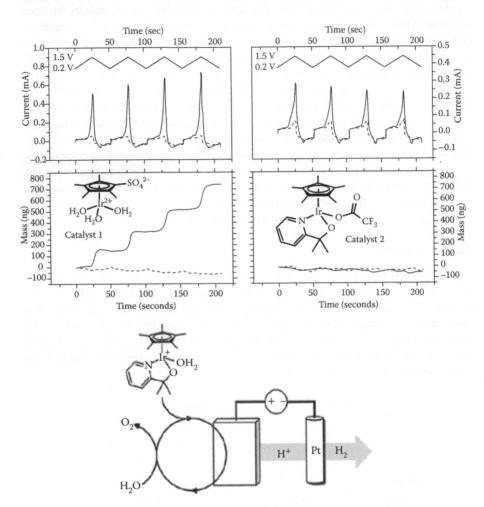

FIGURE 6.14 Comparison of current and mass responses of solutions containing complexes 1 (left panels) or 2 (right panels) as a function of time. In left/right panels: solid lines represent mass/current response in each experiment, while dashed lines correspond to the catalyst-free backgrounds, under the following conditions: 0.1 M KNO_3 in air-saturated deionized water; 3 mM 1; ~2.5 mM 3. Scan rate: 50 mV/s. (From Schley, N.D. et al., *J. Am. Chem. Soc.*, 133, 10473–10481, 2011.)

rotating ring-disk electrode (RRDE) and Clark electrode studies. Athough complexes (1) and (2) are oxidized at a similar potential, their behavior could barely be different. Complex (1) displays a prominent catalytic wave in voltammetry and is shown by EQCN to be a precursor for a heterogeneous catalyst (Schley et al. 2011).

6.7 ROLE OF METALLIC/METALLIC HYDROXIDE COCATALYST IN HYDROGEN EVOLUTION REACTION/OXYGEN EVOLUTION REACTION

The efficiency of the first two steps (generation and separation of photocarriers, i.e., photoelectrons and photoholes) in photocatalytic water splitting can be enhanced in the presence of an H_2-evolution or O_2-evolution cocatalyst (Tong et al. 2012; Zhou et al. 2012; Qu and Duan 2013). The process of reduction and oxidation of water splitting has been promoted by these carriers. Cocatalysts channelize the segregation of the photogenerated charge carriers and improve the efficiency of the catalytic HER or OER by suppressing the photocorrosion of the supporting photocatalyst. Consequently, cocatalysts play a significant role in improving both the activity and stability of semiconductor photocatalysts by improving the charge separation and transfer rate and lowering the activation energy (Lv et al. 2010; Dai et al. 2011; Wang et al. 2012; Meng et al. 2013; Zhang et al. 2013a) at the semiconductor heterojunction. Therefore, a highly active and selective cocatalyst is selected to promote the forward HER or OER and inhibit the backward reactions. Presently, most of the developed photocatalyst systems utilize noble metal–based cocatalysts to achieve high photocatalytic activity due to their large work function and low overpotential for H_2/O_2 evolution (Yang et al. 2013).

6.7.1 METALLIC COCATALYST

A few of the vital cocatalysts that enhance the activity of HER are Ru (Hara et al. 2003; Tsuji et al. 2005), Rh (Sasaki et al. 2008; Maeda et al. 2010a), Pd (Sayed et al. 2012), Pt (Yan 2009; Maeda et al. 2010a; Yu et al. 2010; Li et al. 2011; Xiang et al. 2011; Wu et al. 2012; Lingampalli et al. 2013), Au (Lin et al. 2011; Murdoch et al. 2011), Ag (Korzhak et al. 2008; Onsuratoom et al. 2011), RuO_2 (Sato et al. 2005; Liu et al. 2008b), $Rh_{2-y}Cr_yO_3$ (Maeda et al. 2006a), IrO_2 (Abe et al. 2009; Youngblood et al. 2009; Ma et al. 2010), and core–shell-structured $Rh–Cr_2O_3$ (Maeda et al. 2006b, 2010b). Unfortunately, most of the noble metal–based cocatalysts are too sparse and expensive to be used for large-scale energy production. Therefore, the research on the noble metal–free cocatalysts with high efficiency and low cost is highly desirable. In the name of as precious metal (Pt, Au, Ag,etc)-free cocatalysts, few of the metal hydroxides such as ($Cu(OH)_2$, (Yu and Ran 2011) $Ni(OH)_2$ clusters, (Yu et al. 2011), are studied. With optimal $Ni(OH)_2$ loading content of 0.23 mol% on TiO_2, gives it a H_2-production rate of 3056 μmol/h/g with QE of 12.4% that exceeds 223 times more than the TiO_2 (Ran et al. 2011). The optimal 23 mol%, $Ni(OH)_2$ loading on TiO_2 5085 μmol/h/g H_2 produced with corresponding 28% QE at 420 nm, which exceedsing from CdS and 1wt.% Pt-loaded CdS nanorods by a factor of 145 and 1.3 times (Peng et al. 2011; Yu et al. 2012, 2013). Various semiconductors, for example, TiO_2, CdS, and g-C_3N_4, and nanocarbon materials, such as carbon NTs and graphene nanosheets (Zhang et al. 2012) also exhibited a strong capability to enhance the photocatalytic H_2-production activity. Therefore, cocatalysts with excellent activity and selectivity, which could both promote forward H_2 or O_2 evolution and inhibit backward reactions, are essential for achieving high efficiency in photocatalytic pure water splitting. Up to now, the most active cocatalysts for assisting H_2 evolution in overall water splitting are $Rh_{2-y}Cr_yO_3$ (Maeda et al. 2006a) and core–shell-structured Ni–NiO (NiO_x) (Domen et al. 1986; Kudo et al. 1988; Kim et al. 2000, 2002; Takahara et al. 2001; Kondo et al. 2004; Miseki et al. 2009; Hu et al. 2010; Husin et al. 2011; Wang et al. 2012) and $Rh–Cr_2O_3$ (Maeda et al. 2010b). The reverse reaction of H_2 and O_2 as well as photoreduction of O_2 could be suppressed by the shell. Moreover, the shell of core–shell structures involved in the protection of the metal core from photocorrosion also improved the robustness of cocatalysts. The acquaintance of this type of structure is beneficial in leading the direction to design and fabricate efficient and stable cocatalysts for photocatalytic water splitting. Therefore, the development of noble metal–free cocatalysts with high efficiency and low cost is highly desirable.

A novel metal-free O_2-evolution cocatalyst, $B_2O_{3-x}N_x$ nanocluster, showed the capability of apparently improving the photocatalytic activity of WO_3 (Xie et al. 2012). Bioinspired molecular cocatalysts, such as the $[Mo_3S_4]^{4+}$ cluster, were reported to be effective in photocatalytic overall water splitting (Seo et al. 2012). In some cases, the smaller particle size of cocatalysts results in lowering of energy barriers for the smooth interfacial charge transfer from semiconductors to cocatalysts (e.g., electron-tunneling effect). Copious amount of studies show that the small size and high dispersion of the loaded cocatalysts lead to significantly enhanced photocatalytic efficiency.[64,77,78,96,100,107–126] Furthermore, artificial molecular cocatalysts with very small dimensions like $[Mo_3S_4]^{4+}$ are composed of surface atoms/edge sites and possess an optimized number of active sites, which could considerably increase the photocatalytic activity of semiconductors (Seo et al. 2012).

The OER is a four-ET process coupled with the removal of four protons from water molecules to form an oxygen–oxygen bond. Therefore, this half water-splitting reaction is referred to as the most challenging step. Thus, the cocatalysts are utilized as follows:

1. Lower the O_2-evolution overpotential that could remarkably boost up the efficiency of photocatalytic overall water splitting for the OER and HER process.
2. Assist in photoelectron–hole separation at the cocatalyst/semiconductor interface.
3. Suppress photocorrosion and increase the stability of semiconductor photocatalysts.
4. Improve the photostability of the catalysts by timely consuming of the photogenerated charges, particularly the holes/electrons.
5. Most importantly, a cocatalyst catalyzes the reactions by lowering the activation energy.
6. A cocatalyst can provide trapping sites for the photogenerated charges and promote the charge separation, thus enhancing the QE.
7. Facilitate oxidation and reduction reactions by providing the active sites/reaction sites while suppressing the charge recombination and reverse reactions.

The formation of an appropriate heterojunction between the cocatalyst and the semiconductor is the key factor for promoting the charge separation and transfer from the semiconductor to the cocatalyst. The relative energy level at the heterojunction interface dictates the direction and efficiency of the carrier separation and transportation. The heterostructure with a short distance between the semiconductor and the cocatalyst reduces the necessary charge transfer distance and can effectively inhibit the bulk recombination and enhance the overall efficiency. Figure 6.15 demonstrates the relationship between the photocatalytic activity and amount of cocatalyst loaded onto the semiconductor. This volcanic curve explains that at the initial state the addition of a cocatalyst onto a semiconductor could gradually increase the photocatalytic water-splitting activity because it facilitates the charge collection and gas evolution reactions. When the amount of a cocatalyst loaded on the semiconductor reaches the optimal value, the cocatalyst/semiconductor system achieves the highest photocatalytic activity. However, further loading of the cocatalyst would severely decrease the photocatalytic activity. Excessive loading of cocatalyst often decreases the activity due to the following factors:

1. This could result in blocking of the surface active sites of the semiconductor that may hinder its contact with sacrificial reagents or water molecules.
2. It could shield the incident light, and thus prevent light absorption and generation of photo carriers inside the semiconductor.
3. Large particle size deteriorates the catalytic properties of cocatalysts or leads to the disappearance of surface effects.
4. That could compel the surface to act as a charge recombination center and decrease the photocatalytic activity.

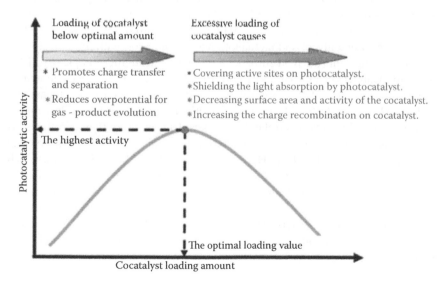

FIGURE 6.15 A graph of the relationship between the loaded amount of a cocatalyst and the photocatalytic activity of the cocatalyst-loaded semiconductor photocatalyst. (From Ran, J. et al., *Chem. Soc. Rev.*, 43, 7787–7812, 2014.)

A few more important factors that influence the catalytic activity of cocatalysts are their size (smaller sizes have a larger surface area and more active sites and charge carriers are less likely to recombine), stability, structure (core shell or different shape), amount, compatibility of cocatalyst with supporting semiconductor, optimum number of active sites, and many more.

Besides the high cost, the main disadvantage of metallic electrodes over the semiconductor electrode is that the metal has a very large impedance mismatch with water and tends to reflect a lot of light. Therefore, noble metal–free cocatalyst materials are getting more attention. One more important category of metal-free cocatalysts is the visible light–responsive semiconductors, such as sulfides, and (oxy) sulfides (Yu et al. 2012).

6.7.2 Roles of Hydroxyl Cocatalysts in Photocatalytic Water Splitting

Cocatalysts are vital for the separation of the charges on the main semiconductor surface. But most of the studied cocatalysts are either noble metals or noble metal oxides, which are not only costlier but also are scarce. Therefore, the search for low-cost active cocatalyst is on to find a superior catalyst with high activity. Hydroxides with low cost and high efficiency are far ahead in this race but are limited in number. Therefore, much research work needs to be done in this field. Recently, $Cu(OH)_2$ (Yu and Ran 2011; Dang et al. 2013; Zhang et al. 2013b), $Ni(OH)_2$ (Jang et al. 2009; Ran et al. 2011; Yu et al. 2011, 2013; Rocha et al. 2013), and $Co(OH)_2$ (Wender et al. 2013) have been developed as efficient and low-cost cocatalysts that compete with semiconductors to generate photocatalytic H_2. A $Cu(OH)_2$-loaded TiO_2 nanoarray was fabricated by different methods such as the precipitation method (Yu and Ran 2011), chemical bath deposition method (Zhang et al. 2013b), and hydrothermal-precipitation process (Dang et al. 2013). The corresponding H_2 production rate was observed as 3418 μmol/h/g (QE of 13.9%), 6.5 μmol/h/cm and 14.94 mmol/h/g catalyst, in sacrificial electrolyte (aqueous ethylene glycol and methanol), under irradiation with ultraviolet light. Figure 6.16i(a) represents the field emission scanning electron microscopic (FESEM) image of the $Cu(OH)_2/TiO_2$. Yu and Ran (2011) found that the loaded amount of $Cu(OH)_2$ in the $Cu(OH)_2–TiO_2$ composite significantly influenced its photocatalytic H_2 evolution (Figure 6.16i(b)). The optimal $Cu(OH)_2$ loading amount was determined to be 0.29 mol%, which exceeded the rate on bare TiO_2 by more than 205 times, as shown in Figure 6.16i(c) and (d). The enhanced activity of the device is attributed to the potential of $Cu(OH)_2/Cu$ ($Cu(OH)_2 + 2e^- = Cu + 2OH^-$, $E_o = -0.224$ V), which was slightly lower than the CB (-0.26 V) of anatase TiO_2 that promotes the ET from the CB of TiO_2, to $Cu(OH)_2$ in $Cu/Cu(OH)_2$ clusters. These $Cu/Cu(OH)_2$ clusters can serve as a cocatalyst to assist the separation and transport of photoinduced electrons from TiO_2 to $Cu/Cu(OH)_2$ clusters, where H^+ is reduced to H_2 molecules (Yu and Ran 2011). The mechanism involved in the water reduction is described by the following photocatalytic steps:

$$Cu(OH)_2 + 2e- = Cu + 2OH^- \qquad (6.57)$$

$$Cu + 2e - + 2H^+ = Cu + H_2 \qquad (6.58)$$

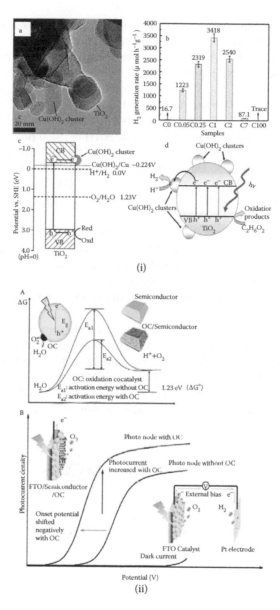

FIGURE 6.16 (i) (a) High-magnification transmission electron microscopy (TEM) images of the Cu(OH)$_2$–TiO$_2$ composite. (b) Comparison of the photocatalytic activity of the Cu(OH)$_2$–TiO$_2$ composite with different contents of Cu(OH)$_2$ for the photocatalytic H$_2$ production in ethylene glycol aqueous solution under UV-light-emitting diode (LED) irradiation. (c) Proposed mechanism for photocatalytic H$_2$ production under UV light irradiation. (d) Schematic illustration of the charge transfer and separation in the Cu(OH)$_2$ cluster-modified TiO$_2$ system under UV-LED irradiation. (ii) (a) Schematic description of the role of oxidation cocatalyst in (a) photocatalytic water oxidation and (b) PEC water oxidation. (From Yu, J.G. and Ran, J.R., *Energy Environ. Sci.*, 4, 1364–1371, 2011; Yang, J. et al., *Acc. Chem. Res.*, 46, 1900–1909, 2013.)

Another popular and efficient cocatalyst-supported system is $Ni(OH)_2$-modified semiconductors. A few significant cases are $Ni(OH)_2$-modified CdS nanorods (Ran et al. 2011) with a visible-light H_2-production rate of 5085 mmol/h/g (QE = 28%), and $Ni(OH)_2$ loaded on Ni-doped δ-FeOOH photocatalysts (Yu et al. 2013) with a light H_2-production rate of 5746 µmol/h/g. $Ni(OH)_2$-modified graphitic carbon nitride ($Ni(OH)_2$–g-C_3N_4) composite photocatalysts (Yu et al. 2013) of 7.6 µmol/h (with an apparent QE of 1.1% at 420 nm), $Ni(OH)_2$ cluster-modified TiO_2 ($Ni(OH)_2$/ TiO_2) nanocomposite photocatalysts (Yu et al. 2011) of 3056 µmol/h/g with QE of 12.4% are also known. Enhanced activity of the $Ni(OH)_2$-modified photocatalyst was a result of Lower negative reduction potential of the photocataytic materials such as Ni^{2+}/Ni ($Ni^{2+} + 2e^- = Ni$, $E° = -0.23$ V) than the CB of CdS = -0.7 V, CB of anatase TiO_2 = -0.26 V, CB of FeOOH = -5.8 V, and CB of g-C_3N_4 = -0.67 V. Moreover, these potentials are more negative than the reduction potential of H^+/H_2 ($2H^+ + 2e^- = H_2$, $E° = 0.0$ V), which favors the transfer of CB electrons from photocatalysts to $Ni(OH)_2$ that promotes the reduction of Ni^{2+} to Ni^0 and H_2 production. The role of Ni^0 is to help the charge separation and to act as a cocatalyst for water reduction, thus enhancing the photocatalytic H_2-production activity. Wender et al. (2013) presented $Co(OH)_2$ as an efficient cocatalyst on the Fe_2O_3 nanoring for photocatalytic H_2 production.

In 2013, Yang et al. illustrated the water-splitting phenomena in the absence and presence of cocatalyst (Figure 6.16ii(a) and (b)). Domen et al. demonstrated the synthetic way to prepare $Ni(OH)_2$-loaded $SrTiO_3$. Initially, NiO-loaded $SrTiO_3$ reduced to Ni-loaded $SrTiO_3$ at 500°C. This reduced shell/core NiO/Ni-loaded $SrTiO_3$ was further oxidized to shell/core $Ni(OH)_2$/Ni-loaded $SrTiO_3$, which was used to reduce and oxidize water to H_2 and O_2, respectively, as shown in Figure 6.17 (Domen et al. 1986).

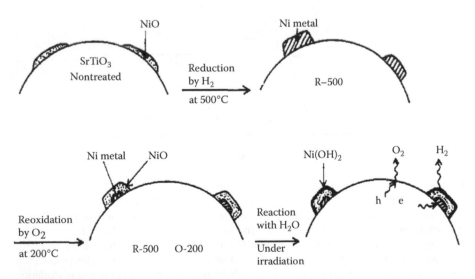

FIGURE 6.17 NiO–$SrTiO_3$ photocatalyst after reduction and oxidation treatments of cocatalyst. (Reprinted with permission from Domen, K. et al., *J. Phys. Chem.*, 90, 292–295, 1986. Copyright [1986] American Chemical Society.)

Appropriate cocatalysts are vital for achieving high efficiency in photocatalytic and PEC water splitting. By providing the active sites for reduction or oxidation, cocatalysts can catalyze the surface reactions by lowering the activation energies, trapping the sunlight-generated electrons and holes, and restraining the recombination of the photocarriers. Noble metals, biomimetic material such as hydrogenase, transition-metal oxides, chalcogenides, metal hydroxides, sulfides, and nitrides can serve as either reduction or oxidation cocatalysts for photocatalytic reactions. The photocatalytic performance largely depends on the nature of the light-harvesting semiconductor and their synergy with the functionality of cocatalysts. In simple words, we can say that a photocatalyst should be composed of the materials with three necessary functions, that is, light-harvesting component (semiconductor), reduction, and oxidation reaction component (dual cocatalysts). An efficient cocatalyst should be in harmony with the semiconductors in terms of compatible lattice, energy levels (band levels or Fermi levels), and electronic structures. The charge transport should proceed in the right direction through the built-in electric field at the interface between the semiconductor and cocatalysts. The exact and precise mechanism of the dual cocatalysts is yet to be known, but practically it was proven that the loading of the suitable dual cocatalysts on semiconductors surface can appreciably increase the photocatalytic activities of the HER and OER and even make the overall water-splitting reaction feasible. Therefore, dual cocatalysts are necessary for developing highly efficient photocatalysts for water-splitting reactions.

6.8 NATURE/ROLE OF THE ACTIVE SITES ON A CATALYST'S SURFACE

Heterogeneous photocatalytic reactions often occur on the catalytic surface. These catalytic surfaces have active sites in from of the atoms or crystal faces where the reaction really occurs. The active sites may exist either as an exposed planar surface or a crystal edge with imperfect cation/anion valence or a complex amalgamation of the two. Therefore, most of the volume of the heterogeneous catalyst including the surface may be catalytically inactive. The presence of surface catalytic active sites is a fundamental issue in the rate of the reaction determination and it needs to be clarified on an experimental basis. It demonstrates the important contributions of the photocatalyst surface properties to the fundamental molecular/electronic structure–photoactivity relationships. The phases of surfaces of single crystals photocatalysts such as TiO_2, ZnO, WO_3, and Ga_2O_3 have been studied for decades as models for metal oxide. Phivilay et al. (2013) illustrated the relationship between the nature of catalytic active sites present onto the surface and photocatalytic activity of the system by taking an example of LaO_x-doped $NaTO_3$ (Figure 6.18a). Tracing out the nature of the catalytically active site and the mechanism is a technically challenging area of the research. Thus, pragmatic research for the discovery of the appropriate catalytic material is going on. The surface stability is an important aspect that appreciably affects the competence of catalytic reactions. The surface stability can be computed in terms of the surface energy to identify the most stable surface orientation (Lazzeri et al. 2001). High temperatures are required to prepare clean surfaces for chemical reactions, which has rendered their study by surface science methods

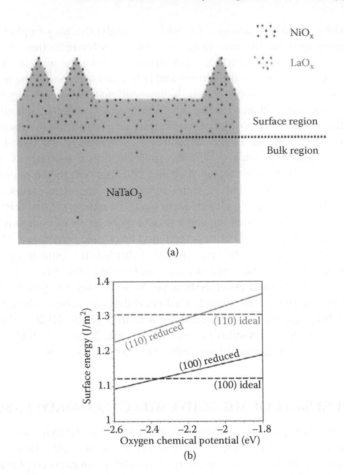

FIGURE 6.18 (a) Schematic diagram of the LaO_x-doped $NaTO_3$ surface region with highly dispersed cocatalyst NiO_x on nanosteps in highly active 0.2% $NiO/NaTaO_3$:2% La photocatalyst. (b) Energies of reduced and ideal {110} and {100} surfaces of Ag_3PO_4 as a function of oxygen chemical potential. (Reprinted with permission (a) from Phivilay, S.P. et al., *ACS Catal.*, 3, 2920–2929, 2013; Bi, Y.P. et al., *J. Am. Chem. Soc.*, 133, 6490–6492, 2011. Copyright [2011] American Chemical Society.)

more complex; most studies are conducted by computation modeling (Lazzeri and Selloni 2001; Tanner et al. 2002; Ruzycki et al. 2003; Bonapasta, and Filippone 2005; Czekaj et al. 2006). Equation 6.59 represents the formula used for computing the surface energy γ:

$$\gamma = \frac{E_{slab} - nE_{bulk}}{2A} \tag{6.59}$$

where E_{slab} is the total energy of the slab/surface, E_{bulk} is the total energy of the bulk per unit cell, n is the number of bulk unit cells possessed in the surface, and A is the surface area of each side of the surface/slab. Similarly, the stability of each surface under a reduced oxygen atmosphere, the surface energies of reduced surfaces can be calculated as follows:

$$\gamma_R = \frac{(E_{surf} + \mu O)}{A} \tag{6.60}$$

$$E_{surf} = \frac{(E_{Rslab} - nE_{bulk})}{2} \tag{6.61}$$

where E_{Rslab} is the total energy of the reduced slab and μO is the chemical potential of oxygen, which is directly related to the partial pressure of the O_2 in atmosphere. Figure 6.18b refers to the plot between the surface energy of {110} plane and {100} plane as a function of μO for ideal and reduced systems, which was referenced to the total energy of a single oxygen molecule. The structures with lower surface energy are more real to be realized. The ideal surface was found to be stable at higher μO, whereas the reduced surface predominated at lower μO. The intersection of the two lines corresponding to the ideal and reduced surfaces gives the transition value of μO, which is higher for the {110} plane than for {100}. This indicates that oxygen vacancies start to form on the {110} surface even under a relatively high O_2 partial pressure. Therefore, the abundance of oxygen vacancies as well as the higher surface energy of the {110} plane is a possible cause of the high reactivity of Ag_3PO_4 rhombic dodecahedral crystals (Bi et al. 2011).

Semiconductor photocatalysts based on d^0 (Ti, Zr, Nb, Ta, and W) transition metal oxides and d^{10} (Ga, In, Ge, Sn, and Sb) main group metal oxides have emerged as good candidates to be used in heterogeneous photocatalytic systems. Their advantageous electronic configurations with empty/filled d-orbitals minimized the trapping centers for excited electrons and holes at their surface. Therefore, there are more possibilities for possessing active sites on their surface. Some of the strategies are adopted to increase the active sites on the photocatalyst surface, that is, addition of a cocatalyst such as NiO, Pt, Rh_2O_3, or RuO_2 as well as doping of the photocatalysts with metal/nonmetal ions and by making morphological changes. The nature and number of the catalytic active sites on the surface are the controlling factors for the efficiency and stability of the photocatalytic system.

One of the key factors in determining surface stability is the coordination number of surface atoms. The closer the coordination number of a given surface atom to that of the bulk, the more stable the surface. In both the rutile and anatase structures of TiO_2, the coordination number is six for Ti atoms in the bulk and three for oxygen. The (110) rutile surface contains alternating rows of Ti and O atoms five- and two-fold coordinated, respectively. Ti atoms underneath the twofold coordinated oxygen atoms are sixfold coordinated. Due to the high coordination numbers of Ti atoms in the (110) rutile surface, this surface is the most stable, as evidenced by the value of its surface energy. The (100) surface is the second-most stable surface, where all surface Ti atoms are fivefold coordinated with twofold O atoms. It was found that doping of metal ions could expand the photoresponse of TiO_2 into the visible spectrum. The chemistry of the metal ion incorporated-TiO_2 lattice that induces the impurity energy levels in the band gap of TiO_2 is indicated as follows:

$$M^{n+} + h\nu \rightarrow M^{(n+1)+} + e^- \text{ (CB)} \tag{6.62}$$

$$M^{n+} + h\nu \rightarrow M^{(n-1)+} + h^+ (VB) \tag{6.63}$$

where M and M^{n+} represent metal and the metal ion dopants, respectively. Furthermore, electron (hole) transfer between metal ions and TiO_2 can alter electron–hole recombination as follows:

$$\text{Electron trap:} \quad M^{n+} + e^-(CB) \rightarrow M^{(n-1)+} \tag{6.64}$$

$$\text{Hole trap:} \quad M^{n+} + h^+(VB) \rightarrow M^{(n+1)+} \tag{6.65}$$

The energy level of $M^{n+}/M^{(n-1)+}$ should be less negative than that of the CB edge of TiO_2, while the energy level of $M^{n+}/M^{(n+1)+}$ should be less positive than that of the VB edge of TiO_2. For photocatalytic reactions, if the trapped electron and hole are transferred to the surface, then reactions can occur. Therefore, metal ions should be doped near the surface of TiO_2 particles for better charge transferring. The deeply doped metal ions likely behave as recombination centers and make electron/hole transferring to the interface more difficult. Furthermore, there exists an optimum concentration of doped metal ion, above which the photocatalytic activity decreases due to the increase in recombination. Twenty-one metal ions were studied, out of which Cu Fe, Mo, Ru, Os, Re, V, and Rh ions can increase photocatalytic activity, while dopants Co and Al ions cause detrimental effects (Choi et al. 1994). The effects of doping transitional metal ions (Cr, Mn, Fe, Co, Ni, and Cu) on photocatalytic activity of TiO_2 were studied by Wu et al. and this was called metal ion doping. As Cu, Mn, and Fe ions can trap both electrons and holes; doping of these metal ions may work better than doping of Cr, Co, and Ni ions. Xu et al. compared photocatalytic activities of different rare earth metal ions (La, Ce, Er, Pr, Gd, Nd, and Sm) doped into TiO_2. Enhanced photocatalytic activities and red shift of photoresponse were observed at certain doping content. Dopant Gd ions were found to be the most effective in enhancing the photocatalytic activity due to their highest ability to transfer charge carriers to the interface (surface of TiO_2). Dye sensitization and composite semiconductors are the two promising surface modification methods to expand light response of TiO_2 to the visible region.

6.9 CONCEPTUAL ADVANCEMENT (MODEL) OF THE ACTIVE MATERIALS FOR HYDROGEN GENERATION THROUGH WATER SPLITTING

Water splitting is the core reaction of the photosynthesis processes, where nature uses sunlight to generate energy. We also want to dissociate water by utilizing solar energy. For this, we need an optimal material with the ability to dissociate the water molecules, suppress the inverse ion recombination reaction, and utilize light in the IR and visible (VS) range and to remain stable in contact with water. But until now we have not been able to discover such an abundant, stable, and effective photocatalyst for water splitting and technology that utilizes the whole range of solar light (7% ultraviolet, 50% visible, and 43% IR light) capable of producing water-splitting reactions (Kudo and Miseki 2009). To address the problems associated with the making of the ultimate material for hydrogen generation through water splitting, various methods have been practiced to infiltrate these restrictions. A new mechanism and many materials are needed. Photocatalytic materials are synthesized with maximum functionality by engineering their electronic structure through modification of structure and composition, which is well exemplified using the MoS_2-layered structure in Figure 6.19a (Hinnemann et al. 2005). After identification of the catalytic system for a particular reaction (here we can take either overall splitting of water or water oxidation or water reduction), one can try to optimize the influential factors that control its catalytic performance. Usually, the critical reaction parameters involve the stability of photocatalyst, content of active surface material (cocatalyst/noble metal or nonprecious transition metal), the ratio of cocatalyst (active metal) to promoter, concentration of photocatalyst and sacrificial electrolyte (if needed) in water, temperature, pressure, and flow rate of carrier gas, and so on. The "trial and error" method is used to regulate each factor in turn while keeping others constant in order to study how the response diverges respecting that particular factor. However, the synergistic effects/interactions between the factors are neglected in this approach. Therefore, the approach is unable to attain a complete understanding and the performance of the system cannot be fully optimized (Du et al. 2006). Moreover, due to the large number of the factors/variables, extensive data are collected and this can make the approach troublesome/impractical. But by designing the proper experiments (DPE), a comprehensive understanding of the effect of individual factors and their interactions can be easily derived with a small number of experiments.

A few conceptual models of photoactive systems are mentioned in the following.

6.9.1 BINARY-LAYERED METALS WITH EXTENDED LIGHT HARVESTING POWER

This is a new class of photocatalytic materials that include ionic/covalent binary metals with layered graphite-like structures for effective absorption of visible and infrared light, facilitating the reaction of water splitting, suppressing the inverse reaction of ion recombination by separating ions due to internal electric fields existing near alternating layers, providing the sites for ion trapping of both polarities, and finally delivering the electrons and holes required to generate hydrogen and oxygen gases. It consists of alternative layers of metals (Al, Mg, Cr, etc.) and metalloids where the ionic (or covalent) exchange between layers results in alternating charges (electrons and holes) sitting on alternating layers. These arrangements of charges lead to the generation of three key properties. First, the electron and hole plasmas present in binary-layered metal (BLM) interact strongly with light of the IR-VS spectra and there is an effective transfer of solar energy to ions. Second, alternate charges of the layers spatially separate ions by providing strong electric fields (in analogy to an array of p–n junctions) and this effectively suppresses the back reaction of opposite ion recombination. Third, the overall metallic properties allow BLM to deliver the electrons and holes necessary to generate hydrogen and oxygen gases to split water.

MgB_2 electrodes are a good example of the class of BLM electrodes that can replace the expensive Pt CE in photochemical cells and can demonstrate high solar energy conversion efficiency (27%) at small bias voltage $V_{bias} = 0.5$ V for photocatalytic water splitting. Two opposite charges of MgB_2 layers can act as both photoanode and cathode (Kravets and Grigorenko 2015). Their novel reaction mechanism for water splitting in

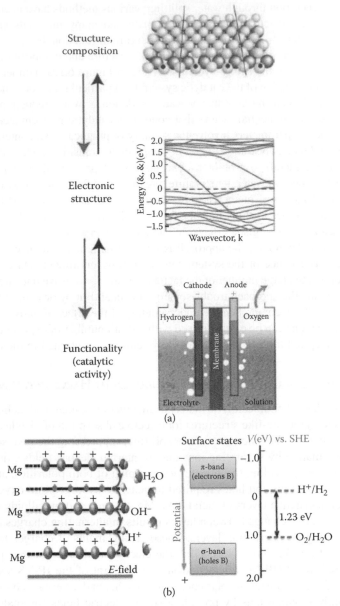

FIGURE 6.19 (a) Schematic illustration of the tailoring process of the materials and their functionality can be engineered by the electronic structure through modification of structure and composition in a MoS$_2$ sheet, which is a few atoms wide, where new electronic states at the edges across the Fermi level increase the catalytic activity, for instance in electrochemical hydrogen evolution. (b) Schematic view of the charge distribution in MgB$_2$ and electric fields responsible for ion separation along with an overview of the energy level positions of surface electrons in MgB$_2$ film with respect to the redox potential of H$^+$/H$_2$ (0 eV vs. standard hydrogen electrode [SHE]). (From Hinnemann, B. et al., *J. Am. Chem. Soc.*, 127, 5308–5309, 2005; Kravets, V.G. and Grigorenko, A.N., *Opt. Express.*, 23, A1651–A1663, 2015.)

BLM is illustrated by Figure 6.19b. Materials of this class such as AlB_2, NbB_2, MoB_2, TaB_2, TiB_2, HfB_2, and CrB_2 with typical graphite-like structures can be engineered to optimal materials for solar water splitting. Such visible–IR light–driven water-splitting devices are capable of opening a new arena for solar energy conversion.

6.9.2 BRIDGING STRUCTURES FOR WATER SPLITTING

Photosynthesis is a vital chemical reaction that sustains our biosphere by producing oxygen and carbohydrates. Photosynthetic oxidation of water to dioxygen is cata- lyzed by the active site of the OEC that consists of a $CaMn_4O_5$ cuboidal structure cluster in the PSII. PSII is one of the most important enzymes in nature (Cox et al. 2013). It is a multisubunit pigment protein–protein complex which is embedded in the thylakoid membrane of photosynthetic organisms (Umena et al. 2011). There have been many extensive theoretical and experimental studies related to determin- ing its electronic structure, properties, oxidative activation, and mechanism of water oxidation. The overall PSII reaction is a visible light-driven water-splitting reaction that is mentioned in Equation 6.66:

$$2H_2O + PQ \xrightarrow{4Av} O_2 + PQH_2 + 2H^-$$ (6.66)

In PSII, light absorbed by the reaction center, a multipigment array of chlorophyll/ pheophytin molecules, initiates a single ET across the thylakoid membrane. Successive ETs facilitate the oxidation of water to oxygen and reduction of plastoquinone to plas- toquinol (PQH_2), thereby storing sunlight energy in chemical bonds. Plastoquinol (PQH_2) is subsequently used to generate reduced nicotinamide adenine dinucleotide phosphate hydrogen (NADPH), one of the two major energy carriers of biology.

Chouhan et al. also proposed one model for hydrogen production by using one solid solution on the basis of surface studies (x-ray diffraction [XRD], differential reflectance spectroscopy [DRS], x-ray photoelectron spectroscopy [XPS], and x-ray absorption spectroscopy [XAS]) of an atomic framework for Cd–ZnGeON with bridging oxygen and nitrogen. Figure 6.20 illustrates the water-splitting mechanism on active sites of the catalytic surface.

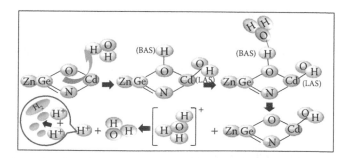

FIGURE 6.20 Schematic mechanism of water cleavage for hydrogen generation within the cluster network of Cd–ZnGeON. (From Chouhan, N. et al., *J. Mater. Chem. A*, 1, 7422–7432, 2013.)

The model comprises a network of tetrahedral units that consist of oxygen and nitrogen atoms at the apex and metallic aggregation at the center. The oxygen and nitrogen atoms bridge the metallic species Zn–Ge with Cd atoms (supported by XPS, Zn K edge, Ge K-edge, O K-edge, and N K-edge XAS studies). The local cluster framework of the tetrahedrally coordinated groups of multivalent metal cations (Zn, Ge, and Cd) and anions (NO) generates a strong local electrostatic field inside the tetrahedra, as is confirmed by the XRD, XPS, and XAS results. Residual water molecules are captured by the strong local electrostatic fields of the molecular device Cd–ZnGeON. These water molecules attract the bridging oxygen through the protonic side and the extra metallic cation, that is, Cd through the hydroxyl side bridging oxygen with a hydroxyl proton and a Cd metallic side with a hydroxyl group that function as Lewis acid sites (LAS), which create strong electron withdrawing centers for neighboring bridging O–H groups. These withdrawing centers can act as superacidic Brønsted acid sites (BAS) with a highly negative cluster framework. BAS can be attacked by another water molecule and form sterically bulky species. H_3O^+ detaches from BAS to release the tension of the bulky species and generates a H^+ ion. These H^+ ions react with the photoelectrons of the solid solutions and produce nascent H that couples with another H. Thus, hydrogen gas is generated. These cations have to neutralize two or three negatively charged frameworks at a significant distance and two OH^- ions per three valences of the cation formed by the cation exchange framework (Chouhan et al. 2013).

6.9.3 Oxygen Activity and Active Surface Sites for Water Splitting

One conceptual model (Figure 6.21i and ii) for the mechanism of TiO_2 catalytic performance in solar-to-chemical energy conversion is illustrated by Bak et al., which gives the close relationship of the catalytic performance between oxygen activity with the band gap, the charge transport, the position of Fermi level, and the concentration of the surface active sites (titanium vacancies). Its photoreactivity with water leading either to the total or partial oxidation may be expressed by the following respective anodic reactions:

$$2H_2O \leftrightarrow O_2 + 4H^+ + 4e^- \tag{6.67}$$

$$2H_2O \leftrightarrow 2HO* + 2H^+ + 2e^- \tag{6.68}$$

where e^- denotes a quasi-free electron in the oxide lattice. The charge compensation requires that these reactions are accompanied by the respective cathodic reactions:

$$4H^+ + e^- \leftrightarrow H_2 \tag{6.69}$$

$$O_2 + e^- \leftrightarrow O_2^- \tag{6.70}$$

Equations 6.67 and 6.68 correspond to the cycle of the water-splitting process. Equations 6.69 and 6.70 represent the formation of active radicals, which could be involved in subsequent reactions with organic molecules in water. As seen from Equations 6.69 and 6.70, water oxidation requires transfer of multielectrons from the

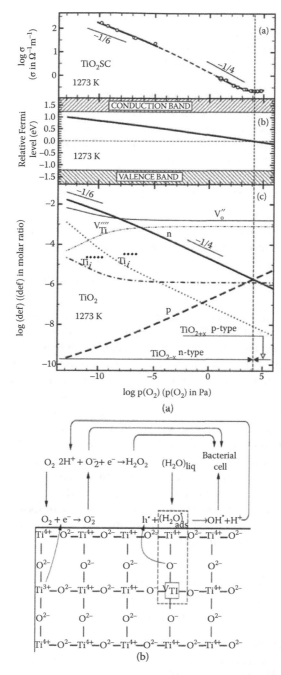

FIGURE 6.21 (a) Effect of oxygen activity on the electrical conductivity, Fermi level, and the concentration of electronic and ionic defects for pure TiO_2 at 1273 K. (Bak, T. Research Reports on Oxide Semiconductors, 2010.) (b) Model representing the role of structural imperfections in the reactivity of TiO_2 with water leading to partial oxidation of water and the related charge transfer (From Bak, T. et al., *J. Phys. Chem. A*, 119, 9465–9473, 2015.).

water molecules. In this process, the increase in oxygen activity of the oxide lattice results in a downward shift in the Fermi level and modification in the concentration of titanium vacancies, which all together enhanced the rate of the aforementioned reactions. A change of oxygen activity leads to the following reaction:

$$O_2 \leftrightarrow 2O_O^x + V_{Ti}'''' + 4h^+ \tag{6.71}$$

Titanium vacancies V_{Ti}'''' are the active sites for adsorption of water molecules that leads to the formation of an active complex, which may be represented by the following reaction (Nowotny et al. 2006):

$$2H_2O + V_{Ti} \leftrightarrow \left\{2H_2O^{2+} - V_{Ti}''''\right\}* \tag{6.72}$$

Decomposition of the preceding complex leads to formation of the hydroxyl radical species and regenerates the titanium vacancies:

$$\left\{2H_2O^{2+} - V_{Ti}''''\right\}* \leftrightarrow V_{Ti}'''' + 2HO* + 2h^+ \tag{6.73}$$

The formation and transportation of titanium vacancies is an extremely slow process (Nowotny et al. 2006). It is essential to note that an increase of oxygen activity leads to the formation of titanium vacancies (Equation 6.73), and results in a decrease of the Fermi level and the conversion of n-type TiO_2 to p-type. This analysis indicates that two key catalytic activity performance controlling factors for rutile phase of TiO_2 are the concentration of surface active sites and Fermi level. Furthermore, the group study is the same system with two other parameters, that is, band gap and charge transport, but they found a minor effect on the performance. Although the outcomes of the present work are shocking, the findings are important and lead to the important result that the photocatalytic activity must be considered in terms of specific surface sites. In the case of TiO_2 photocatalysis, such surface active sites are titanium vacancies and they are able to effectively remove electrons from the water molecules. Availability of local sites is essential for effective charge transfer and is required to perform water splitting and other chemical reactions at the rutile surface.

6.9.4 INTRINSIC KINETIC REACTOR MODEL FOR PHOTOCATALYTIC HYDROGEN PRODUCTION USING CADMIUM ZINC SULFIDE CATALYST IN SULFIDE AND SULFITE ELECTROLYTE

This model (Tambago et al. 2015) reflects the influence of sulfide and sulfite concentrations and photon absorption in the form of the local volumetric rate of photon absorption (LVRPA). The term LVRPA communicates the amount of incident photons on a region of the reactor that is absorbed by the photocatalyst. Suspension of the photocatalyst (the absorbing species) causes the attenuation of light. Therefore, the radiation field and the LVRPA are not uniform throughout the reactor and are dependent on several system properties such as the intensity and spectral irradiance of the light source, the geometrical configuration of the reactor, the amount of the catalyst, and the absorption and scattering properties of the catalyst and the presence of the other species in the system. The kinetic model is used to design the photoreactor and scale up the hydrogen production by optimizing the model parameters by keeping experimental configuration independent. The photocatalytic hydrogen production on semiconductor particles involves the absorption of photons with energy higher than the catalyst band gap, generating electron hole pairs, and the reduction by electrons and oxidation by holes of water and sacrificial species, respectively, on the surface of the photocatalyst. A competing step is the recombination of electrons and holes. Equations 6.74 through 6.88 show the reaction mechanism for hydrogen production on a cadmium zinc sulfide catalyst (Tambago et al. 2015):

Absorption of photon and electron–hole pair generation:

$$CdZnS + h\nu \rightarrow CdZnS + e^- + h^+$$
(6.74)

Recombination of electron and hole: $e^- + h + \rightarrow \phi$ (6.75)

Adsorption of reactants: $CdZnS + H_2O \rightarrow CdZnS - H_2O$ (6.76)

$$S^{2-} + H_2O \rightarrow HS^- + OH^-$$
(6.77)

$$SCdZn + HS^- \rightarrow SCdZn - HS^-$$
(6.78)

$$SCdZn + SO_3^{2-} \rightarrow SCdZn - SO_3^{2-}$$
(6.79)

Reactions on the catalyst surface

Reduction of water to hydrogen:

$$CdZnS - H_2O + SCdZn + e^- \rightarrow CdZnS - H + SCdZn - OH^-$$
(6.80)

$$2\,CdZnS - H^\bullet \rightarrow 2\,CdZnS + H_2$$
(6.81)

Oxidation of sulfide: $SCdZn - HS^- + h^+ \rightarrow SCdZn - HS^\bullet$ (6.82)

$$SCdZn - HS^\bullet + SCdZn - OH^- \rightarrow CdZnS + SCdZn - S^{\bullet-} + H_2O$$ (6.83)

$$SCdZn - HS^\bullet + SCdZn - S^{\bullet-} \rightarrow SCdZn - HS_2^- + CdZnS$$
(6.84)

Desorption of products: $SCdZn - HS_2^- + CdZnS + HS_2^-$ (6.85)

Reaction of disulfide with sulfite in the liquid phase:

$$HS_2^- + SO_3^{2-} \rightarrow S_2O_3^{2-} + HS^-$$
(6.86)

Adsorption of other species in the system: (6.87)

$$SCdZn + OH^- \rightarrow SCdZn - OH^-$$

$$SCDZn + SO_3^{2-} \rightarrow SCdZ_n - SO_3^{2-} \qquad (6.88)$$

The reactions are assumed to take place among adsorbed species on the photocatalyst, following the Langmuir–Hinshelwood mechanism. For conciseness, the $Cd_xZn_{1-x}S$ photocatalyst site is represented as CdZnS. The symbol CdZnS is used to denote the site on the negatively charged S atom while SCdZn is used for the site on the positively charged Cd and Zn atoms. The expression for the rate of hydrogen production is derived from the mechanism with the following assumptions:

1. Reactions take place among species adsorbed on the catalyst surface.
2. Reactions are elementary and irreversible.
3. Concentrations of electrons, holes, and radicals are at steady state.
4. Species adsorbed on the photocatalyst are in equilibrium with those in the bulk solution.
5. Sulfide, sulfite, and hydroxide ions compete for the same adsorption site.
6. Concentration of water on the catalyst surface is constant.
7. Rate of electron–hole generation is proportional to $\gamma \alpha, \nu$, the LVRPA or the rate of photon absorption per unit volume of catalyst suspension, by the average QE ϕ.

A modification to the rate expression was introduced such that the rate retains its dependence on species concentrations when it is reduced to a form having linear dependence with the LVRPA at low intensities. The resulting kinetic rate expression is as follows:

$$r_{H_2}^v = k \left\{ \frac{K_{SH} - [SH^-]}{(1 + K_{SH} - [SH^-] + K_{SO_3^{2-}}[SO_3^{2-}] + K_{OH-}[OH]^-)^2} \right\} \gamma^{av} \qquad (6.89)$$

where $\gamma_{H_2}^v$ is the rate of hydrogen production per unit volume of the suspension, K_{SH^-}, $K_{SO_3^{2-}}$, and K_{OH^-} are the adsorption equilibrium constants for SH⁻, SO_3^{2-}, and OH⁻, respectively, and the parameter k is represented in Equation 6.90:

$$k = \left\{ k_{rd}[H_2O]k_{ox}[SCdZn]_{tot}^2 \right\} \phi \qquad (6.90)$$

where k_{rd} is the rate constant for reduction of water to hydrogen, k_{ox} is the rate constant for oxidation of sulfide, $[H_2O]$ is the concentration of water on the catalyst surface, and $[SCdZn]_{tot}$ represents the total amount of SCdZn sites (Figure 6.22a–d) (Tambago et al. 2015).

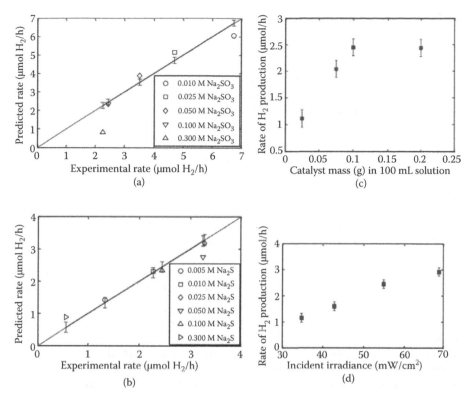

FIGURE 6.22 Initial rates of hydrogen production at various amounts of (a) Na_2SO_3, (b) Na_2S, and (c) catalyst (100 mL solution with 0.1 M Na_2S, 0.1 M Na_2SO_3, 54.95 mW/cm² incident irradiance), and (d) various incident irradiances (0.1 g catalyst in 100 mL solution with 0.1 M Na_2S, 0.1 M Na_2SO_3). (From Tambago, H.M.G. and de Leon, R.L., *J. Chem. Eng. Appl.* 6, 220–227, 2015.)

The model shows Langmuir–Hinshelwood characteristics in the concentration-dependent term and linear dependence of the rate on the LVRPA valid at low intensities. The model, with four parameters—a kinetic parameter and the adsorption equilibrium constants of sulfide, sulfite, and hydroxide ions—estimated from experimental data, shows good fit to the experimental data. The estimated model parameters may be used to provide information for the photoreaction kinetics component in photocatalytic reactor design and scale-up using the $Cd_{0.4}Zn_{0.6}S$ catalyst for the photocatalytic production of hydrogen (Tambago et al. 2015). For the validation of the computed absorption and scattering coefficients of the cadmium zinc sulfide particles in water, the developed intrinsic kinetic model may be applied to other photoreactors. The optical properties and the model parameters may be further improved by solving the complete form of the radiative transfer equation (RTE) applied and by taking into account the influence of the products such as thiosulfate and the participation of sulfite in the oxidation reactions at high sulfite concentrations.

6.9.5 Remedial Treatment for Improving Efficiency by Improvement in Catalytic Activity of the Nanoparticles by Synthesizing them in Ionic Liquids

NPs are important in enhancing catalytic activities their catalytic activities may be further improved by improving synthesis technique. IL are a good choice as a medium for many of the metallic/metal oxide NP syntheses. The main advantages of using ILs in synthesis and immobilization of metal NP are their dual roles as both reaction solvent and NP stabilizer. Besides this, they have tunable properties by virtue of having a synthetically accessible carbon backbone, for rationally controllable catalytic performance of NPs to modify their surface (Ariga et al. 2010; Datta et al. 2010). ILs enable the facial recycling of these intrinsic green natured solvents. Stable suspensions of metal NPs can be synthesized in ILs because the cationic and anionic species act as electrostatic stabilizers. However, NP immobilization and stabilization is not the only benefit of ILs in catalysis; of greater importance are the positive effects of ILs on catalytic rates and selectivity (Lee et al. 2010) due to the unique microenvironment created by the anions/cations of ILs. ILs have some limitations, like poor dispersion of some NPs in ILs, and low catalytic activity and/or selectivity for certain reactions; precipitation/agglomeration of NPs upon recycling and remarkable strategies/efforts have been made to address these problems (Dupont and Scholten 2010; Luska and Moores 2012; Scholten et al. 2012). A few of the effective ILs are imidazolium-based salts, 1,3-bis(9-anthracenylmethyl)imidazolium chloride ([Bamim][Cl]), 1,3-bis(1-naphthalenylmethyl)imidazolium chloride ([Bnmim][Cl]), and 1,3-bis(benzyl methyl)imidazolium chloride ([Bbmim][Cl]), 1-butyl-3-methylimidazolium chloride ([C_4mim]Cl), 1-ethyl-3-methylimidazolium boron tetrafluoride ([C_2mim][BF_4]), 1-butyl-3-methylimidazolium [C4mim][Tf_2N] and [C_2OHmim][Tf_2N],1-(2-hydroxyethyl)-3-methylimidazoliumboron tetrafluoride ([C_2OHmim][BF_4]), functionalized ILs based on pyridinium ([C_2OHmpy][BF_4]) and pyrrolidinium ([C_2OHmpip][BF_4]), and so on. It is desirable that future studies concentrate on the development of IL-supported catalysts based on inexpensive metals (e.g., Fe, Zn, Cu, Ni, Co), NPs for the production of highly value-added chemicals in a one-pot manner. Figure 6.23 exhibits the stepwise formation of small-sized Pd NPs in the presence of ILs by using oxidative addition, chemical etching, metathetic exchange, transmetalation, reductive elimination, and second oxidative addition reactions (Yan et al. 2010).

6.9.6 Addition of Carbonate Salts to Suppress Backward Reaction

Sayama et al. reported that the addition of carbonate salts (Na_2CO_3, $NaHCO_3$, Li_2CO_3, etc.) to aqueous medium was found to be effective in a significant enhancement of the stoichiometric production of hydrogen and oxygen by splitting of water (Sayama and Arakawa 1994). Semiconductor photocatalysts including TiO_2 (Sayama and Arakawa 1992), $A_4Ta_xNb_{6-x}O_{17}$ (A = K; Rb) (Sayama et al. 1996), Ta_2O_5 (Arakawa and Sayama 2000a), and ZrO_2 (Arakawa and Sayama 2000b) were tested for water splitting in the presence of Na_2CO_3 and it was found to be very beneficial for hydrogen and oxygen production. The IR spectrum of exposed catalytic surface announced that the surface of the photocatalyst was covered by many types of carbonate-species, such as

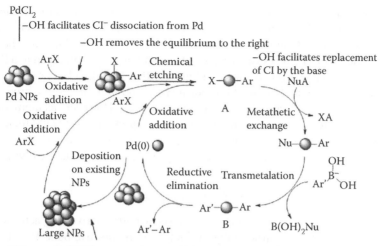

FIGURE 6.23 The catalytic cycle for Suzuki coupling and the roles of a hydroxyl group-functionalized ionic liquid (IL) in the synthesis of Pd nanoparticles. (Reprinted from *Coord. Chem. Rev.*, 254, Yan, N., Transition metal nanoparticle catalysis in green solvents, 1179–1218, Copyright (2010), with permission from Elsevier.)

HCO_3^-, CO_3^{-*}, HCO_3^*, and $C_2O_6^{2-}$, in the presence of carbonates. These carbonate species were formed during the following photocatalytic reactions:

$$CO_3^{2-} + H^+ \leftrightarrow HCO_3^- \tag{6.91}$$

$$HCO_3^- + h^+ \rightarrow HCO_3^* \tag{6.92}$$

$$HCO_3^* \leftrightarrow H^+ + CO_3^{-*} \tag{6.93}$$

$$2CO_3^{-*} \rightarrow C_2O_6^{2-} \tag{6.94}$$

Here, the photogenerated holes were consumed by reacting with carbonate species to form bicarbonate radicals, which facilitate the photoexcited electron/hole separation. The reactions (Equation 6.94) finally result in the product, that is, peroxycarbonates, which were easily decomposed into O_2 and CO_2, as follows:

$$C_2O_6^{2-} + 2h^+ \rightarrow O_2 + 2CO_2 \tag{6.95}$$

The evolution of CO_2 and O_2 in Equation 6.95 supported the desorption of O_2 from the photocatalytic surface that suppresses the formation of H_2O through the backward reaction of H_2 and O_2. Consequently, the desorbed CO_2 was also dissolved and converted into HCO_3^-, resulting in higher H_2 production. Since the company of CO_3^{2-} with Pt-TiO$_2$ was the main factor responsible for higher H_2 photocatalytic production, addition of the same amount of Na_2CO_3 and K_2CO_3 should exhibit comparable photocatalytic activity. The reason behind the greater effectiveness of the Na_2CO_3 than K_2CO_3 is still not clear. Similar to the effectiveness of CO_3^{2-}, addition of iodide was also found to be advantageous for hydrogen production (Abe et al. 2003b). Iodide anion (I^-) in a suspension can be adsorbed preferentially onto a Pt surface, forming an iodine layer, as shown in Figure 6.1b. The iodine layer can thus suppress backward reaction of H_2 and O_2 to form H_2O. And in harmony, the production of hydrogen and oxygen was enhanced very significantly. However, accumulation of too much carbonate salt or iodide anion beyond the optimum level could reduce the beneficial effects and adsorption of these species at the active sites of the catalyst surface could decrease light harvesting capacity (Sayama and Arakawa 1996).

6.9.7　Design of Active and Stable Chalcogels

Inorganic solids are an important class of catalysts that often derive their activity from meager active sites that are structurally distinct from the inactive bulk. Rationalized optimum activity is consequently viewed in light of these active sites at molecular level, which is the grand challenge in the study. Triangular active edge site fragments of molybdenum disulfide (MoS_2) are a guaranteed low-cost alternative to platinum for electrocatalytic hydrogen production at the industrial level. By controlling the strong coordination environment of a pentapyridyl ligand, Karunadasa et al. (2012) synthesized and characterized a well-defined MoIV-disulfide complex ($[(Py_5Me_2)MoS_2]$ $(CF_3SO_3)_2$) that can catalytically generate hydrogen in organic acidic media as well as in acidic water, under electrochemical reduction. Molecular analogs of the active components of inorganic solids have been broadly implicated for designing and optimizing the density of these functional metallic sites. Electronic structure calculations were conducted on nanoparticulate MoS_2 that revealed only a quarter of the edge sites are useful in hydrogen production (Hinnemann et al. 2005; Karunadasa et al. 2012). The number of active edge sites per unit volume can be increased by the tailoring the structures at the nanolevel or changing the electronics of the system to increase the enthalpy of hydrogen adsorption, which are the major challenges in inorganic materials and nanoscience. Thus, an alternative strategy was developed using discrete molecular units, which in principle can be tailored to give a high density of catalytically active metal sites without the rest of the inactive bulk material.

Researchers also designed good functionalized materials by developing some specific properties like equal activity/efficiency in any pH solution by proper combination of the substances to create more active sites at the surface. CoS_x materials are more active but less stable than MoS_x. Therefore, Jirkovský et al. combined highly active CoS_x and highly stable MoS_x units to prepare a compact and tough $CoMoS_x$ chalcogen that elucidates the low cost alternative to noble metal catalysts for efficient electrocatalytic production of hydrogen by splitting of water in both alkaline and acidic environments (Figure 6.24a and b). Although the computational insights about the mechanism are required to be fully understood, it was suggested on an experimental basis that the Transition metal (TM^{n+}) cations play a significant role in accelerating the slow rate of the Volmer step and establishing the activity-stability trends that are of the foremost importance for elucidating the role of TM^{n+} cations in the HER on transition metal sulphide (TMS_x).

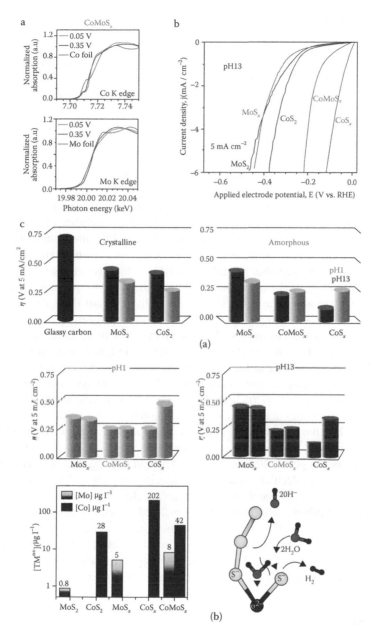

FIGURE 6.24 (a) *In situ* characterization (XAS at Co and Mo K-edge, current-voltage plot) of TMS_x and trends in activity for the hydrogen evolution reaction (HER) on TMS_x reveal that $CoMoS_x$ is a pH universal catalyst. Polarization curves for TMS_x in alkaline solutions, recorded during initial potential sweeps, provide clear trends in activity with a unique activity for CoS_x. (b) Comparison between trends in activities for the HER, expressed as overpotential required for 5 mA/cm^2 current densities, in 0.1 M HClO$_4$ (pH 1) and 0.1 M KOH (pH 13) for both crystalline and amorphous TMS_x. For crystalline chalcogenides and their water splitting mechanism. (From Jirkovský, J.S. et al., *Nat. Mater.*, 15, 197–203, 2016.)

6.10 SUMMARY

Carbon neutral fuels (hydrogen or oxygen) obtained from catalytic water splitting using sunlight would offer an attractive solution for cleaner and greener energy. A low cost, simple and efficiency-wise superior catalytic system for efficient water cleavage is still considered to be one of the holy grails. These catalytic systems are to be exploited in electrochemical and photoelectrochemical devices for light-driven water splitting for production of chemical energy in the form of hydrogen or oxygen. Catalytic systems such as metal oxides, sulfide, nitride, or their modified versions such as oxynitride, oxysulfide, composite materials, noble metal complexes, transition metal organometallics with multinuclear to monosite catalysts, and many more are considered good catalysts. But most of them are either lacking stability or efficiency or both and their manufacturing processes are too expensive. Therefore, this chapter gives us a good account of the reasons for the instability or corrosion and solutions to avoid corrosion. This chapter also discussed the mechanism behind the photocatalytic cleavage of water in electrolytes (electron scavenger and hole scavenger) and role of scavengers/sacrificial electrolytes, which is necessary to develop a better understanding of the nature/mechanism of the catalytic process. On the basis of the mechanism and nature of a catalyst, it is divided into three categories: heterogeneous catalysis, homogeneous catalysis, and mixed/synergic catalysis. All are discussed in detail with examples. Besides having a lot of advantageous qualities, heterogeneous catalysis still suffers from repetitive design, restricted modification in materials via synthetic and limited experimental access to surface mechanisms. On the contrary, molecular (homogeneous) catalysts are always readily available for synthetic modifications with the help of a variety of experimental techniques, available for monitoring rates and mechanism in solution but suffer from low recyclability and stability. By sharing the mutual properties of the homogeneous molecular catalysts (selectivity and reactivity) with the heterogeneous catalysts (recyclability and stability), tricks are given to bridge the gap between heterogeneous electrocatalysis and homogeneous molecular catalysis with specific examples. This could open the door to applications in the field of fuel cells, solar cells, dye-sensitized photoelectrochemical cells, multiphase industrial reactions, and so on, of course with improved efficiency. Furthermore, the role of metallic/metallic hydroxide cocatalysts in HER/OER catalysis is conferred by shedding light on the roles of hydroxyl cocatalysts in photocatalytic water splitting. Lastly, a few conceptual models of active materials were specified for hydrogen production through water splitting.

REFERENCES

Abe, R., K. Sayama, and H. Arakawa. 2003a. Significant effect of iodide addition on water splitting into H_2 and O_2 over Pt-loaded TiO_2 photocatalyst: Suppression of backward reaction. *Chem. Phys. Lett.* 371: 360–4.

Abe, R., K. Sayama, and H. Arakawal. 2003b. Significant influence of solvent on hydrogen production from aqueous I^{3-}/I^- redox solution using dye-sensitized Pt/TiO_2 photocatalyst under visible light irradiation. *Chem. Phys. Lett.* 379: 230–35.

Abe, R., K. Shinmei, K. Hara, and B. Ohtani. 2009. Robust dye-sensitized overall water splitting system with two-step photoexcitation of coumarin dyes and metal oxide semiconductors. *Chem. Commun.* (24): 3577–79.

Antonietti, M., D. B Kuang, B. Smarsly, and Z. Yong. 2004. Ionic liquids for the convenient synthesis of functional nanoparticles and other inorganic nanostructures. *Angew. Chem. Int. Ed.* 43: 4988–92.

Appel, A. M., D. L. DuBois, and M. R. DuBois. 2005. Molybdenum-sulfur dimers as electrocatalysts for the production of hydrogen at low overpotentials. *J. Am. Chem. Soc.* 127: 12717–26.

Arakawa, H. and K. Sayama. 2000a. Solar hydrogen production: significant effect of Na_2CO_3 addition on water splitting using simple oxide semiconductor photocatalysts. *Catal. Surv. Jpn.* 4:75–80.

Arakawa, H. and K Sayama. 2000b. Oxide semiconductor materials for solar light energy utilization. *Res. Chem. Intermed.* 26:145–52.

Ardalan, P., T. P. Brennan, H. B. R. Lee, et al. 2011. Effects of self-assembled monolayers on solid-state CdS quantum dot sensitized solar cells. *ACS Nano* 5(2): 1495–504.

Ariga, K., A. Vinu, Y. Yamauchi, Q. Ji, and J. P. DuBois. 2012. Nanoarchitectonics for mesoporous materials. *Bull. Chem. Soc. Jpn.* 85: 1–32.

Armor, J. N. 1999. The multiple roles for catalysis in the production of H_2. *Appl. Catal. A Gen.* 176: 159–76.

Avrami, M. 1940. Kinetics of phase change. II Transformation-time relations for random distribution of nuclei. *J. Chem. Phys.* 8(2): 212–24.

Autrey, T., S. Xantheas, L. Li, et al. 2002. Hydrogen from water: Photocatalytic splitting of water with visible light? *Fuel Chem.* 47(2): 752–54.

Baffert, C., V. Artero, and M. Fontecave. 2007. Cobaloximes as functional models for hydrogenases. 2. Proton electroreduction catalyzed by difluoroborylbis(dimethylglyoximato) cobalt(II) complexes in organic media. *Inorg. Chem.* 46:1817–24.

Bak, T., W. Li, J. Nowotny, A. J. Atanacio, and J. Davis. 2015. Photocatalytic properties of TiO_2: Evidence of the key role of surface active sites in water oxidation. 2015. *J. Phys. Chem. A* 119: 9465–73.

Beard, M. C., A. G. Midgett, M. Law, O. E. Semonin, R. J. Ellingson, and A. J. DuBois. 2009. Variations in the quantum efficiency of multiple exciton generation for a series of chemically treated PbSe nanocrystal films. *Nano Lett.* 9(2): 836–45.

Bhanage, B., M. Y. Ikushima, M. Shirai, and M. Arai. 1999a. Heck reactions using water-soluble metal complexes in supercritical carbon dioxide. *Tetrahedron Lett.* 40: 6427–30.

Bhanage, B. M., Y. Ikushima, M. Shirai, and M. Arai. 1999b. Multiphase catalysis using water-soluble metal complexes in supercritical carbon dioxide. *Chem.Commun.* (14): 1277–78.

Bhattacharjee, A. 2012. Combined experimental–theoretical characterization of the hydrido-cobaloxime [$HCo(dmgH)_2(PnBu_3)$]. *Inorg. Chem.* 51: 7087–93.

Bi, Y. P., S. X. Ouyang, N. Umezawa, J. Y. Cao, and J. H. Ye. 2011. Facet effect of single-crystalline Ag_3PO_4 sub-microcrystals on photocatalytic properties. *J. Am. Chem. Soc.* 133: 6490–92.

Blakemore, J. D., N. D. Schley, G. W. Olack, C. D. Incarvito, G. W. Brudvig, and R. H. Crabtree. 2011. Anodic deposition of a robust iridium-based water-oxidation catalyst from organometallic precursors. *Chem. Sci.* 2: 94–98.

Blanchard, L. A., D. Hancu, E. J. Beckman, and J. F. Brennecke. 1999. Green processing using ionic liquids and CO_2. *Nature* 399: 28–29.

Bonapasta, A. A. and F. Filippone. 2005. Photocatalytic reduction of oxygen molecules at the (100) TiO_2 anatase surface. *Surf. Sci.* 577: 59–68.

Bonilla, R. J., B. R. James, and P. G. Jessop. 2000. Colloid-catalysed arene hydrogenation in aqueous/supercritical fluid biphasic media. *Chem. Commun.* 941–42.

Brown, P. and P. V. V. 2008. Quantum dot solar cells. Electrophoretic deposition of CdSe–C_{60} composite films and capture of photogenerated electrons with nC_{60} cluster shell. *J. Am. Chem. Soc.* 130(18): 8890–91.

Brown, R. A., P. Pollet, E. McKoon, C. A. Eckert, C. L. Liotta, and P. G. G. 2001. Asymmetric hydrogenation and catalyst recycling using ionic liquid and supercritical carbon dioxide. *J. Am. Chem. Soc.* 123: 1254–55.

Busch, M., E. Ahlberg, and I. Panas. 2011. Electrocatalytic oxygen evolution from water on a Mn(III–V) dimer model catalyst—A DFT perspective. *Phys. Chem. Chem. Phys.* 13: 15069–76.

Cape, J. L., W. F. Siems, and J. K. Hurst. 2009. Pathways of water oxidation catalyzed by ruthenium "blue dimers" characterized by ^{18}O-isotopic labeling. *Inorg. Chem.* 48: 8729–35.

Castelli, P., S. Trasatti, F. H. Pollak, and W. E. O'Grady. 1986. Single crystals as model electrocatalysts: Oxygen evolution on RuO_2 (110). *J. Electroanal. Chem.* 210:189–94.

Chao, T. H. and J. H. Espenson. 1978. Mechanism of hydrogen evolution from hydridocobaloxime. *J. Am. Chem. Soc.* 100: 129–33.

Choi, J. J., Y. F. Lim, M. B. Santiago-Berrios, et al. 2009. PbSe nanocrystal excitonic solar cells. *Nano Lett.* 9: 3749–55.

Choi, W. Y., A. Termin, and M. R. Hoffmann. 1994. The role of metal ion dopants in quantum-sized TiO_2: Correlation between photoreactivity and charge carrier recombination dynamics. *J. Phys Chem.* 84:13669–79.

Chouhan, N., R.-S. Liu, and S-F Hu. 2013. Cd-ZnGeON solid solution: The effect of local electronic environment on the photocatalytic water cleavage ability. *J. Mater. Chem. A* 1: 7422–32.

Collin, J. P. and J. P Sauvage. 1986. Synthesis and study of mononuclear ruthenium(II) complexes of sterically hindering diimine chelates. Implications for the catalytic oxidation of water to molecular oxygen. *Inorg. Chem.* 25: 135–41.

Connolly, P. and J. H. Espenson. 1986. Cobalt-catalyzed evolution of molecular-hydrogen. *Inorg. Chem.* 25: 2684–88.

Cox, N., D. A. Pantazis, F. Neese, and W. Lubitz. 2013. Biological water oxidation. *Acc. Chem. Res.* 46: 1588–96.

Crossley, S., J. Faria, M. Shen, and D. E. Resasco. 2010. Solid nanoparticles that catalyze biofuel upgrade reactions at the water/oil interface. *Science* 327: 68–72.

Cuenya, B. R. 2010. Synthesis and catalytic properties of metal nanoparticles: Size, shape, support, composition, and oxidation state effects. *Thin Solid Films* 518: 3127–50.

Czekaj, I., G. Piazzesi, O. Krocher, and A. Wokaun. 2006. DFT modeling of the hydrolysis of isocyanic acid over the TiO_2 anatase (101) surface: Adsorption of HNCO species. *Surf. Sci.* 600: 5158–67.

da Silva, J. G., M. O. F. Goulart, and M. Navarro. 1999. Electrocatalytic hydrogenation of diethyl fumarate. A simple system development. *Tetrahedron* 55: 7405–10.

Dai, G. P., J. G. Yu, and G. Liu. 2011. Synthesis and enhanced visible-light photoelectrocatalytic activity of p–n junction $BiOI/TiO_2$ nanotube arrays. *J. Phys. Chem. C* 15: 7339–46.

Damjanovic, A. 1969. *Modern Aspects of Electrochemistry* (eds. Bockris, J. O. M., and B. E. Conway), Vol. 5. London, UK: Butterworth, p. 369.

Dang, H. F., X. F. Dong, Y. C. Dong, Y. Zhang, and S. Hampshire. 2013. TiO_2 nanotubes coupled with nano-Cu(OH)$_2$ for highly efficient photocatalytic hydrogen production *Int. J. Hydrogen Energy*, 38: 2126–35.

Datta, K. K. R., B. V. S Reddy, K. Ariga, and A. Vinu. 2010. Gold nanoparticles embedded in a mesoporous carbon nitride stabilizer for highly efficient three-component coupling reaction. *Angew. Chem. Int. Ed.* 49: 5961–65.

Dempsey, J. L., J. R. Winkler, and H. B. Gray. 2010. Mechanism of H_2 evolution from a photogenerated hydridocobaloxime. *J. Am. Chem. Soc.* 132: 16774–76.

Dickinson, A., D. James, N. Perkins, T. Cassidy, and M. Bowker. 1999. The photocatalytic reforming of methanol. *J. Mol. Catal. A Chem.* 146: 211–21.

Dicks, A. L. 1996. Hydrogen generation from natural gas for the fuel cell systems of tomorrow. *J. Power Sources* 61: 113–24.

Didden, A., P. Hillebrand, B. Dam, and R. V. D. Krol. 2015. Photocorrosion mechanism of TiO_2-coated photoanodes. *Int. J. Photoenergy* 2015: 457980 (8 p.).

Domen, K., A. Kudo, T. Onishi, N. Kosugi, and H. Kuroda. 1986. Photocatalytic decomposition of water into hydrogen and oxygen over nickel(II) oxide-strontium titanate (SrTiO3) powder. 1. Structure of the catalysts. *J. Phys. Chem.* 90: 292–95.

Du, G., Y. Yang, W. Qiu, S. Lim, L. Pfefferle, and G. L. Haller. 2006. Statistical design and modeling of the process of methane partial oxidation using V-MCM-41 catalysts and the prediction of the formaldehyde production *Appl. Catal. A* 313: 1–13.

DuBois, M. R. and D. L. DuBois. 2009. The roles of the first and second coordination spheres in the design of molecular catalysts for H_2 production and oxidation. *Chem. Soc. Rev.* 38: 62–72.

Dupont, J. and J. D. Scholten. 2010. On the structural and surface properties of transition-metal nanoparticles in ionic liquids. *Chem. Soc. Rev.* 39: 1780–804.

Fierro, S., T. Nagel, H. Baltruschat, and C. Comninellis. 2007. Investigation of the oxygen evolution reaction on Ti/IrO_2 electrodes using isotope labelling and on-line mass spectrometry. *Electrochem. Comm.* 9: 1969–74.

Flannigan, D. J. 2002. Seminar lecture on fluorous biphasic catalysis. Inorganic Chemistry Department, Illinois. University, 21 November 2002 pp. 49–52.

Fujishima, A. and K. Honda. 1972. Electrochemical photolysis of water at a semiconductor electrode. *Nature* 238: 37–38.

Gao, J. B., J. M. Luther, O. E. Semonin, R. J. Ellingson, A. J. Nozik, and M. C. Beard. 2011. Quantum dot size dependent *J–V* characteristics in heterojunction ZnO/PbS quantum dot solar cells. *Nano Lett.* 11: 1002.

Gersten, S. W., G. J. Samuels, and T. J. Meyer. 1982. Catalytic oxidation of water by an oxo-bridge ruthenium dimer. *J. Am. Chem. Soc.* 104: 4029–30.

Ghouse M, H. Abaoud, and A. Al-Boeiz. 2000. Operational experience of a 1 kW PAFC stack. *Appl. Energy* 65: 303–14.

Gloaguen, F. and T. B. Rauchfuss. 2009. Small molecule mimics of hydrogenases: Hydrides and redox. *Chem. Soc. Rev.* 38: 100–108.

Gorlin, Y. and T. F. Jaramillo. 2010. A bifunctional nonprecious metal catalyst for oxygen reduction and water oxidation. *J. Am. Chem. Soc.* 132: 13612–14.

Guo, Z., B. Liu, Q. H. Zhang, W. P. Deng, Y. Wang, and Y. H. Yang. 2014. Recent advances in heterogeneous selective oxidation catalysis for sustainable chemistry. *Chem. Soc. Rev.* 43: 3480–24.

Hara, M., J. Nunoshige, T. Takata, J. N. Kondo, and K. Domen. 2003. Unusual enhancement of H_2 evolution by Ru on TaON photocatalyst under visible light irradiation. *Chem. Commun.* (24): 3000–3001.

Heldebrant D. J. and P. G. Jessop. 2003. Liquid Poly (ethylene glycol) and supercritical carbon dioxide: A benign biphasic solvent system for use and recycling of homogeneous catalysts. *J. Am. Chem. Soc.* 125: 5600–01.

Helm, M. L., M. P. Stewart, R. M. Bullock, M. R. DuBois, and D. L. DuBois. 2011. A synthetic nickel electrocatalyst with a turnover frequency above 100,000 s^{-1} for H_2 production. *Science.* 333: 863–66.

Heyduk, A. F. and D. G. Nocera. 2001. The formation of H_2 from irradiation of a mixed valence metal dimer using a sacrificial donor was recently reported. *Science* 293: 1639.

Hinnemann, B., P. G. Moses, J. Bonde, et al. 2005. Biomimetic hydrogen evolution: MoS_2 nanoparticles as catalyst for hydrogen evolution. *J. Am. Chem. Soc.* 127: 5308–09.

Hodes, G., J. Manassen, and D. Cahen. 1977. Photo-electrochemical energy conversion: Electrocatalytic sulphur electrodes. *J. Appl. Electrochem.* 7: 181–82.

Hodes, G., J. Manassen, and D. Cahen. 1980. Electrocatalytic electrodes for the polysulfide redox system. *J. Electrochem. Soc.* 127: 544–49.

Hodes. G. 2008. Comparison of dye- and semiconductor-sensitized porous nanocrystalline liquid junction solar cells. *J. Phys. Chem. C* 112: 17778–87.

Horváth, I. and J. Rába. 1994. Facile catalyst separation without water: Fluorous biphase hydroformylation of olefins. *Science* 266: 72–75.

Hu, C. and H. S. Teng. 2010. Structural features of p-type semiconducting NiO as a co-catalyst for photocatalytic water splitting. *J. Catal.* 272: 1–8.

Husin, H., W. Su, H. Chen, et al. 2011. Photocatalytic hydrogen production on nickel-loaded $La_xNa_{1-x}TaO_3$ prepared by hydrogen peroxide-water based process. *Green Chem.* 13: 1745–54.

Huynh, M.H.V. and T. J. Meyer. 2007. Proton-coupled electron transfer. *Chem. Rev.* 107: 5004–64.

Ikeda, S., H. Nur, T. Sawadaishi, K. Ijiro, M. Shimomura, and B. Ohtani. 2001. Direct observation of bimodal amphiphilic surface structures of zeolite particles for a novel liquid-liquid phase boundary catalysis. *Langmuir* 17: 7976–79.

Ishizuka, T., A. Watanabe, H. Kotani, et al. 2016. Homogeneous photocatalytic water oxidation with a dinuclear Co[III]–pyridylmethylamine complex. *Inorg. Chem.* 55: 1154–64.

Iyer, A., J. Del-Pilar, C. K. King'ondu, et al. 2012. Water oxidation catalysis using amorphous manganese oxides, octahedral molecular sieves (OMS-2), and octahedral layered (OL-1) manganese oxide structures. *J. Phys. Chem. C* 116: 6474–83.

Jacobsen, G. M., J. Y. Yang, B. Twamley, et al. 2008. Hydrogen production using cobalt-based molecular catalysts containing a proton relay in the second coordination sphere. *Energy Environ. Sci.* 1: 167–74.

Jacobson, G. B., C. T. Lee, K. P. Johnston, and W. Tumas. 1999. Enhanced catalyst reactivity and separations using water/carbon dioxide emulsions. *J. Am. Chem. Soc.* 121: 11902–03.

Jacques, P-A, V. Artero, J. Pecaut, and M. Fontecave. 2009. Cobalt and nickel diimine-dioxime complexes as molecular electrocatalysts for hydrogen evolution with low overvoltages. *Proc. Natl. Acad. Sci. USA.* 106: 20627–24.

Jang, J. S., S. H. Choi, D. H. Kim, J. W. Jang, K. S. Lee, and J. S. Lee. 2009. Enhanced photocatalytic hydrogen production from water–methanol solution by nickel intercalated into titanate nanotube. *J. Phys. Chem. C* 113:8990–96.

Jin, L., B. AlOtaibi, D. Benetti, et al. 2016. Near-infrared colloidal quantum dots for efficient and durable photoelectrochemical solar-driven hydrogen production, *Adv. Sci.* 3(3): 1500345.

Jirkovský, J. S., C. D. Malliakas, P. P. Lopes, et al. 2016. Design of active and stable Co–Mo–Sx chalcogels as pH-universal catalysts for the hydrogen evolution reaction. *Nat. Mat.* 15: 197–203.

Jun, H. K., M. A. Careem, and A. K. Arof. 2013. A suitable polysulfide electrolyte for cdse quantum dot-sensitized solar cells. *Int. J. Photoenergy* 2013: 942139.

Kamat, P. V., K. Tvrdy, D. R. Baker, and J. G. Radich. 2010. Beyond photovoltaics: Semiconductor nanoarchitectures for liquid-junction solar cells. *Chem. Rev.* 110: 6664–88.

Kaneco, S., K. Iiba, N.Hiei, K. Ohta, T. Mizuno, and T. Suzuki. 1999. Electrochemical reduction of carbon dioxide to ethylene with high Faradaic efficiency at a Cu electrode in CsOH/methanol. *Electrochim Acta.* 44: 4701–4706.

Karunadasa, H. I., C. J. Chang, and J. R. Long. 2010. A molecular molybdenum-oxo catalyst for generating hydrogen from water. *Nature* 464: 1329–33.

Karunadasa, H. I., E. Montalvo, Y. Sun, M. Majda, J. R. Long, and C. J. Chang. 2012. A molecular MoS_2 edge site mimic for catalytic hydrogen generation. *Science* 335: 699–702.

Katakis, D., C. Mitsopoulou, and E. Vrachnou. 1994. Photocatalytic water splitting: Increase in conversion and energy storage efficiency. *J. Photchem. Photobio A.* 81: 103–06.

Kaur-Ghumaan, S., L. Schwartz, R. Lomoth, M. Stein, and S. Ott. 2010. Catalytic hydrogen evolution from mononuclear iron(II) carbonyl complexes as minimal functional models of the [FeFe] hydrogenase active site. *Angew. Chem. Int. Ed.* 49: 8033–36.

Kim D. W. H., H. G. Kim, J. Kim, K. Y. Cha, Y. G. Kim and J. S. Lee. 2000. Photocatalytic water splitting over highly donor-doped (110) layered perovskites. *J. Catal.* 193:40–48.

Kim, J., D. W. Hwang, H. G. Kim, S. W. Bae, S. M. Ji, and J. S. Lee. 2002. Nickel-loaded $La_2Ti_2O_7$ as a bifunctional photocatalyst. *Chem. Commun.* 2488–89.

Kondo, J. N., M. Uchida, K. Nakajima, D. Lu, M. Hara, and K. Domen. 2004. Synthesis, mesostructure, and photocatalysis of a highly ordered and thermally stable mesoporous Mg and Ta mixed oxide. *Chem. Mater.* 16: 4304–10.

Korzhak, A. V., N. I. Ermokhina, A. L. Stroyuk, et al. 2008. Photocatalytic hydrogen evolution over mesoporous TiO_2/metal nanocomposites. *J. Photochem. Photobiol. A* 198: 126–34.

Kravets, V. G. and A. N. N. 2015. A new class of photocatalytic materials and a novel principle for efficient water splitting under infrared and visible light: MgB_2 as unexpected example. *Opt. Express.* 23: A1651–63.

Kudo, A., A. Tanaka, K. Domen, K. Maruya, K. Aika, and T. Onishi. 1988. Photocatalytic decomposition of water over $NiO-K_4Nb_6O_{17}$ catalyst. *J. Catal.* 111: 67–76.

Kudo A. and Y. Miseki. 2009. Heterogeneous photocatalyst materials for water splitting. *Chem. Soc. Rev.* 38: 253–78.

Law, M., M. C. Beard, S. Choi, J. M. Luther, M. C. Hanna, and A. J. Nozik. 2009. Determining the internal quantum efficiency of PbSe nanocrystal solar cells with the aid of an optical model. *Nano Lett.* 8(11): 3904–10.

Lazarides,T., T. McCormick, P. Du, G. Luo, B. Lindley, and R. Eisenberg. 2009. Making hydrogen from water using a homogeneous system without noble metals. *J. Am. Chem. Soc.* 131: 9192–94.

Lazzeri, M. and A. Selloni. 2001. Stress driven reconstruction of an oxide surface: The anatase TiO_2 (001) (1x4) surface. *Phys. Rev. Lett.* 87: 266105.

Lazzeri, M., A. Vittadini, and A. Selloni. 2001. Structure and energetics of stoichiometric TiO_2 anatase surfaces. *Phy. Rev. B* 63: 155409 and Erratum *Phys. Rev. B* 65: 119901 (2002).

Lee, J. W., J. Y. Shin, Y. S. Chun, H. B. Jang, C. E. Song, and S.-G. Lee. 2010. Toward understanding the origin of positive effects of ionic liquids on catalysis: Formation of more reactive catalysts and stabilization of reactive intermediates and transition states in ionic liquids. *Acc. Chem. Res.* 43: 985–94.

Lee, S. J., S. Mukerjee, E. A. Ticianelli, and J. McBreen. 1999. Electrocatalysis of CO tolerance in hydrogen oxidation reaction in PEM fuel cells. *Electrochim. Acta.* 44: 3283–93.

Leschkies, K. S., R. Divakar, J. Basu, et al. 2007. Photosensitization of ZnO nanowires with CdSe quantum dots for photovoltaic devices. *Nano Lett.,* 7: 1793–98.

Leschkies, K. S., T. J. Beatty, M. S. Kang, D. J. Norris, and E. S. Aydil. 2009. Solar cells based on junctions between colloidal PbSc nanocrystals and thin ZnO films. *ACS Nano* 3: 3638–48.

Li, Q., B. D. Guo, J. G. Yu, et al. 2011. Highly efficient visible-light-driven photocatalytic hydrogen production of CdS-cluster-decorated graphene nanosheets. *J. Am. Chem. Soc.* 133: 10878–84.

Lichterman, M. F., A. I. Carim, M. T. McDowell, et al. 2014. Stabilization of n-cadmium telluride photoanodes for water oxidation to $O_2(g)$ in aqueous alkaline electrolytes using amorphous TiO_2 films formed by atomic-layer deposition, *Energy Environ. Sci.* 7: 3334–37.

Lin, H. Y., H. C. Yang, and W. L. Wang. 2011.Synthesis of mesoporous Nb_2O_5 photocatalysts with Pt, Au, Cu and NiO cocatalyst for water splitting. *Catal. Today* 174: 106–113.

Lingampalli, S. R., U. K. Gautam, and C. N. R. Rao. 2013. Highly efficient photocatalytic hydrogen generation by solution-processed ZnO/Pt/CdS, ZnO/Pt/Cd$_{1-x}$Zn$_x$S and ZnO/Pt/CdS$_{1-x}$Se$_x$ hybrid. *Energy Environ. Sci.* 6: 3589–94.

Liu, C., P. C. Wu, T. Sun, et al. 2009. Synthesis of high quality *n*-type CdSe nanobelts and their applications in nanodevices. *J. Phys. Chem. C* 113: 14478–81.

Liu, F., M. B. Abrams, R. T. Baker, and W. Tumas. 2001. Phase-separable catalysis using room temperature ionic liquids and supercritical carbon dioxide. *Chem. Commun.* (5): 433–434.

Liu, F., J. J. Concepcion, J. W. Jurss, T. Cardolaccia, J. L. Templeton, and T. J. Meyer. 2008a. Mechanisms of water oxidation from the blue dimer to photosystem II. *Inorg Chem* 47:1727–1752.

Liu, H., J. Yuan, W. Shangguan, and Y. Teraoka. 2008b. Visible-light-responding BiYWO$_6$ solid solution for stoichiometric photocatalytic water splitting. *J. Phys. Chem. C* 112: 8521–23.

Liu, T. and M. Y. Darensbourg. 2007. A mixed-valent, Fe(II)Fe(I), diiron complex reproduces the unique rotated state of the [FeFe]hydrogenase active site. *J. Am. Chem. Soc.* 129: 7008–09.

Lodi, G., E. Sivieri, A. D. Battisti, and S. Trasatti. 1978. Ruthenium dioxide-based film electrodes III. Effect of chemical composition and surface morphology on oxygen evolution in acid solutions. *J. Appl. Electrochem.* 8: 135–43.

Lu, G.-Q. and A. Wieckowski. 2000. Heterogeneous electrocatalysis: A core field of interfacial science. *Curr. Opin. Colloid Interface Sci.* 5(1–2): 95–100.

Lu, H. and C. Li. 2007. Oxidative desulfurization of dibenzothiophene with molecular oxygen using emulsion catalysis. *Chem. Commun.* (2): 150–52.

Lu, H., J. Gao, Z., Jiang, et al. 2006. Ultra-deep desulfurization of diesel by selective oxidation with [C$_{18}$H$_{37}$N(CH$_3$)$_3$]$_4$[H$_2$NaPW$_{10}$O$_{36}$] catalyst assembled in emulsion droplets. *J. Catal.* 239: 369–75.

Luo, J., N. P. Rath, and L. M. Mirica. 2011. Dinuclear Co(II)Co(III) mixed-valence and Co(III) Co(III) Complexes with N- and O-donor ligands: Characterization and water oxidation studies. *Inorg. Chem.* 50: 6152–57.

Luska, K. L. and A. Moores. 2012. Functionalized ionic liquids for the synthesis of metal nanoparticles and their application in catalysis. *Chem. Cat. Chem.* 4: 1534–46.

Lv, J., T. Kako, Z. S. Li, et al. 2010. Synthesis and photocatalytic activities of NaNbO$_3$ rods modified by In$_2$O$_3$ nanoparticles. *J. Phys. Chem. C* 114: 6157–62.

Ma, B., J. Yang, H. Han, et al. 2010. Enhancement of photocatalytic water oxidation activity on IrO$_x$–ZnO/Zn$_{2-x}$GeO$_{4-x-3y}$N$_{2y}$ Catalyst with the solid solution phase junction. *J. Phys. Chem. C* 114: 12818–22.

Maeda, K., K. Teramura, D. L. Lu, et al. 2006a. Photocatalyst releasing hydrogen from water. *Nature* 440: 295.

Maeda, K., K. Teramura, D. Lu, N. Saito, Y. Inoue, and K. Domen. 2006b. Noble-metal/Cr$_2$O$_3$ core/shell nanoparticles as a cocatalyst for photocatalytic overall water splitting. *Angew. Chem. Int. Ed.* 45: 7806–09.

Maeda, K., M. Higashi, D. L. Lu, et al. 2010a. Efficient nonsacrificial water splitting through two-step photoexcitation by visible light using a modified oxynitride as a hydrogen evolution photocatalyst. *J. Am. Chem. Soc.* 132: 5858–68.

Maeda, K., N. Sakamoto, T. Ikeda, et al. 2010b. Preparation of core–shell-structured nanoparticles (with a noble-metal or metal oxide core and a chromia shell) and their application in water splitting by means of visible light. *Chem. Eur. J.* 16: 7750–59.

Mahapatra P. K. and A. R. Dubey. 1994. Photoelectrochemical behaviour of mixed polycrystalline n-type CdS-CdSe electrodes. *Sol. Energy Mater. Sol. Cells* 32: 29–35.

Marinescu, S. C., J. R. Winkler, and H. B. Gray. 2012. Molecular mechanisms of cobalt-catalyzed hydrogen evolution. *Proc. Natl. Acad. Sci. USA* 109: 15127–31.

Mathur, A., S. Bali, M. Balakrishnan, R. Perumal, and V. S. Batra. 1999. Demonstration of coal gas run molten carbonate fuel cell concept. *Int. J. Energy Res.* 23: 1177–1185.

McCrory, C. C. L., S. Jung, J. C. Peters, and T. F. Jaramillo. 2013. Benchmarking heterogeneous electrocatalysts for the oxygen evolution reaction. *J. Am. Chem. Soc.* 135: 16977–87.

McNicol, B. D., D. A. J. Rand, and K. R.Williams. 1999. Direct methanol-air fuel cells for road transportation. *J. Power Sources* 83: 15–31.

Meissner, D., R. Memming, and B. Kastening. 1998. Photoelectrochemistry of cadmium sulfide. 1. Reanalysis of photocorrosion and flat-band potential. *J. Phys. Chem.* 92(12): 3476–83.

Meng, F., J. T. Li, S. K. Cushing, M. J. Zhi, and N. Q. Wu. 2013. Solar hydrogen generation by nanoscale *p–n* junction of *p*-type molybdenum disulfide/*n*-type nitrogen-doped reduced graphene oxide. *J. Am. Chem. Soc.* 135: 10286–89.

Meng, N., M. K. H. Leung, D. Y. C. Leung, and K. Sumathy. 2007. A review and recent developments in photocatalytic water-splitting using TiO_2 for hydrogen. *Renew. Sustain. Energy Rev.* 11(3): 401–25.

Mills, A. 1989. Heterogeneous redox catalysts for oxygen and chlorine evolution. *Chem. Soc. Rev.* 18: 285–316.

Miseki, Y., H. Kato, and A. Kudo. 2009. Water splitting into H_2 and O_2 over niobate and titanate photocatalysts with (111) plane-type layered perovskite structure. *Energy Environ. Sci.* 2: 306–14.

Mohtadi, R.W., K. Lee, S. Cowan, J. W. Van Zee, and M. Murthy. 2003. Effects of hydrogen sulfide on the performance of a PEMFC. *Electrochem. Solid-State Lett.* 6: A272–74.

Muckerman, J. T. and E. Fujita. 2011. Theoretical studies of the mechanism of catalytic hydrogen production by a cobaloxime. *Chem. Commun.* 47: 12456–58.

Murdoch, M., G. I. N. Waterhouse, M. A. Nadeem, et al. 2011. The effect of gold loading and particle size on photocatalytic hydrogen production from ethanol over Au/TiO_2 nanoparticles. *Nat. Chem.* 3: 489–92.

Nada, A. A., M. H. Barakat, H. A. Hamed, N. R. Mohamed, and T. N. Veziroglu. 2005. Studies on the photocatalytic hydrogen production using suspended modified TiO_2 photocatalysts. *Int. J. Hydrogen Energy* 30(7): 687–91.

Nakagawa, T., C. A. Beasley, and R. W. Murray. 2009. Efficient electro-oxidation of water near its reversible potential by a mesoporous IrO_x nanoparticle film. *J. Phys. Chem. C* 113: 12958–61.

Naruta, Y., M. Sasayama, and T. Sasaki. 1994. Oxygen evolution by oxidation of water with manganese porphyrin dimers. *Angew. Chem. Int. Ed. Engl.* 33: 1839–41.

Nørskov, J. K., T. Bligaard, J. Rossmeisl, and C. H. Christensen. 2009. Towards the computational design of solid catalysts. *Nat. Chem.* 1: 37–46.

Nowotny, M. K. T. Bak, and J. Nowotny. 2006. Electrical properties and defect chemistry of TiO_2 single crystal. IV. Prolonged oxidation kinetics and chemical diffusion. *J. Phys. Chem. B* 110: 16302–308.

Nur, H., S. Ikeda, and B. Ohtani. 2001. Phase-boundary catalysis of alkene epoxidation with aqueous hydrogen peroxide using amphiphilic zeolite particles loaded with titanium oxide. *J. Catal.* 204: 402–08.

Ohko, Y., S. Saitoh, T. Tatsuma, and A. Fujishima. 2001. Photoelectrochemical anticorrosion and self-cleaning effects of a TiO_2 coating for type 304 stainless steel. *J. Electrochem. Soc.* 148: B24–28.

Onsuratoom, S., T. Puangpetch, and S. Chavadej. 2011. Comparative investigation of hydrogen production over Ag-, Ni-, and Cu-loaded mesoporous-assembled TiO_2–ZrO_2 mixed oxide nanocrystal photocatalysts. *Chem. Eng. J.* 173: 667–75.

Paal, Z., K. Matusek, and M. Muhler. 1997. Sulfur adsorbed on Pt catalyst: its chemical state and effect on catalytic properties as studied by electron spectroscopy and n-hexane test reactions. *Appl. Catal. A* 149: 113–32.

Park S. D., J. M.Vohs, and R. J. Gorte. 2000. Direct oxidation of hydrocarbons in a solid-oxide fuel cell. *Nature* 404: 265–67.

Park, S., Y. Seo, M. S. Kim, and S. Lee. 2013. Solar energy conversion by the regular array of TiO₂ nanotubes anchored with ZnS/CdSSe/CdS quantum dots formed by sequential ionic bath deposition. *Bull. Korean Chem. Soc.* 34(3): 856–62.

Patzke, G., R. Alberto, P. Hamm, and K. Baldridge. Homogenous photocatalytic water splitting. SNF Sinergia Project Report. Duration of project is from January 2011 to October 2014. Available at http://www.research-projects.uzh.ch/p16425.htm. Accessed on 1.11.2014.

Peng, T. Y., P. Zeng, D. N. Ke, et al. 2011. Hydrothermal preparation of multiwalled carbon nanotubes (MWCNTs)/CdS nanocomposite and its efficient photocatalytic hydrogen production under visible light irradiation. *Energy Fuels* 25: 2203–10.

Phivilay, S. P., A. A. Puretzky, K. Domen, and I. E. Wachs. 2013. Nature of catalytic active sites present on the surface of advanced bulk tantalum mixed oxide photocatalysts. *ACS Catal.* 3: 2920–29.

Plass, R., S. Pelet, J. Krueger, M. Gratzel, and U. Bach. 2002. Quantum dot sensitization of organic–inorganic hybrid solar cells. *J. Phys. Chem. B* 106(31): 7578–80.

Pushkar, Y., J. Yano, K. Sauer, A. Boussac, and V. K. Yachandra. 2008. Structural changes in the Mn₄Ca cluster and the mechanism of photosynthetic water splitting. *Proc. Natl. Acad. Sci. USA* 105: 1879–84.

Qu Y. Q. and X. F. Duan. 2013. Progress, challenge and perspective of heterogeneous photocatalysts. *Chem. Soc. Rev.* 42: 2568–80.

Ran, J. R., J. G. Yu, and M. Jaroniec. 2011. Ni(OH)₂ modified CdS nanorods for highly efficient visible-light-driven photocatalytic H₂ generation. *Green Chem.* 13: 2708–13.

Ran, J., J. Zhang, J. Yu, M. Jaroniec, and S. Z. Qiao. 2014. Earth-abundant cocatalysts for semiconductor based photocatalytic water splitting. *Chem. Soc. Rev.* 43: 7787–812.

Robel, I., B. A. Bunker, and P. V. Kamat. 2005. Single-walled carbon nanotube–CdS nanocomposites as light-harvesting assemblies: Photoinduced charge-transfer interactions. *Adv. Mater.* 17: 2458–63.

Robel, I., V. Subramanian, M. Kuno, and P. V. Kamat. 2006. Quantum dot solar cells. Harvesting light energy with CdSe nanocrystals molecularly linked to mesoscopic TiO₂ films. *J. Am. Chem. Soc.* 128(6): 2385–93.

Rocha, T. S., E. S. Nascimento, A. C. Silva, et al. 2013. Enhanced photocatalytic hydrogen generation from water by Ni(OH)₂loaded on Ni-doped δ-FeOOH nanoparticles obtained by one-step synthesis. *RSC Adv.* 3: 20308–14.

Ruzycki, N., G. S., Herman, L. A., Boatner, and U. Diebold. 2003. Scanning tunneling microscopy study of the anatase (1 0 0) surface. *Surf. Sci.* 529: L239–44.

Sasaki, Y., A. Iwase, H. Kato, and A. Kudo. 2008. The effect of co-catalyst for Z-scheme photocatalysis systems with an Fe³⁺/Fe²⁺ electron mediator on overall water splitting under visible light irradiation *J. Catal.* 259: 133–37.

Sato, J., N. Saito, Y. Yamada, et al. 2005. RuO₂-Loaded β-Ge₃N₄ as a non-oxide photocatalyst for overall water splitting. *J. Am. Chem. Soc.* 127: 4150–51.

Sauer, K., J. Yano, and V. K. Yachandra. 2005. X-ray spectroscopy of the Mn₄Ca cluster in the water-oxidation complex of photosystem II. *Photosynth. Res.* 85: 73–86.

Sayama, K. and H. Arakawa. 1994. Effect of Na₂CO₃ addition on photocatalytic decomposition of liquid water over various semiconductors catalysis. *J. Photochem. Photobiol. A Chem.* 77: 243–7.

Sayama, K. and H. Arakawa. 1996. Effect of carbonate addition on the photocatalytic decomposition of liquid water over a ZrO₂ catalyst. *J. Photochem. Photobiol. A Chem.* 94: 67–76.

Sayama, K. and H. Arakawa. 1992. Significant effect of carbonate addition on stoichiometric photodecomposition of liquid water into hydrogen and oxygen from platinum-titanium (IV) oxide suspension. *J. Chem. Soc. Chem. Commun.* 2: 150–2.

Sayama, K., H. Arakawa, and K. Domen. 1996. Photocatalytic water splitting on nickel inter-calated $A_4Ta_xNb_{6-x}O_{17}$ (A = K; Rb). *Catal. Today* 28: 175–82.

Sayama, K., K. Yase, H. Arakawa, et al. 1998. Photocatalytic activity and reaction mechanism of Pt-intercalated $K_4Nb_6O_{17}$ catalyst on the water-splitting in carbonate salt aqueous solution. *J Photochem. Photobiol. A Chem.* 114: 125–35.

Sayed, F. N., O. D. Jayakumar, R. Sasikala, et al. 2012. Photochemical hydrogen generation using nitrogen-doped TiO_2–Pd nanoparticles: Facile synthesis and effect of Ti^{3+} incor-poration. *J. Phys. Chem. C* 116: 12462–67.

Schley, N. D., J. D. Blakemore, N. K. Subbaiyan, et al. 2011. Distinguishing homogeneous from heterogeneous catalysis in electrode-driven water oxidation with molecular irid-ium complexes. *J. Am. Chem. Soc.* 133: 10473–81.

Schlögl, R. and S. B. Abd Hamid. 2004. Nanocatalysis: Mature science revisited or something really new?. *Angew. Chem. Int. Ed.* 43: 1628–37.

Scholten, J. D., B. C. Leal, and J. Dupont. 2012. Transition metal nanoparticle catalysis in ionic liquids. *ACS Catal.* 2: 184–200.

Schrauzer, G. N. and R. J. Holland. 1971. Hydridocobaloximes. *J. Am. Chem. Soc.* 93: 1505–06.

Sellin, M. F., P. B. Webb, and D. J. Cole-Hamilton. 2001. Continuous flow homogeneous catalysis: hydroformylation of alkenes in supercritical fluid–ionic liquid biphasic mix-tures. *Chem. Commun.* (8): 781–82.

Semagina, N. and L. Kiwi-Minsker. 2009. Recent advances in the liquid-phase synthesis of metal nanostructures with controlled shape and size for catalysis. *Catal. Rev. Sci. Eng.* 51: 147–217.

Seo, S. W., S. Park, H. Jeong, et al. 2012. Enhanced performance of $NaTaO_3$ using molecular co-catalyst $[Mo_3S_4]^{4+}$ for water splitting into H_2 and O_2. *Chem. Commun.* 48: 10452.

Sheeney-Haj-Khia, L., B. Basnar, and I. Willner. 2005. Efficient generation of photocurrents by using CdS/carbon nanotube assemblies on electrodes. *Angew. Chem. Int. Ed.*, 44: 78–83.

Shen, M. and D. E. Resasco. 2009. Emulsions stabilized by carbon nanotube–silica nanohy-brids. *Langmuir* 25(18): 10843–51.

Shi, R., J. Lin, Y. Wang, J. Xu, and Y. Zhu. 2010. Visible-light photocatalytic degradation of $BiTaO_4$ photocatalyst and mechanism of photocorrosion suppression. *J. Phys. Chem. C* 114(14): 6472–77.

Shimazaki, Y., T. Nagano, H. Takesue, B.-H. Ye, F. Tani, and Y. Naruta. 2003. Characterization of a dinuclear Mn^V=O complex and its efficient evolution of O_2 in the presence of water. *Angew. Chem. Int. Ed.* 43: 98–100.

Sirjoosingh, A. and S. Hammes-Schiffer. 2011. Proton-coupled electron transfer versus hydro-gen atom transfer: Generation of charge-localized diabatic states. *J. Phys. Chem. A* 115: 2367–77.

Solis, B. H. and S. Hammes-Schiffer. 2011. Substituent effects on cobalt diglyoxime catalysts for hydrogen evolution. *J. Am. Chem. Soc.* 133: 19036–39.

Solis, B. H. and S. Hammes-Schiffer. 2011a. Theoretical analysis of mechanistic pathways for hydrogen evolution catalyzed by cobaloximes. *Inorg. Chem.* 50: 11252–62.

Stubbert, B. D., J. C. Peters, and H. B. Gray. 2011. Rapid water reduction to H_2 catalyzed by a cobalt bis (iminopyridine) complex. *J. Am. Chem. Soc.* 133: 18070–73.

Suh, W. H., Y. H. Suh, and G. D. Stucky. 2009. Multifunctional nanosystems at the interface of physical and life sciences. *Nano Today* 4: 27–36.

Tachan, Z., M. Shalom, I. Hod, S. Ruhle, S. Tirosh, and A. Zaban. 2011. PbS as a highly cata-lytic counter electrode for polysulfide-based quantum dot solar cells. *J. Phys. Chem. C* 115: 6162–66.

Takahara, Y., J. N. Kondo, T. Takata, D. L. Lu, and K. Domen. 2001. Mesoporous tantalum oxide. 1. Characterization and photocatalytic activity for the overall water decomposi-tion. *Chem. Mater.* 13: 1194–99.

Tambago, H. M. G. and R. L. de Leon. 2015. Intrinsic kinetic modeling of hydrogen production by photocatalytic water splitting using cadmium zinc sulfide catalyst. *Int. J. Chem. Eng. Appl.* 6: 220–27.

Tanaka, K., H. Isobe, S. Yamanaka, and K. Yamaguchi. 2012. Similarities of artificial photosystems by ruthenium oxo complexes and native water splitting systems. *Proc. Natl. Acad. Sci. USA* 109: 15600–05.

Tanner, R. E., A. Sasahara, Y. Liang, E. I. Altman, and H. Onishi. 2002. Formic acid adsorption on anatase TiO_2 (001) (1x4) thin films studied by NC AFM and STM. *J. Phys. Chem. B* 106: 8211–22.

Tong, H., S. X. Ouyang, Y. P. Bi, N. Umezawa, M. Oshikiri, and J. H. Ye. 2012. Nanophotocatalytic materials: Possibilities and challenges. *Adv. Mater.* 24: 229–51.

Trasatti, S. 1992. *Advances in Electrochemical Science and Engineering*, (eds. Gerischer, H., and C. W. Tobias), Vol. 2, Weinheim, Germany: Weily VCH, pp. 1–85.

Trasatti, S. 1999. Interfacial electrochemistry of conductive oxides for electrocatalysis. In: *Interfacial Electrochemistry: Theory, Experiment, and Applications* (ed. Wieckowski, A). New York, NY: Marcel Dekker, pp. 769–92.

Trasatti, S. 2000. Electrocatalysis: Understanding the success of DSA. *Electrochem. Acta* 45: 2377–85.

Tsuji, I., H. Kato, and A. Kudo. 2005. Visible-light-induced H_2 evolution from an aqueous solution containing sulfide and sulfite over a $ZnS–CuInS_2–AgInS_2$ solid-solution photocatalyst. *Angew. Chem. Int. Ed.* 44: 3565–68.

Tsuji, I., H. Kato, H. Kobayashi, and A. Kudo. 2004. Photocatalytic H_2 evolution reaction from aqueous solutions over band structure-controlled $(AgIn)_xZn_{2(1-x)}S_2$ solid solution photocatalysts with visible-light response and their surface nanostructures. *J. Am. Chem. Soc.* 126: 13406–13.

Umena, Y., K. Kawakami, J.-R.Shen, and N. Kamiya. 2011. Crystal structure of oxygen-evolving photosystem II at a resolution of 1.9 Å. *Nature* 473: 55–60.

Ungureanu, S., H. Deleuze, M. I. Popa, C. Sanchez, and R. Backov. 2008. First Pd@Organo–Si(HIPE) open-cell hybrid monoliths generation offering cycling heck catalysis reactions. *Chem. Mater.* 20: 6494–500.

Vijayaraghavan, K., T. K., Ramanujam, and N. Balasubramanian. 1999. In situ hypochlorous acid generation for the treatment of syntan wastewater. *Waste Manage.* 19: 319–323.

Wang, X., Q. Xu, M. R. Li, et al. 2012. Photocatalytic overall water splitting promoted by an α–β phase junction on Ga_2O_3. *Angew. Chem. Int. Ed.* 51: 13089–92.

Wang, C.-M. and Wang, C.-Y. 2014. Photocorrosion of plasmonic enhanced Cu_xO photocatalyst. *J. Nanophoton.* 8(1): 084095.

Wasielewski, M. R. 2009. Self-assembly strategies for integrating light harvesting and charge separation in artificial photosynthetic systems. *Acc. Chem. Res.* 42: 1910–21.

Wei, W., X. Y. Bao, C. Soci, Y. Ding, Z. L. Wang, and D. Wang. 2009. Direct heteroepitaxy of vertical InAs nanowires on Si substrates for broad band photovoltaics and photodetection. *Nano Lett.* 9: 2926–33.

Wender, H., R. V. Goncalves, C. S. B. Dias, et al. 2013. Photocatalytic hydrogen production of $Co(OH)_2$ nanoparticle-coated α-Fe_2O_3 nanorings. *Nanoscale* 5: 9310–16.

Wu, K. F., H. M. Zhu, Z. Liu et al. 2012. Ultrafast charge separation and long-lived charge separated state in photocatalytic CdS–Pt nanorod heterostructures. *J. Am. Chem. Soc.* 134: 10337–40.

Wu, N.-L., M.-S. Lee, Z.-J. Pon, and J.-Z. Hsu. 2004. Effect of calcination atmosphere on TiO_2 photocatalysis in hydrogen production from methanol/water solution. *J. Photochem. Photobiol. A: Chem.* 163(1–2): 277–80.

Wydrzynski, T. and S. Satoh. 2005. *Photosystem II: The Light-Driven Water: Plastoquinone Oxidoreductase*. Berlin, Germany: Springer, Dordrecht.

Xiang, Q. J., J. G. Yu, and M. Jaroniec. 2011. Preparation and enhanced visible-light photo-catalytic H_2-production activity of graphene/C_3N_4 composites. *J. Phys. Chem. C* 115: 7355–63.

Xie, Y. P., G. Liu, G. Q. Lu, and H. M. Cheng. 2012. Boron oxynitride nanoclusters on tung-sten trioxide as a metal-free cocatalyst for oxygen evolution from photocatalytic water splitting. *Nanoscale* 4: 1267–70.

Xu, A. W., Y. Gao, and H. Q. Liu. 2002. The preparation characterization and their photocata-lytic activities of rareearthdoped TiO_2 nanoparticles. *J. Catal.* 207: 151–7.

Yan, H. J., G. J. Ma, G. P. Wu, et al. 2009. Visible-light-driven hydrogen production with extremely high quantum efficiency on Pt–PdS/CdS photocatalyst. *J. Catal.* 266: 165–68.

Yan, N., C. Xiao, and Y. Kou. 2010. Transition metal nanoparticle catalysis in green solvents. *Coord. Chem. Rev.* 254: 1179–1218.

Yanagida, S., K. Mizumoto, and C. J. Pac. 1986. Semiconductor photocatalysis. Part 6. Cis-trans photoisomerization of simple alkenes induced by trapped holes at surface states. *J. Am. Chem. Soc.* 108: 647–54.

Yang, J. H., D. Wang, H. X. Han, and C. Li. 2013. Roles of cocatalysts in photocatalysis and photoelectrocatalysis. *Acc. Chem. Res.* 46: 1900–09.

Yang, Z. S., C. Y. Chen, C. W. Liu, and H. T. T. 2010. Electrocatalytic sulfur electrodes for CdS/CdSe quantum dot-sensitized solar cells. *Chem. Commun.* 46: 5485–87.

Youngblood, W. J., S. H. A. Lee, Y. Kobayashi, et al. 2009. Photoassisted overall water split-ting in a visible light-absorbing dye-sensitized photoelectrochemical cell. *J. Am. Chem. Soc.* 131: 926–27.

Yu J. G. and J. R. Ran. 2011. Facile preparation and enhanced photocatalytic H_2-production activity of $Cu(OH)_2$ cluster modified TiO_2. *Energy Environ. Sci.* 4: 1364–71

Yu, J. G., B. Yangm and B. Cheng. 2012. Noble-metal-free carbon nanotube-$Cd_{0.1}Zn_{0.9}S$ com-posites for high visible-light photocatalytic H_2-production performance. *Nanoscale* 4: 2670–77.

Yu, J. G., L. F. Qi, and M. Jaroniec. 2010. Hydrogen production by photocatalytic water split-ting over Pt/TiO_2 nanosheets with exposed (001) facets. *J. Phys. Chem. C* 114: 13118–25.

Yu, J. G., S. H. Wang, B. Cheng, Z. Lin, and F. Huang. 2013. Noble metal-free $Ni(OH)_2$–g-C_3N_4 composite photocatalyst with enhanced visible-light photocatalytic H_2-production activity. *Catal. Sci. Technol.* 3: 1782–89.

Yu, J. G., Y. Hai, and B. Cheng. 2011. Enhanced photocatalytic H_2-production activity of TiO_2 by $Ni(OH)_2$ cluster modification. *J. Phys. Chem. C* 115: 4953–58.

Yu, P. R., K. Zhu, A. G. Norman, S. Ferrere, A. J. Frank, and A. J. Nozik. 2006. Nanocrystalline TiO_2 solar cells sensitized with InAs quantum dots. *J. Phys. Chem. B* 110: 25451–54.

Yu, Z. R., J. H. Du, S. H. Guo, H. Y. Zhang, and Y. Matsumoto. 2002. CoS thin films prepared with modified chemical bath deposition. *Thin Solid Flims* 415: 173–76.

Yuan, Y., N. Yan, and P. J. Dyson. 2012. Advances in the rational design of rhodium nanopar-ticle catalysts: Control via manipulation of the nanoparticle core and stabilizer. *ACS Catal.* 2: 1057–69.

Zaban, A., O. I. Micic, B. A. Gregg, and A. J. Nozik. 1998. Photosensitization of nanoporous TiO_2 electrodes with InP quantum dots. *Langmuir* 14: 3153–56.

Zhang, J., J. G. Yu, M. Jaroniec, and J. R. Gong. 2012. Noble metal-free reduced graphene oxide-$Zn_xCd_{1-x}S$ nanocomposite with enhanced solar photocatalytic H_2-production per-formance. *Nano Lett.* 12: 4584–89.

Zhang, J., S. Z. Qiao, L. F. Qi et al. 2013a. Fabrication of NiS modified CdS nanorod p–n junc-tion photocatalysts with enhanced visible-light photocatalytic H_2-production activity. *Phys. Chem. Chem. Phys.* 15: 12088–94.

Zhang, Q. X., Y. D. Zhang, S. Q. Huang, et al. 2010. Application of carbon counterelectrode on CdS quantum dot-sensitized solar cells (QDSSCs). *Electrochem. Commun.* 12: 327–30.

Zhang, S. S., H. J. Wang, M. S. Yeung, Y. P. Fang, H. Yu, and F. Peng. 2013b. $Cu(OH)_2$-modified TiO_2 nanotube arrays for efficient photocatalytic hydrogen production *Int. J. Hydrogen Energy*. 38: 7241–45.

Zhou, H. L., Y. Q. Qu, T. Zeida, and X. F. Duan. 2012. Towards highly efficient photocatalysts using semiconductor nanoarchitectures. *Energy Environ. Sci.* 5: 6732–43.

7 Nanostructured Semiconducting Materials for Water Splitting

7.1 INTRODUCTION

The world economy is at a threshold point, where we need to change not only our energy resources from conventional to renewable but also our present energy carrier to develop an energy efficient, low price, nontoxic, stable, eco-friendly fuel at large scale. Hydrogen is one of the potential candidates that fulfills all of the aforementioned requirements. Therefore, hydrogen is known as the "fuel of the future." It has been recognized that hydrogen is playing a critical role in the development of the green energy economy since it is an ultimate form of clean energy and can be used in fuel cells and other processes. However, currently hydrogen is primarily formed by steam reforming, in which fossil fuels are consumed and carbon dioxide is generated (Kudo and Miseki 2009). Simultaneously, water is an abundant renewable source of hydrogen. Sunlight has attracted considerable interest and it is an inexpensive, nonpolluting, abundant (10,000 times the total energy that is consumed on this planet), and endlessly renewable source of clean energy. With respect to solar energy conversion, four major technologies are currently available for converting sunlight into useful forms of energy—electrical and chemical (hydrogen). The most feasible technology involves converting the electrical energy into some form of chemical fuel such as hydrogen or other small, low carbon organics, which can be easily stored, transported, and used. Photocatalysis or photoelectrolysis (PE) can be utilized to split water into hydrogen and oxygen without any harmful emission of by-products. This is because the only combustion product of hydrogen, that is, water, can be recycled. Researchers from industry and academic institutes have put forward their best efforts in developing photocatalytic material that can quickly generate hydrogen from water using sunlight, potentially creating a clean and renewable source of energy. Usually, solar or photovoltaic (PV) cells, thin film (TF) PV cells, wet chemical photosynthesis cells, and PE are the prominent technologies that are used for converting sunlight into energy (electricity/chemical). We will detail these technologies/strategies in Section 7.2. We will discuss nanomaterial structuring, energetic transport dynamics, and material design for advancement in materials with reference to nanocrystalline materials, TF materials, mesoporous materials, and advanced nanostructures for water splitting.

7.2 NANOMATERIAL STRUCTURE, ENERGETIC TRANSPORT DYNAMICS, AND MATERIAL DESIGN

PE of water requires two electrodes—a photoanode and photocathode (counter electrode) dipped in water with a reference electrode. These photoelectrodes need photoactive material (or semiconducting material) to make photoelectrodes active, which is irradiated by a light beam during water splitting. In order to drive the photoelectrode reactions, the photoactive material of the photoelectrode, band edges (conduction band [CB] minima and valence band [VB] maxima) must straddle between the redox potential of the water, that is, 1.23 V, so the water molecule can be oxidized to form O_2 and protons (H^+) and reduced to form H_2 and hydroxyl ions OH^-. These photoreactions are used to convert solar energy into useful work, that is, electricity or chemical energy by means of different transport dynamics. They are categorized as follows.

7.2.1 DEVICES WITH DIFFERENT ENERGETIC TRANSPORT DYNAMICS

7.2.1.1 Solar or PV Cell

This is an electrical device that directly converts the energy of light into electricity by the PV effect, which is a physicochemical phenomenon and is used to convert sunlight into electric or chemical energy at a large scale. It comprises two steps, photon-to-electric energy conversion and electric energy-to-chemical energy conversion. During the generation of electricity from a Si solar cell, the first process is the absorption of the photons by crystalline silicon wafers (semiconductor/photocatalyst) of 180–300 mm thick film causes the migration of excitons to a p–n junction and leads to the charge separation that results in electron–hole pair generation. This device is used to drive and direct the current of electrons in the form of electricity and also used to reduce chemicals and holes used to oxidize chemicals. Unfortunately, these systems suffer the problem of high cost ($0.62/watt in 2012; two to five times higher than that of conventional energy technologies; material costing and associated processing are expensive) that is attributed to making and processing of the high purity silicon film (99.999%). It is necessary to know the purity of the Si material that is required to promote efficient photon absorption process and to suppress recombination, which are the major challenges faced by the current technology. Impurity causes defects, which are popular centers for recombination of carriers (Atwater and Polman 2010). To overcome the associated challenges, various technologies (dye- or quantum dots–sensitized solar cells, provskite, polymer, and organic PVs) are introduced for developing a visible light harvesting low-cost material for this technique (Gratzel 2001; Yang et al. 2007).

7.2.1.2 Thin-Film PVs

To reduce the cost of expensive silicon-based photoactive materials in PV cells, TF PVs have been explored. These TFs are made of low-cost photoactive material such as GaAs, CdTe, $CuInSe_2$, as well as amorphous/polycrystalline silicon as a substitute for expensive conventional crystalline silicon. On cheap substrates, 1- to 2-mm thick films of aforementioned materials such as glass, stainless steel, or plastic are fabricated (Atwater and Polman 2010).

7.2.1.3 Wet-Chemical Photosynthesis

In order to explore new technology inspired by nature's photosynthesis that converts sunlight into chemicals, wet-chemical photosynthesis (sometimes called artificial photosynthesis) has been investigated. The goal is to achieve an efficient and stable energy conversion device, which utilizes only Earth-abundant materials and operates under mild conditions. Production of organics by the reduction of CO_2 and water or splitting of water, in the presence of photocatalytic systems and sunlight, is one of the forms of wet-chemical photosynthesis. The harvesting and storing of solar energy via chemical methods, which occurs naturally in photosynthesis, is an attractive strategy for meeting the challenges of solar energy applications.

7.2.1.4 Photoelectrolysis

Amalgamation of the concepts of PVs and wet-chemical photosynthesis generates a new technique, that is, PE. It is an ability to split water through solar and electric energy to produce hydrogen (a chemical fuel). PE is an electrochemically assisted process, which has a solid photoelectrode, used to convert the energy of sunlight into splitting of water for generating chemical energy. PE devices can be divided into three categories:

Type 1: Schottky, single absorber
Type 2: Tandem, dual absorber
Type 3: Multiple junction (a) subcell separated and (b) monolith devices
(Figure 7.1a–c) (Walker et al. 2010; Krol 2012).

In type 1 PE, light absorption and charge separation occur at a single absorber particle connected to one or more cocatalysts to complete the circuit for water electrolysis. The ideal maximum achievable solar-to-hydrogen (STH) efficiency for this Schottky-type configuration is $\eta = 14.4\%$, which would be achieved by using a light absorber of 2.0 eV band gap and conversion losses of $E_{Loss} = 0.8$ eV per electron (Bolton et al. 1985; Varghese and Grimes 2008). But in acutal usage type 1 photocatalysts such as NiO-modified La:KTaO$_3$ (Kato et al. 2003) and Cr/Rh-modified GaN:ZnO (quantum efficiency [QE] = 2.5%, pure water, visible light) (Ohno 2012) show less than 0.1% STH due to low visible sunlight absorption (E_g of La:KTaO$_3$ is 3.6 eV) and recombination losses at the surfaces of the photocatalyst particles. Although a large number of other catalysts have been reported, including In$_{1-x}$Ni$_x$TaO$_4$ ($x = 0$–0.2) (Zou et al. 2001; Zou and Arakawa 2003), CoO (Liao et al. 2014), and Cu$_2$O (Hara et al. 1998), their reproducibility is still in question (De Jongh et al. 1999; Malingowski et al. 2012). The type 2 photocatalyst is based on the tandem (or Z-scheme) concept, shown in Figure 7.1b. Here, two (or more) separate light absorbers are connected in series. Because the device voltage is divided into several contributions, semiconductors with smaller and large band gaps can be used simultaneously, which absorb a much greater fraction of the solar flux (Bard 1979; Bard and Faulkner 2001). The ideal limiting STH efficiency of a dual absorber configuration is 24.4% (for a combination of light-absorber systems with band gaps of

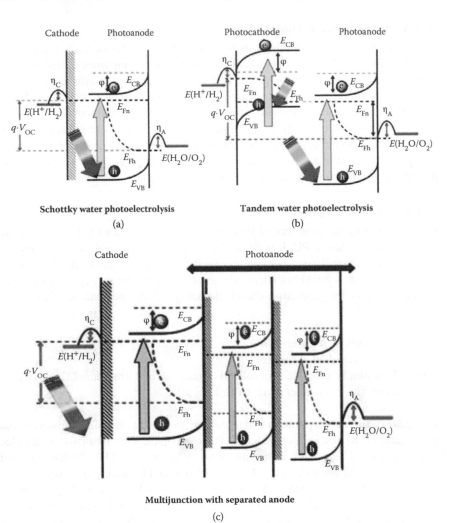

FIGURE 7.1 Schematic representation of the energetic transport dynamics of suspended photocatalysts for water photoelectrolysis (PE): (a) type 1 (single absorber) and (b) type 2 (dual absorber, tandem) and (c) type 3 (multiple absorber, multijunction separated anode). E_{CB} and E_{VB} are conduction and valance band positions, E_{Fn} and E_{Fh} are the quasi Fermi levels of the illuminated catalysts that need to be above and below the water redox potentials, φ = degree of band bending, qV_{OC} is maximum possible energy output, q = charge, V_{OC} = open circuit current, and η_A and η_C are anodic and cathodic electrochemical overpotentials, respectively.

2.25 and 1.77 eV). This is nearly twice that of the Schottky junction photocatalysts (Bolton et al. 1985; Varghese and Grimes 2008).

However, because of the greater complexity (two absorbers instead of one with mediator), functional tandem photocatalysts have only been fabricated for about 15 years (Maeda 2013). Of these, the combination of Rh:SrTiO$_3$ and BiVO$_4$ with a Fe$^{3+/2+}$ redox shuttle mediator gives the highest STH efficiency (0.1%) (Kato et al. 2013). A multijunction device can reach STH efficiency up to 30%, by utilizing several small band gap absorbers of band gap 1.25–1.45 eV in a series (Peharz et al. 2007). These kind of devices are categorized in two types: (1) three terminals or a typical three-junction amorphous hydrogen evolution reaction (HER)/Si–Ge on a glass/tin-doped indium oxide (ITO) cathode with separate anode oxygen evolution reaction (OER) displayed by Figure 7.2a and a (2) two-terminal monolith triple junction (energy conversion device) on stainless steel substrate, coated with CoMo HER material and OER film (FeNiO$_x$ coated) at the backside (Malingowski et al. 2012). A typical example of the three-junction amorphous Si–Ge on glass/ITO with separate anode is displayed by Figure 7.2a. A monolith triple junction (energy conversion device) on stainless steel substrate is displayed in Figure 7.2b.

7.2.2 INTERFACIAL ELECTRON-TRANSFER REACTIONS BY NANOMATERIAL

Nanostructuring is used to exploit scaling laws and specific effects at the nanoscale to overcome the limiting problems of metal oxide absorbers, such as their short electron–hole lifetimes and low mobility. Nanostructuring has gained significant interest in the last 20 years and is used for smooth electronic transport. It is widely believed that nanostructured photocatalysts can improve the efficiency of the process

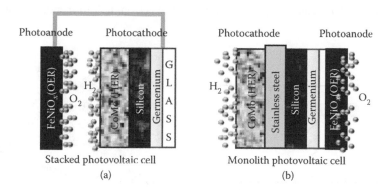

FIGURE 7.2 Multijunction solar cell for water splitting: (a) a stacked or three-terminal cell, where the two subcells (anode and cathode) are electrically separated. (b) A monolithic or two-terminal cell (with oxygen evolution reaction (OER) and hydrogen evolution reaction (HER) materials at opposite faces), where the single cells are electrically connected in series. (From Malingowski, A.C. et al., *Inorg. Chem.*, 51(11), 6096–6103, 2012.)

and lower the costs. Nanostructured photocatalysts have several advantages over traditional bulk materials like shorter charge transport pathways and larger redox active surface areas. Moreover, it is also possible to adjust the energetics of the nanoparticles (NPs) via the quantum size effect or with adsorbed ions.

The effects of nanostructuring of the light absorber on free energy conversion are illustrated well with the circuit diagram for a PV cell in Figure 7.3. In the diagram, the light-absorbing component corresponds to a photon-driven current source (I_L), and the rectifying (charge separating, I_D) component is shown as a diode connected in parallel to it. After photon-driven charge carrier (electron and hole) generation at the photocatalyst surface, generated electrons move toward the cocatalyst site (electron-rich noble metals, working as a cathode) and the innocent photocatalyst surface becomes hole rich (worked as an anode). In total, the photocatalyst surface loaded with cocatalyst works as a diode (or rectifier, a device having two electrodes that direct electron flow in one direction). In addition, there are parallel R_{SH} (shunt resistance; think of as an unwanted short circuit) and serial resistances R_S. The former is associated with the nonselective charge transfer that leads to leakage or shunting, whereas the latter is produced by the transport of charge carriers from the absorber interior to the interface. In PVs, R_{SH} is attributed to the recombination of the photogenerated electrons and holes at surface defects. In H_2-producing photocatalyst, R_S is described as a load and attributed to water electrolysis (Würfel 2005). In the circuit I_{SH}, R_S, and R_{SH}, are dependent upon the physical size of the junction. The surface area of the nanosubstances is larger than their bulk counterparts. If we take twice the surface area of the nanomaterial, then it is expected to have half the shunt resistance and double the leakage current I_{SH} because the current can leak across the doubled junction area. It also has half the series resistance R_S because it has twice the cross-sectional area through which current can flow. If the leakage current exceeds the generation current IL, water electrolysis comes to a stop. On the other hand, the decrease of R_S caused by the reduction in electron–hole transport resistances

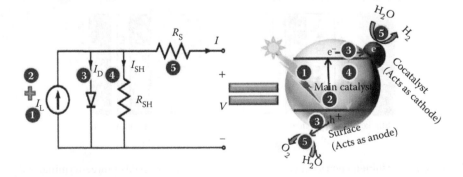

FIGURE 7.3 Equivalent circuit diagram for a *photovoltaic* (PV) cell with a photocatalytic system. I_L represents the photon-driven current generation; the I_D diode current represents the rectifying component due to charge separation; R_{SH} is a shunt resistance due to the shunting or leakage of current, and R_S is a serial resistance, described as a load and attributed to the water electrolysis. 1 = light absorption, 2 = charge carrier formation 3 = charge separation, 4 = recombination, and 5 = redox reactions at the photocatalyst and cocatalyst surface, respectively.

improves the photocurrent. Finally, at the nanoscale of the semiconductor particles approach, which reduces the space charge layer effectiveness and diminishes the rectifying properties. This corresponds to reduction of R_{SH}.

7.2.3 Aspects of the Material Design

In photoelectrochemical (PEC) cells, low performance of the photoelectrodes is the limiting factor. The performance is controlled by the three main driving forces for PEC cells, that is, efficiency, stability, and cost of the material used in above techniques. To design ideal photoactive materials that can fulfill the above criteria effort should be made in two directions: light absorbance and charge carrier transportation. To facilitate the above two points, the following strategies are illustrated for improvement of material design.

7.2.3.1 Surface Passivation

In this technique, the traditional absorber materials that are used in PVs (e.g., III/V, II–VI, and IV semiconductors with small band gap) have to be encapsulated by the protecting layers to inhibit the photocorrosion process. Recently, this strategy has been applied to a number of semiconductors, which have been used as promising photocathodes and photoanodes, with reasonable success, including GaAs, gallium phosphide (GaP) photoanodes, carbon-coated CdS (Han et al. 2016), p-InP (Rh-H alloy), p-InP (Re-H alloy) (Aharon-Shalom and Heller 1982; Lee et al. 2012a; Muñoz et al. 2013), Pt/Ru–coated WSe_2 (McKone et al. 2012), ultrathin film–coated Cu_2O (Paracchino et al. 2012), Zn-doped GaP (Liu et al. 2012), and CuIn-GaSe$_2$ (Moriya et al. 2013). PEC properties of CuGaSe$_2$ were modified by deposition of a thin CdS layer, which forms a p–n junction on the electrode surface. It is a good example of the balance between the charge separation effect and light absorption by CdS that results in a high stability system that evolved hydrogen continuously for more than 10 days.

Some of these photoelectrodes have shown pretty good energy conversion efficiency that exceeded 10%. This was attributed to the stability, light-harvesting character, and effective electron transfer in the photocatalyst by improving carrier dynamics and by suppressing surface recombination velocity, surface coating of metalloxide, nonmetals and metals to photocatalyst material, which have been employed to protect the photoelectrodes. The intimate and large contact interfaces between the main light absorber photocatalyst and the coated layer result in effective electron transfer. Ultrathin carbon shell layer–coated CdS composites exhibited good efficiency for water splitting because of the good electrical conductivity of the carbon layer that ensures a nice electron transfer between CdS and the carbon layer in all directions and also protects from photocorrosion of CdS. Similarly, corrosion protected TiO_2 TF-encrusted p-InP photocathodes have been used for solar hydrogen generation. The large valence band offset between TiO_2 and InP creates an energy barrier for holes to reach the surface. But the conduction band of TiO_2 is well-aligned with that of InP. The combination of these two effects creates an electron-selective contact with low interface recombination (Lin et al. 2015). Chlorine-passivated Cd surface atoms of CdSe nanocrystals (NCs) (larger than 2.7 nm) show enhanced photocatalytic H_2O reduction as compared with the nonpassivated samples. Moreover, passivated samples show increased photo-luminescence quantum yield, for example, from 9% to 22% because of untreated CdSe NCs of the same size as Cl treated sample but Cl-treatment removes trap states with energy below the H_2O reduction potential (Figure 7.4d) (Kim et al. 2016).

7.2.3.2 Development of New Oxide or Nonoxide or Semioxide Materials

Usually, metal oxides are wide band semiconductors, having great stability and good electron-transfer properties. Therefore, they are able to exploit a lesser portion of the light (i.e., ultraviolet [UV]), and to extend their light-harvesting capacity, band gap engineering is incorporated by formation of mixed metal oxides, doping, introducing metal at the valance band side, codoping, or formation of oxynitrides, oxysulfides, or solid solutions. Another approach involves the development of a low-cost and stable new metal oxide or nonoxide materials that acquire suitable properties (visible light–sensitive band gap, high carrier mobilities, long carrier lifetimes) with better chemical stability for PEC water splitting. In the designing of the ideal photoactive materials for the efficient conversion of solar energy, one has to modify photoactive material at three fronts: (1) improved light absorbance, (2) life of charge carrier, and (3) charge carrier transportation, for efficient water splitting. Such materials can be made by direct synthesis, sometimes guided by theory, or they can be made by combinatorial approaches, as described by Woodhouse and Parkinson, Eric McFarland (Jaramillo et al. 2005), and Nathan Lewis (Katz et al. 2009).

Recently, a tandem pair of solar cells with bifunctional electrocatalyst, that is, layered NiFe double hydroxide grown on nickel foam, was developed by Luo et al. (2014) for an impressive 12.3% overall water photolysis efficiency; the group was also able to attain 10 mA/cm² photocurrent in 1 M NaOH solution at 1.7 V (iR uncorrected) with a 470 mV overpotential from the equilibrium. Furthermore, lithium-induced ultrasmall $NiFeO_x$ NPs were used as active bifunctional catalysts that exhibit high activity and stability for overall water splitting in 1 M KOH solution. They show better performance than the benchmark catalysts (combination of iridium and platinum), that is, 10 mA/cm² water-splitting current at only 1.51 V for over 200 hours without degradation in a two-electrode configuration (Wang et al. 2015). Cobalt(II) oxide (CoO) NPs (5–8 nm, $E_g = 2.6$ eV) synthesized from nonactive CoO micropowders (805 nm, $E_g = 2.6$ eV) using two distinct methods (femtosecond laser ablation and mechanical ball milling) can carry out overall water splitting (without any cocatalysts or sacrificial reagents) with a STH efficiency ~5% (Liao et al. 2014). This is a result of a shifting of the band edges of microcrystalline CoO to a nanocrystalline CoO photocatalyst at a more favorable redox potential, demonstrated by Figure 7.5a. However, despite progress in the past decade, semiconductor water-splitting photocatalysts (such as $(Ga_{1-x}Zn_x)(N_{1-x}O_x)$) do not exhibit good activity beyond 440 nm and water-splitting devices that can harvest visible light typically have a low STH efficiency of around 0.1% (effect of postcalcination on photocatalytic activity) (Meada et al. (2008).

A solid solution of perovskite-type $BaTaO_2N$ ($E_g = 1.8–1.9$ eV) with $BaZrO_3$ improves (six- to ninefold) the photocatalytic activity for nonsacrificial H_2 evolution from water under visible light ($\lambda > 420$ nm) (Figure 7.5b). The improvement is attributable to the increased driving force for the redox reactions and the reduced density of defects, which minimizes undesirable electron–hole recombinations

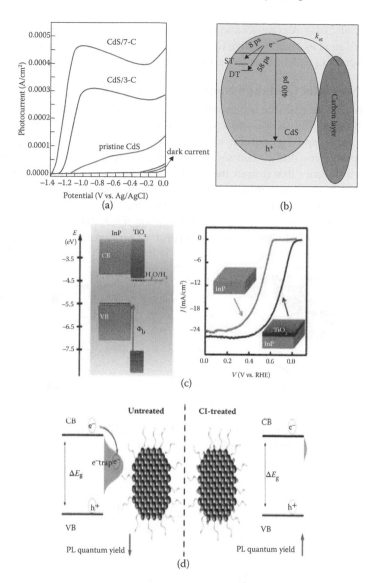

FIGURE 7.4 (a) Photocurrent density–voltage curves in the dark and under light illumination with pristine CdS, CdS-3-nm-thick C-layer, and CdS-73-nm-thick C-layer; (b) schematic illustration of charge carrier dynamics in CdS/7-C composites. The electrons can be trapped into shallow trap (ST) and deep trap (DT) states with a 6 ps lifetime and a 58 ps lifetime, respectively, and electrons of the conduction band (CB) can undergo nonradiative recombination with holes of the valence band (VB) on the 400 ps time scale. (c) Band positions of p-InP-coated TiO_2 and J–V plots of bare p-InP and p-InP/TiO_2 photocathodes measured in 1 M $HClO_4$ solution under simulated solar light of intensity 100 mW/cm². and (d) chlorine-passivated CdSe nanocrystals for photocatalytic hydrogen generation. [(a and b) From Han, S. et al., *J. Mater. Chem. A*, 4, 1078–1086, 2016. (c) From Lin, Y. et al., *J. Phys. Chem. C*, 119(5), 2308–2313, 2015. (d) From Kim, W.D. et al., *Chem. Mater.*, 28(3), 962–968, 2016.]

(Matoba et al. 2011). A compact and robust pH universal catalyst $CoMoS_x$, a chalcogel, was designed as an alternative to noble metal catalysts for efficient electrocatalytic production of hydrogen at low cost by Staszak-Jirkovsky et al. (2016).

7.2.3.3 Nonmetal Oxide and Nonoxide Metals

The majority of our research is presently based on metal oxide photocatalysts owing to their photostability in aqueous solution. The inefficient light absorption due to their large band gap (O_{2p} orbital is located at ca. +3.0 eV or higher) and poor optoelectronic properties, that is, their short electron–hole lifetimes and low mobility, are the major disabilities that restrict them from achieving a practical STH efficiency, that is, 10%. To overcome these limitations, a number of strategies were adopted including band engineering and nanostructuring, metal/nonmetal ion doping (e.g., Na, Li, Ni, C, N, and S), making of a solid solution (GaN:ZnO, quantum efficiency $\eta = 5.9\%$) (Maeda et al. 2006), elemental substitution, sensitization, combinatorial

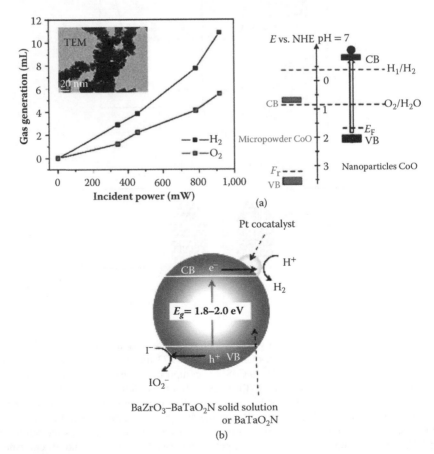

FIGURE 7.5 (a) Nanocrystalline CoO photocatalyst with more favorable redox potentials for water in comparison to their microcounterpart and (b) solid solution of $BaTaO_2N$ (E_g = 1.8–1.9 eV) with $BaZrO_3$ (From Liao, L. et al., *Nat. Nanotechnol.*, 9, 69–73, 2014; Matoba, T. et al., *Chem. Eur. J.* , 17(52), 14731–14735, 2011.)

approaches, for example, with η = 5% from CoO, 2% from C dots C_3N_4 (Liu et al. 2015), and 1.8% from p-GaN/p-InGaN, and so on (Kibria and Mi 2016). But their long-term instability and high cost remain the key concerns for commercialization. Research on metal/nonmetal-nitride photocatalysts and photoelectrodes (e.g., GaN, InGaN, C_3N_4, T_3N_5, Ge_3N_4, W_2N, InN, BCN) for water splitting has drawn considerable attention. Recently, low-cost, eco-friendly and highly stable catalysts, made up of Earth-abundant materials, that is, metal-free carbon nanodot–carbon nitride (C_3N_4) nanocomposites, demonstrated a remarkable performance against photocatalytic water splitting. They exhibited quite high quantum efficiencies under visible light exposure, such as 16% at λ = 420 ± 20 nm, 6.29% at λ = 580 ± 15 nm, and 4.42% at λ = 600 ± 10 nm and an overall solar energy conversion efficiency of 2.0% (Liu et al. 2015). NCs of CdSe (1.75–4.81 nm) in the presence of aqueous sodium sulfite show good photocatalytic hydrogen production activity, controlled by the degree of quantum confinement (Holmes et al. 2012) (Figure 7.6b).

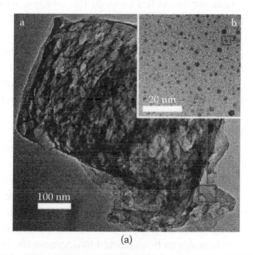

(a)

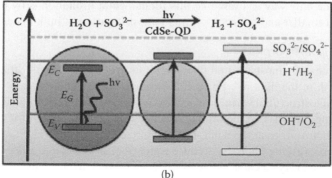

(b)

FIGURE 7.6 (a) Metal-free carbon nanodot–carbon nitride (C_3N_4) nanocomposites and (b) quantum confinement, controlled photocatalytic water splitting of suspended CdSe. (From Liu, J. et al., *Science*, 347(6225), 970–974, 2015; Holmes, M.A. et al., *Chem. Commun.*, 48, 371–373, 2012.)

7.2.3.4 Nanostructuring

Nanostructures mystify us with their intensely interesting properties. The advantages of the nanostructured materials are unique with respect to their bulk counterpart. Worldwide researchers have been attracted to nanomaterials due to their ability to reduce fabrication and installation cost as well as improve cell power conversion efficiency (PCE), along with low temperature processing, flexible, lightweight substrates (Forrest 2005), tunability of their optical and electronic properties, module cost, lifetime (10–15 years) of a module, and so on. Besides their unbeatable properties and ability to synthesize new forms of matter by varying synthetic parameters at low temperature, with remarkable economy, ease, and safety, nanostructures are more ordered aggregates of molecules that lead to new phenomena, and are the basis of nanoelectronics and nanophotonics (Al'tshuler and Lee 1988). Moreover, the small size of the nanostructures is in a range where quantum phenomena occurs (Hey and Walters 2003). Their size and morphology match well with the functional biochemicals that perform the most sophisticated tasks of the biology, which can help us to understand their reaction mechanisms. Their potential confirmed that nanostructural photoelectrode materials would attract considerable attention in the current molecular devices and open the path for the future development of the solar splitting of water. For this reason, they can be used for the application of third-generation concepts such as multiple exciton generation (MEG), singlet fission, and stacked tandem. A few of the important properties of nanostructuring are listed below.

7.2.3.4.1 Surface Area

The larger specific surface area of nanomaterials promotes charge transfer across the material interfaces (solid–solid and solid–liquid), which allows water redox reactions to occur at lower current densities and lower overpotentials (Bard and Faulkner 2001). But on increasing the junction area, the rate for reverse charge transfer increases, which opposes the electron drift from the surface and reduces the rectifying character of the junction (Lewis et al. 2005). Every decadic increase of the reverse saturation current J_0 can be expected to decrease the open circuit voltage by 59 mV. The only way to overcome this fundamental limitation is by making the junction area smaller and by incorporating nanostructuring of bulk material.

7.2.3.4.2 Light Absorption

The dimensions of nanostructured photocatalysts are usually smaller than the penetration length of the light α^{-1}. So each NP can absorb a small fraction of the incident light. Nanostructured surfaces also reduce the reflection losses and increase the light scattering. This has an important consequence for the ability of the particles to generate a photovoltage at the solid–liquid interface and to engender the necessary thermodynamic driving force required for water electrolysis. Photovoltage decreases by 0.059 V for every decadic decrease in j_{photo} (photocurrent density) or decadic increase in junction area. Therefore, minimize the solid–liquid junction area through inert coatings or replace the solid–liquid junction with localized solid–solid junctions on the surface of the light absorber (Osterloh 2016a) for maximizing the water electrolysis.

7.2.3.4.3 Charge Carrier Transport

Photoexcitation of the semiconductor produces charge carriers with finite mobility and lifetime, depending on the light intensity, the type of the material and carrier. When these charge carriers arrive at the material interfaces with electrolyte and cocatalysts, it compels redox reactions of water. In the absence of an external field, charge carriers move by diffusion. The extent of diffusion depends on the mean free diffusion length L that depends upon the carrier diffusion constant D, the carrier lifetime τ, and a dimensionality factor $q = 2$, 4, or 6 for one-, two-, or three-dimensional diffusion, as shown by the following equation:

$$<L^2> = qD\tau \tag{7.1}$$

Photoelectrons catalyze hydrogen evolution in aqueous methanol media under UV light illumination, but hydrogen evolution rates for the nanosheets are consistently higher than those for the bulk particles, even in the presence of cocatalysts. The electronic rate of the catalysts is fitted to the kinetic model shown in Figure 7.7c and expressed as a sum of the inverse rates of charge generation R_G, charge and mass transport J_{CT}, chemical conversion $J_{OX/RED}$, and charge recombination R_R:

$$E_R = \frac{1}{R_G - R\dfrac{L}{R} - R\dfrac{S}{R}} - \frac{r}{\displaystyle\int_{CT}^+} - \frac{r}{\displaystyle\int_{CT}^-} - \frac{d_{OX}}{\displaystyle\int OX} - \frac{d_{RED}}{\displaystyle\int RED} - \frac{1}{R_{OX}} - \frac{1}{R_{RED}} \tag{7.2}$$

Upon doping, the concentration of the majority carriers increases, with their τ and L values. On the other hand, the lifetime and diffusion length of the minority carriers decrease. For optimum collection of both carrier types at the interface, the semiconductor film thickness d has to be in the same range as L_e and L_h (Figure 7.7d) (Osterloh 2016b). This can be achieved by increasing the surface roughness, as shown in Figure 7.7d. Nanostructuring of the surface is a useful tool for the first-row transition metal oxides (Fe_2O_3), which suffer from low hole mobility and short hole lifetimes. Ideal electron–hole collection is possible with suspended NPs if their particle size $d < L_e$, L_h. The impact of the absorber size on charge collection has been experimentally verified with NPs and bulk particles of $KCa_2Nb_3O_{10}$ (Sabio et al. 2012). Chemical exfoliation of this layered Dion–Jacobson phase leads to 1- to 2-nm-thick sheet-like tetrabutylammonium stabilized $HCa_2Nb_3O_{10}$ NPs; their sonication leads to the formation of 227- to 202-nm particles referred to as "bulk" (Figure 7.7).

7.2.3.4.4 Multicomponent Oxides/Heterostructures and Tandem Junctions

Multicomponent oxides and their heterostructures are rapidly emerging as promising light absorbers, required to drive many reductive and oxidative reactions. Size and nature of heterojunction, conduction band minimum (CBM) position and compatibility of individual oxides, and direct band gap or indirect band gap

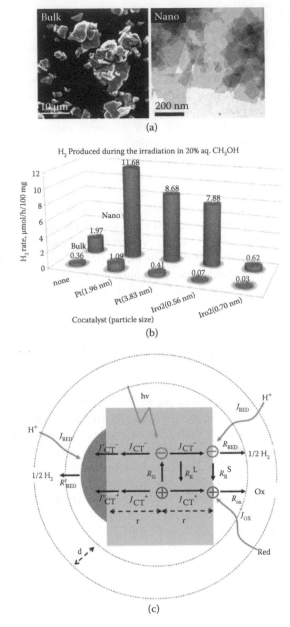

FIGURE 7.7 (a) Electron micrographs of bulk and nanoscale niobate particles, (b) H_2 evolution rates of bare and cocatalyst-modified particles of bulk and nanoscale niobate particles, (c) kinetic rate model of the above system with respect to the electronic rate of the catalyst (mol/s/cm³) to the rates for light-induced electron–hole generation (R_G), electron–hole recombination in the lattice (R_R^L) and on the surface (R_R^S), rates for charge and mass transfer to the catalyst–water interface (J_{CT}, J_{MT}), and to the rates for the redox reactions with the substrates (R_{RED}, R_{OX}). (From Sabio E.M. et al., *J. Phys. Chem. C*, 116(4), 3161–3170, 2012; Osterloh, F.E., *Top Curr. Chem.*, 371, 105–142, 2016.) *(Continued)*

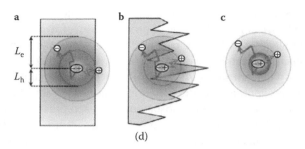

(d)

FIGURE 7.7 (*CONTINUED*) (a) Electron micrographs of bulk and nanoscale niobate particles, (b) H_2 evolution rates of bare and cocatalyst-modified particles of bulk and nanoscale niobate particles, (c) kinetic rate model of the above system with respect to the electronic rate of the catalyst (mol/s/cm³) to the rates for light-induced electron–hole generation (R_G), electron–hole recombination in the lattice (R_R^l) and on the surface (R_R^S), rates for charge and mass transfer to the catalyst–water interface (J_{CT}, J_{MT}), and to the rates for the redox reactions with the substrates (R_{RED}, R_{OX}). (d) Nanoscale effects in water splitting photocatalysis (From Sabio E.M. et al., *J. Phys. Chem. C*, 116(4), 3161–3170, 2012; Osterloh, F.E., *Top Curr. Chem.*, 371, 105–142, 2016.)

play a key role in understanding and directing efficient charge separation/collection, transport, and efficiency in the heterostructured materials. Hematite (α-Fe$_2$O$_3$) has a wide band gap or a suitable band gap (2.0–2.2 eV) for solar water splitting (Lin et al. 2011). However, the carrier-diffusion length in Fe$_2$O$_3$ is shorter than that in other commonly used semiconductors, resulting in difficult photogenerated carrier collection. Many ongoing efforts are focused on addressing this challenge by either doping with foreign ions to increase the carrier-diffusion distance or altering the morphology to reduce its dimension for improving the collection of carriers (Sivula et al. 2011). To fully exploit their overall functionality, the exact tuning of their composition and structure is vital. The combination of transient absorption spectroscopy (picoseconds to second time scales) and PEC measurements is an important tool to reveal photocarrier dynamics and structural tenability of these complex oxide heterostructured films. Assembly of nanostructured bismuth vanadate (BiVO$_4$) with metal oxides (TiO$_2$, WO$_3$, and Al$_2$O$_3$) and layered manganese oxide is a good example of this type, as demonstrated in Figure 7.8a (Loiudice et al. 2015). Lucht and Mendoza-Cortes studied the effect of substituting different intercalated cations (Li$^+$, Na$^+$, K$^+$, Be^{2+}, Mg^{2+}, Ca^{2+}, Sr^{2+}, Zn^{2+}, B^{3+}, Al^{3+}, Ga^{3+}, Sc^{3+}, and Y^{3+}) and the role of waters in the intercalated layer in layered manganese oxide (birnessite) by density function theory (DFT) to find the reason behind the transition of indirect band gap to direct band gap (Figure 7.8b). Birnessite structures with Sr, Ca, B, and Al, and anhydrous B-birnessite are predicted to have a suitable direct band gap for light capturing and to have the potential for water splitting (Lucht and Mendoza-Cortes 2015).

Tandem junction electrodes are more efficient than the simple junction electrode. Many tandem junction devices have been designed, fabricated, and investigated photoelectrochemically for solar-driven water splitting. The tandem devices (n–p$^+$-Si/ITO/WO$_3$/liquid core–shell microwire) with open circuit potentials (E_{oc} = −1.21 V versus E^0(O$_2$/H$_2$O)) demonstrated additive voltages across the individual junctions (n–p⁺-Si E_{oc} = –0.5 V versus solution; WO_3/liquid E_{oc} = –0.73 V versus $E^{0'}$ (O_2/H_2O)) (Shaner et al. 2014). The approach of direct hydrogen generation by PEC water splitting utilized the custom appearance of tandem absorber structures to mimic the Z-scheme of natural photosynthesis. Here a combined chemical surface transformation of a tandem structure and catalyst deposition at ambient temperature yields photocurrents approaching the theoretical limit of the absorber and results in a STH efficiency of 14%. The potentiostatically assisted photoelectrode efficiency is 17%. Present benchmarks for integrated systems are clearly exceeded. Details of the in situ interface transformation, the electronic improvement, and chemical passivation are presented. The surface functionalization procedure is widely applicable

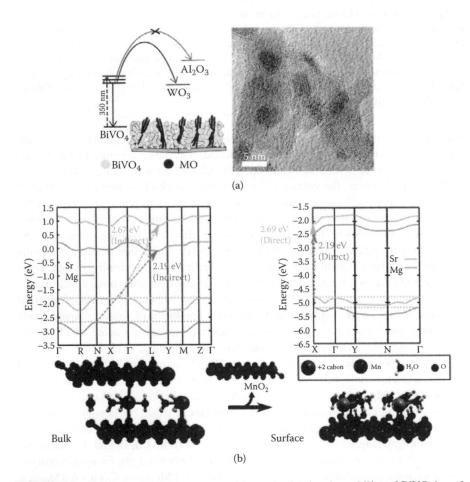

FIGURE 7.8 (a) Structural compatibility with conduction band tunability of $BiVO_4$/metal oxide (TiO_2, WO_3, and Al_2O_3) nanocomposites as a key factor in understanding and directing charge separation, transport, and efficiency in heterostructured films. (b) An indirect to direct band transition is observed when the bulk structure is separated into a single layer oxide in layered manganese oxide. [(a) From Loiudice, A. et al., *Nano Lett.*, 15(11), 7347–7354, 2015. (b) From Lucht, K.P. and Mendoza-Cortes, J.L., *J. Phys. Chem. C*, 119(40), 22838–22846, 2015.]

and can be precisely controlled, allowing further developments of high-efficiency robust hydrogen generators.

7.2.3.4.5 Positioning of the Ligands (Photocatalyst)

In metalloenzymes, the proper positioning of noninnocent pyrazine ligands on a single cobalt center was found crucial in promoting efficient hydrogen evolving catalysis in aqueous media (Figure 7.9a). The findings highlight the worth of electronic structure considerations in designing effective electron–hole reservoirs for multielectron transformations (Jurss et al. 2015).

7.2.3.4.6 Lowering the Reaction Barrier

The hydrogen adsorption energy has been a useful descriptor for screening materials to identify candidates for HER. By considering the protons and electrons separately, the free energies were treated as a function of electrochemical potential and pH. Theoretical barrier energy was calculated from DFT and the experimental from turnover frequency

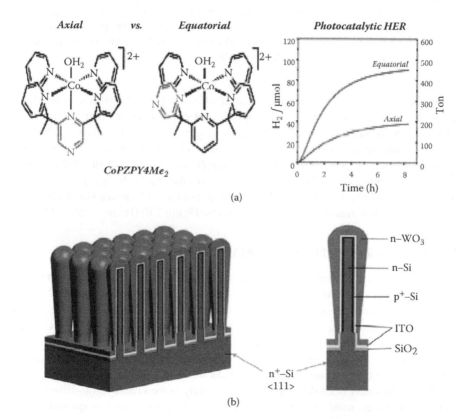

FIGURE 7.9 (a) Role of proper positioning of noninnocent pyrazine ligands on a single cobalt center on enhancing the amount of hydrogen release and turnover number. (b) Tandem junction cells and their performance in terms of photocurrent with respect to simple junction cells. (From Jurss, J.W. et al., *Chem. Sci.*, 6, 4954–4972, 2015; Shaner, M.R. et al., *Energy Environ. Sci.*, 7, 779–790, 2014.)

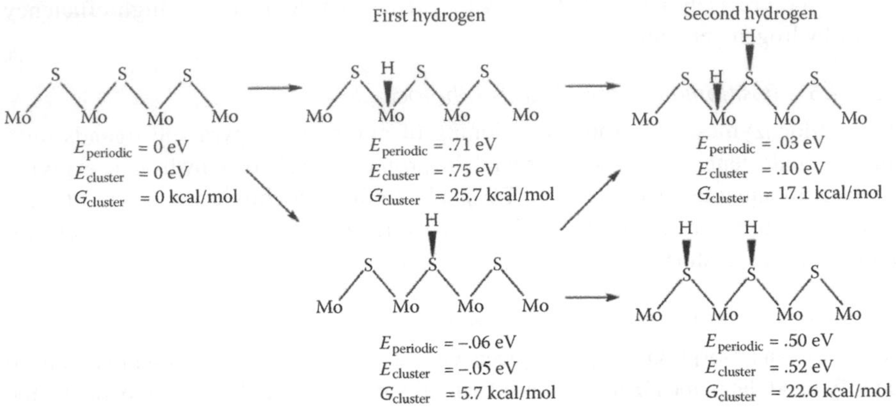

FIGURE 7.10 Hydrogen adsorption energy on the Mo-edge and S-edge. $E_{periodic}$ is the relative electronic energy calculated from periodic calculations using $1/2\ H_2$ as the reference energy for the H atom. $E_{cluster}$ is the relative electronic energy from cluster calculations, and $G_{cluster}$ is the relative free energy from cluster calculations. This indicates that the first H strongly prefers to bind to S by 0.8 eV, but the second H prefers to bind to a Mo. (From Huang, Y. et al., *J. Am. Chem. Soc.*, 137(20), 6692–6698, 2015.)

(TOF). Reaction barriers describe the solvation effects at the Poisson–Boltzmann level and determine the rates. Thus, to design the most efficient HER catalysts, we must determine the reaction barriers for the various expected reaction sequences that can convert protons and electrons to H_2. The pathway will follow the lowest free energy and is expected to be the rate-determining step (RDS), which will rule the reaction. Hydrogen adsorption on the Mo S-edge indicates that the first H strongly prefers to bind to S by 0.8 eV energy, but the second H prefers to bind to a Mo instead of binding to a second S by 0.42 eV reaction rate, as illustrated by Figure 7.10 (Huang et al. 2015).

7.3 NANOCRYSTALLINE MATERIALS

To meet the future demands of the hydrogen-based economy, a copious amount and inexpensive supply of hydrogen from eco-friendly sources is required. Water is always welcomed as a carbon-free/cleaner source of hydrogen but the breaking of water is an energy uphill reaction that needs strong HER and OER active catalysts, supported by solar energy. The use of solar energy to produce molecular hydrogen and oxygen (H_2 and O_2) from overall water splitting is a pledge for renewable energy storage and carrier. For more than 45 years, a variety of inorganic and organic systems have been developed as photocatalysts for visible light–driven water splitting. However, most of the photocatalysts still suffer from low quantum efficiency and/or poor stability. Therefore, nanocrystalline materials are designed and fabricated. Nanocrystalline materials are basically polycrystalline materials with a crystallite size of a few nanometers. These materials fill the gap between amorphous materials without any long range order and conventional coarse-grained materials. Nanocrystalline substances can have perfect local order and stoichiometry but are not easy to quantify. Hence, despite their useful properties such as heat transport and thermoelectric behavior, they have ugly X-ray diffractograms (Bux et al. 2009).

In nanocrystalline films and metals, grain boundaries are as important as their crystal structure and can control material hardness and mechanics (Dao et al. 2007). A good number of the nanocrystalline compounds are being explored as the economical alternative to cathode/anode materials, that is, platinum/ruthenium and iridium oxides with enhanced HER/OER activity in the water-splitting process.

High catalytically active, facile to prepare and low cost, single crystalline $Fe_xNi_{1-x}O$ NPs are promising catalysts for the OER. Fominykh et al. prepared $Fe_{0.1}Ni_{0.9}O$ using solvothermal synthesis in *tert*-butanol. The increase in Fe content (up to 20%) in $Fe_x Ni_{1-x}O$, accompanied by a decrease in particle size, resulted in non-agglomerates of 1.5- to 3.8-nm-sized NCs. They show the highest water oxidation activity in basic media with a TOF = 1.9 s^{-1} at an overpotential = 300 mV; current density = 10 mA/cm^2 at an overpotential of 297 mV with a Tafel slope of 37 mV/dec (Figure 7.11a) (Fominykh et al. 2015).

Active and stable molybdenum compounds (Mo_2C, Mo_2N, and MoS_2) on carbon nanotube (CNT)–graphene hybrid support have been developed via a modified urea glass route. Among the prepared catalysts, Mo_2C/CNT–graphene shows the highest activity (photocurrent density = 21.8 mA/cm^2) for HER with a small onset overpotential of 62 mV and Tafel slope of 58 mV/dec as well as an excellent stability in acid media. Such enhanced catalytic activity may originate from its low hydrogen binding energy and high conductivity. Moreover, the CNT–graphene hybrid support plays a crucial role to facilitate the electron transfer by providing a large area to contact with the electrolyte (Youn et al. 2014). Similarly, hematite (α-Fe_2O_3) is one of the most promising candidates in PEC water-splitting systems, but the low visible light absorption coefficient and short hole diffusion length of pure α-Fe_2O_3 limit its performance. These drawbacks are overcome by the single-crystalline ITO nanowire (NW) core, and α-Fe_2O_3 NC shell (ITO@α-Fe_2O_3) electrodes were fabricated by covering the chemical vapor deposited method, as shown in Figure 7.12a. The J–V curves and incident photon-to-current efficiency (IPCE) of ITO@α-Fe_2O_3 core–shell NW array electrode showed nearly twice as high performance as those of the α-Fe_2O_3 on planar Pt-coated silicon wafers (Pt/Si) and on planar ITO substrates, which was considered to be attributed to more efficient hole collection and more loading of α-Fe_2O_3 NCs in the core–shell structure than with the planar structure and low interface resistance between α-Fe_2O_3 and ITO NW arrays (Yang et al. 2015).

A series of ZnO/TiO_2 was prepared by the sol–gel method followed by supercritical drying by varying the ZnO quantity (1–10%), with high Brunauer–Emmett–Teller (BET) surface area to improve the PEC efficiency of the material. Red shift in the PEC onset induced the less energetic but more intense visible part of the solar spectrum (Indrea et al. 2009). Liao et al. synthesized the cobalt(II) oxide (CoO) NPs (5–8 nm, E_g = 2.6 eV) from nonactive CoO micropowders (805 nm, E_g = 2.6 eV) using two distinct methods, that is, femtosecond laser ablation and mechanical ball milling, respectively. CoO NPs can decompose pure water under visible light irradiation without any cocatalysts or sacrificial reagents with an STH efficiency of around 5%. Electrochemical impedance spectroscopy revealed the significant shift in the position

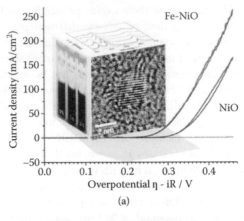

(a)

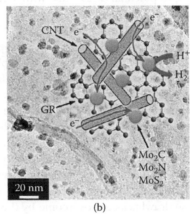

(b)

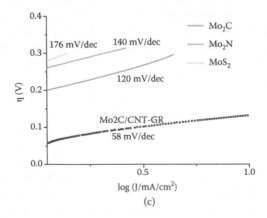

(c)

FIGURE 7.11 (a) Catalytic and characteristic performance of $Fe_{0.1}Ni_{0.9}O$, (b) stable molybdenum compounds (Mo_2C, Mo_2N, and MoS_2) on carbon nanotube (CNT)–graphene hybrid, and (c) Tafel plots of Mo_2C, Mo_2N, MoS_2Mo_2C/CNT–GR catalyst. (a) (From Fominykh, K. et al., *ACS Nano*, 9(5), 5180–5188, 2015.) (b and c) (From Youn, D.H. et al., *ACS Nano*, 8(5), 5164–5173, 2014.)

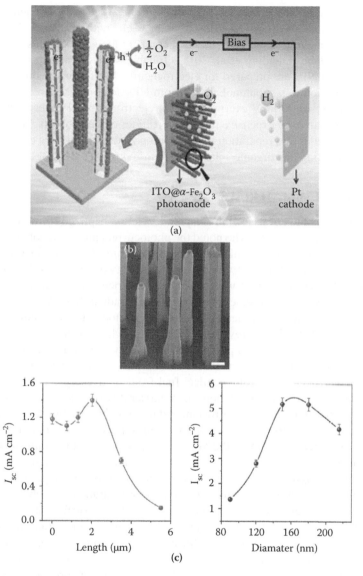

FIGURE 7.12 (a) Core–shell ITO@Fe₂O₃. (b) A close-up of gallium phosphide nanowires (NWs) at the heart of the prototype solar fuel cell. (c) Variation of the saturated photocurrent density length and diameter of GaP NWs. (From Standing, A. et al., *Nat. Commun.*, 6, 7824, 2015.)

of the band edge of the CoO NPs, which is favorable to overall water splitting (Figure 7.5a) (Liao et al. 2014).

Another promising semiconductor material is 500-nm-long and 90-nm-thick ordered arrays of GaP (Figure 7.12b and c), which can convert sunlight into an electrical charge and can also split water. Unfortunately, the material is expensive, but researchers have now used a processed form of gallium phosphide to create a prototype PV cell that not only requires 10,000 times less of the precious material, but also boosts the hydrogen yield by a factor of 10 with efficiency 2.9%, which was an improvement of a factor of 10 when compared to solar cells using GaP with a large flat surface (Standing et al. 2015). Tunable sized ultrasmall (2–5 nm) and crystalline NiO NPs are prepared for electrochemical water oxidation, using a solvothermal reaction in *tert*-butanol. The 3.3 nm NPs demonstrate a remarkably high TOF of ~0.29 s^{-1} at 300 mV overpotential for electrochemical water oxidation that can do better than even expensive rare earth iridium oxide catalysts. NiO NCs also have great potential for preparation of novel composite materials that can be used for PEC water splitting and organic solar cells (Fominykh et al. 2014).

7.4 THIN FILM MATERIALS

The PEC phenomenon described in Chapter 4 was first baptized by the young Henri Becquerel (Becquerel, 1839), who noted the photocurrent and photovoltage produced by sunlight acting on silver chloride–coated platinum electrodes in various electrolytes. But the actual interest was developed after two key events: Fujishima and Honda's (1972) nature paper in which they used this phenomenon in water splitting, and the 1973 gasoline price shock. Fujishima and Honda utilized low-cost polycrystalline semiconductor electrodes n-TiO_2 and Pt photoanodes to sustain the PE of water by renewable solar energy. Since then, the tracking of this and related goals has transformed semiconductor photoelectrochemistry from a specialist domain into a major interdisciplinary subject. PEC water splitting is facing great challenges in generating hydrogen as a clean chemical fuel from solar energy. These challenges, such as inadequate photon absorption, short carrier diffusion length, surface recombination, vulnerability to photocorrosion, and unfavorable reaction kinetics, need to be solved. Highly controlled TF deposition using sophisticated/common techniques are used to overcome some of the key challenges: (1) coating conformal TFs on three-dimensional scaffolds to facilitate the separation and migration of photocarriers and enhance light trapping, as well as realizing controllable doping for band gap engineering and forming homojunctions for carrier separation; (2) achieving surface modification through deposition of anticorrosion layers, surface state passivation layers, and surface catalytic layers; and (3) identifying the main rate limiting steps with model electrodes with highly defined thickness, composition, and interfacial structure (Wang et al. 2014). Nanostructured TFs added the major booster shot to these solar photon conversion processes. This is because they produce very interesting size quantization effects in optoelectronic and other properties: band gaps shifts, increase in carrier lifetimes, high carrier mobility, persuasive catalytic properties, and very high surface-to-volume ratios. Vielstich et al. (2003) and Debe et al. (2006) demonstrate the quality of TF in eliminating or significantly reducing many performance-related problems such as cost, durability, and barriers standing in the way of cathodes and anodes used in PEM cells. This seems to result in significantly higher activities and durability over the conventional carbon-supported Pt catalysts, and their mass activities are closely approaching the 2015 Department of Energy (DOE) targets, that is, current 0.44 A/mg at 900 mV (Debe et al. 2012) and emerge the PV to be exploited with high solar to hydrogen conversion efficiencies. Solar water

PE must employ stable, nontoxic, abundant, and inexpensive visible light absorbers for hydrogen generation. Semiconductor photoelectrodes of many oxide materials, such as $n-TiO_2$, $n-SnO_2$, $n-WO_3$, $n-SrTiO_3$, $n-ZnO$, $p-ZnO$, and so on, are resistant to electrochemical corrosion and often used for solar hydrogen generation from water. Cuprous oxide and $\alpha-Fe_2O_3$, with a band gap excellent for capturing solar spectrum energy, 2.2 eV, generally suffers from electrochemical corrosion (Dai et al. 2014). Usually electrode materials are divided into two parts, that is, HER materials and OER materials. Some classic representatives of the TFs are discussed below.

7.4.1 HEMATITE ($\alpha-Fe_2O_3$) THIN FILMS

Hematite thin film (HTF) $\alpha-Fe_2O_3$ is an OER material that emerges as chemically stable in aqueous environments and as a nontoxic, abundant, inexpensive, and efficient visible light absorber material but with poor transport and optoelectronic properties, which is the present challenge for efficient charge carrier generation, separation, collection, and injection. This leads to both low light-harvesting efficiencies and a large mandatory overpotential for photoassisted water oxidation. It has a band gap of 2.0–2.2 eV and can absorb all ultraviolet (UV) light and most of the visible light, up to 565 nm, which comprises nearly 38% of sunlight photons at air mass (AM) 1.5G. Recently, nanostructuring techniques and advanced interfacial engineering established a landmark in the performance of hematite photoanodes (Sivula et al. 2011). This has resulted in enhanced photocurrent by precise morphology control and reduced overpotential with surface treatments in order to fully exploit the potential of this promising material for solar energy conversion.

Dotan et al. addressed these challenges by means of resonant light trapping in ultrathin films designed as optical cavities. Interference between forward- and backward-propagating waves enhances the light absorption in quarter wave or, in some cases, deeper subwavelength films, amplifying the intensity close to the surface wherein photogenerated minority charge carriers (holes) can reach the surface and oxidize water before recombination takes place. Combining this effect with photon retrapping schemes provides efficient light harvesting in ultrathin films of high internal quantum efficiency, overcoming the trade-off between light absorption and charge collection. A water photooxidation current density of 4 mA/cm^2 was achieved using a V-shaped cell comprising ~26-nm-thick Ti-doped $\alpha-Fe_2O_3$ films on back-reflector substrates coated with silver–gold alloy, as shown in Figure 7.13a and b (Dotan et al. 2013).

Another way to improve the electronic and/or catalytic effects of HTFs is by addition of different dopants in hematite ($\alpha-Fe_2O_3$) photoanodes that induce changes in the layer morphology and microstructure. Doping is an effective way to tailor the electronic and optical properties of metal oxide semiconductors. Incorporation of cations or anions into metal oxide matrices enhanced PEC water-splitting efficiencies through a variety of dopant effects such as electrical conductivity enhancements, intraband gap state formations, and band gap narrowing. This enables the systematic comparison of the effect of different dopants without bogus side effects due to variations in the microstructure and morphology. Observations revealed that the PEC properties and performance vary with type of dopants, dopant concentration, distribution, and the morphology and microstructure of the photoanode in which the

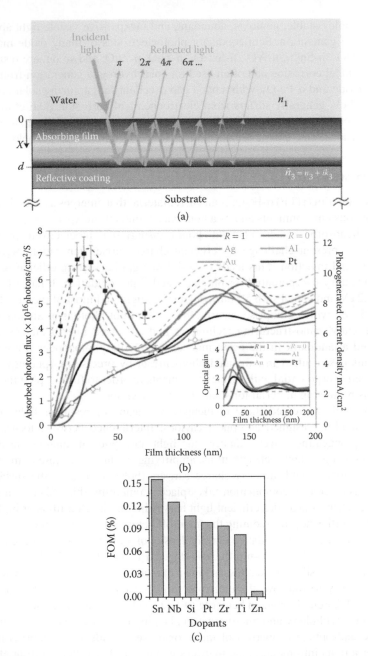

FIGURE 7.13 (a) Schematic illustration of the light propagation in a quarter wave ($d = \lambda/4$, for normal incidence) absorbing film on a back-reflector substrate. The different colors represent the light intensity distribution across the film (red); (b) absorbed photon flux and the corresponding photogenerated current density as a function of film thickness for specimens comprising Ti-doped α-Fe$_2$O$_3$ films on platinized (black squares) or FTO-coated substrates (blue circles). and (c) effect of dopant variation in thin film hematite (α-Fe2O3) photoanodes for solar water splitting in terms of figure of merit (FOM %) [(a and b) From Dotan, H. et al. *Nat. Mater.*, 12, 158–164, 2013. (c) From Malviya, K.D. et al., *J. Mater. Chem. A*, 4, 3091–3099, 2016.]

dopant is incorporated. Malviya et al. examine the effect of the ~1% cation, namely, Sn, Nb, Si, Pt, Zr, Ti, Zn, Ni, and Mn dopants onto the fluorine-doped tin oxide (FTO)-coated glass substrates with Fe_2O_3 targets. PEC properties of the HTF (~50 nm) photoanodes were investigated. In this series, the morphology and microstructure of the films were kept nearly the same and independent of the different dopants in the films. The Sn-doped hematite photoanode outperformed all the other photoanodes that were examined for the photocurrent and photovoltage, achieving the highest photocurrent (~1 mA/cm^2) and the lowest onset potential (~1.1 V_{RHE}). Doped photoanodes ranked in the following order: Sn > Nb > Si > Pt > Zr > Ti > Zn > Ni > Mn, as shown in Figure 7.13c.

Gratzel's group fabricated dendritic silicon-doped Fe_2O_3 TFs by atmospheric pressure chemical vapor deposition (APPVCD) from $Fe(CO)_5$ and tetraethoxysilane (TEOS) on SnO_2-coated glass at 415°C (Figure 7.14a). Under illumination in 1 M NaOH, water is oxidized at the cobalt monolayer decorated and silicon-doped Fe_2O_3 electrodes with higher efficiency (IPCE = 42% at 370 nm and 2.2 mA/cm^2 in AM 1.5G sunlight of 1000 W/m^2 at 1.23 V_{RHE}) than at the best reported single crystalline Fe_2O_3 electrodes. This high efficiency is attributed to the dendritic nanostructure which minimizes the distance photogenerated holes have to diffuse to reach the Fe_2O_3/electrolyte interface while still allowing efficient light absorption (Kay et al. 2006).

Zirconium-doped nanostructured HTFs were successfully prepared by using a electrodeposition method and applied as a photoanode in a PEC cell for hydrogen generation. A maximum photocurrent density of 2.1 mA/cm^2 at 0.6 V/saturated calomel electrode (SCE) was observed for 2.0 at.% Zr^{4+}-doped α-Fe_2O_3 sample with STH conversion efficiency of 1.43% (Figure 7.15a–c). Flat-band potential (−0.74 V/SCE) and donor density (2.6 × 10^{21} cm^{-3}) were found to be maximum for the same sample (Kumar et al. 2011). Similarly, the Pt-doped α-Fe_2O_3 photoanode films were fabricated by coelectrodeposition, followed by high temperature calcinations. Doped photoanodes showed a significant improvement in incident photon to electron conversion efficiency in comparison to undoped samples (Hu et al. 2008).

Nanocrystalline TFs of n-Fe_2O_3 with band gap energy 2.05 eV, flat-band potential of −0.74 V/SCE, and the apparent donor density of 2.2 × 10^{20} cm^{-3} at the alternating current (AC) frequency of 1000 Hz was synthesized using a spray pyrolysis method at the optimum condition of substrate temperature of 350°C, spray time of 60 seconds, and spray solution of 0.11 M FeCl$_3$ in 100% ethanol (Bolts and Wrighton 1976). These films were used for the PEC splitting of water to hydrogen and oxygen gases. The maximum photocurrent density is 3.7 mA/cm^2 at 0.7 V/SCE. The n-Fe_2O_3 films synthesized using the optimum conditions gave rise to a total conversion efficiency of 4.92% and a practical photoconversion efficiency of 1.84% at an applied potential of 0.2 V/SCE at pH 14 (Khan and Akikusa 1999).

7.4.2 TiO$_2$ THIN FILMS FOR WATER SPLITTING

Among all the metal oxide photocatalytic TF HER materials, titanium dioxide has been explored the most due to its superb qualities such as higher electron mobility, high donor concentration (3 × 10^{20}), high photocatalytic activity, more open structures (possibility to incorporate dopants), lower temperature of deposition (<500°C),

FIGURE 7.14 Thin film morphology (a) cobalt monolayer decorated dendritic silicon-doped Fe_2O_3 electrodes and (b) hierarchical nanostructures of SnO_2. (From Kay, A.I.C. and Grätzel, M., *J. Am. Chem. Soc.*, 128(49), 15714–15721, 2006; Wang, H. and Rogach, A.L., *Chem. Mater.*, 26(1), 123–133, 2014.)

and nontoxic, stable, and optoelectronic properties. Moreover, TiO_2 is found in three different natural forms (rutile, anatase, and brookite), as well as in five polymorphs that can be synthesized in a laboratory. TFs of TiO_2 deposited on various supports are very useful photocatalysts for a wide variety of applications, such as solar cells, water splitting (without the need for an external bias), anode materials in batteries, ceramics, pigments, photocatalysis, protective coatings, antireflection coatings, and optoelectronics. Since Fujishima and Honda's use of a crystal wafer of n-type TiO_2 (rutile) as a photoanode for water splitting, a huge pile of work has been done on this material with modification. It is a large band gap (3.2 eV) semiconductor with band positions suitable for reduction and oxidation of water, but often suffers from short exciton diffusion lengths. So it is mainly the carriers generated within the space charge layer which contribute to the photocurrent. TiO_2 is active in the UV region (4%–5%) of solar radiation. In order to extend its absorption spectrum toward longer wavelengths, various modifications of TiO_2 have been proposed such as band gap engineering (Radecka et al. 2008), nonstoichiometry oxygen addition (Zakrzewska et al. 1999; Radecka et al. 2007), and doping of metal and nonmetals, but with limited success. Undoped TiO_2 can be modified by reduction in order to produce non-stoichiometry, which plays a special role in the case of nanomaterials. Doping of anionic or cationic sublattices of TiO_2 has been shown to narrow the band gap and, thus, alter the optical properties of TiO_2. Modification of cationic and anionic sublattices can be achieved by metal type doping (W, Cr, Fe, Nb, Ta, Mn) (Radecka et al. 2004) or nonmetal (N, S, C) type doping (Wong et al. 2006; Henderson 2011; Dong et al. 2011; Chaudhuri and Paria 2014).

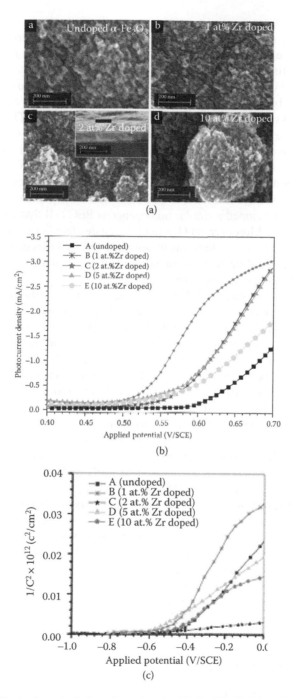

FIGURE 7.15 Electrodeposited zirconium-doped α-Fe₂O₃ thin film for photoelectrochemical (PEC) water splitting: (a) morphology, (b) photocurrent density versus applied potential plot, and (c) Mott–Schottky analysis plot for donor density. (From Kumar, P. et al. *Int. J. Hydrogen Energy*, 36(4), 2777–2784, 2011.)

To optimize the water-splitting efficiency of a TiO_2 photoanode (Fujishima and Honda 1972), a narrower band gap material was added to TiO_2 to exploit the visible light energy region (Sieber et al. 1985) high contact area with the electrolyte to increase the splitting of the e^-–h^+ pairs, and a thick film to enhance the total solar light absorption (Fujishima and Honda 1971). Theoretical quantum efficiency of the TiO_2 surface has been increased to 10% on introduction of more reducible species (like Fe^{3+} ions) to the light-irradiated Pt electrode compartment. The band bending at the TiO_2 surface is responsible for incompetent electron–hole pair separation and the bias voltage is needed for increasing overall solar conversion efficiency of PE (Mavroides et al. 1975). Nozik observed that the pH biased PEC cell containing 0.1 M KOH in anodic and 0.2 M H_2SO_4 in cathodic compartments exhibits 9.5% efficiency at zero bias voltage and 10.1% at a bias voltage of 0.4 V (Nozik 1975). Ohnishi et al. (1975) have developed a similar homogeneous PEC cell that contained mixed electrolyte solutions. Moreover, an O_2 saturated anode chamber causes inhibition in H_2 formation and this can be overcome by applying external bias up to 0.7 V. Besides the morphological modification, sensitization, codoping, and incorporation of the cocatalyst can also improve efficiency.

Recently, Ibadurrohman and Hellgardt successfully synthesized TiO_2 films with modified morphology via a facile spray pyrolysis method in the presence of poly(ethylene glycol) (PEG) as a templating agent. Their photocatalytic properties were assisted by PEC water-splitting measurements. The surface roughness introduced by PEG exceptionally improves PEC responses, revealing a photoconversion efficiency of 1.15% at −0.50 V versus HgO|Hg (in a 1 M NaOH electrolyte under broad spectrum illumination), which is nearly triple that of the unmodified film (0.45% at −0.38 V vs. HgO|Hg) (Figure 7.16a and b) (Ibadurrohman and Hellgardt 2015).

Not only the morphology but also the orientation of the crystal growth can play a crucial role in the photocatalytic activity. A selectively exposed (101) crystal facet engineered TiO_2 photoanode was investigated as a high-efficiency electrode for the hydrogen evolution reaction in comparison to (001) and (100) crystal facet electrodes. Selectively exposed (101), (100), and (001) crystal facet-manipulated TiO_2 TFs are fabricated over presynthesized microcrystals through a three-step strategy: (1) hydrothermal synthesis of microcrystals, (2) positioning of microcrystals via polymer-induced manual assembly, and (3) fabrication of selectively exposed crystal facets of a TiO_2 TF through a secondary growth hydrothermal reaction and exposed for PEC performance. The photocurrent density of the selectively exposed (101) crystal faceted TiO_2 TF photoanode is determined as 0.13 mA/cm^2 and has an 18% conversion efficiency of incident photon-to-current at a 0.65 V Ag/AgCl potential under AM 1.5G illumination. Its PEC hydrogen production reached 0.07 $mmol/cm^2$ for 12 hours, which is higher than those of (100) and (001) faceted photoelectrodes (Kim et al. 2015).

The size of Nb^{5+} (0.69 Å) and Ti^{4+} (0.64 Å) is appropriate for isotropic dissolution and thus permits niobium ions to reduce Ti^{4+} to Ti^{3+}. This leads to an increase in the majority (electron) carrier concentration and therefore the conductivity. The incomplete dissolution as well as large dopant concentrations leads to decrease in photocurrent densities (Salvador 1980). Doping TiO_2 with various metal oxides, (a) TiO_2–SiO_2, (3.4% SiO_2), (b) TiO_2–SiO_2–Al_2O_3 (3.4% SiO_2, 0.2%Al_2O_3),

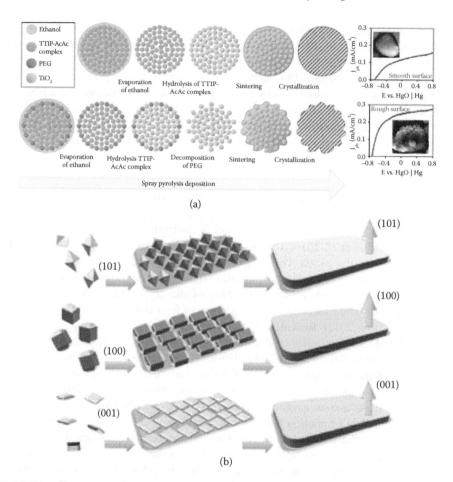

FIGURE 7.16 (a) Morphological modifications of TiO$_2$ thin films (TFs) by using a poly(ethylene glycol) (PEG) template for efficient PEC water splitting and (b) selectively exposed (101), (100), and (001) crystal facet-oriented TiO$_2$ TFs. (From Ibadurrohman, M. and Hellgardt, K., *ACS Appl. Mater. Interfaces*, 7(17), 9088–9097, 2015; Kim, C.W. et al., *Energy Environ. Sci.* 8, 3646–3653, 2015.)

(c) TiO$_2$–SiO$_2$–In$_2$O$_3$ (3.4% SiO$_2$, 0.2% Al$_2$O$_3$), and (d) TiO$_2$–SiO$_2$–RuO$_2$ (3.4% SiO$_2$, 0.2% Al$_2$O$_3$), are reported for PE of water (Nair et al. 1991). Investigation of the mixed oxide materials TiO$_2$–SiO$_2$, TiO$_2$–SiO$_2$–Al$_2$O$_3$, TiO$_2$–SiO$_2$–In$_2$O$_3$, and TiO$_2$–SiO$_2$–RuO$_2$ in regard to the PE.

7.4.3 ZnO Thin Films

The II–IV-type semiconductor ZnO with its inherent properties such as abundancy, nontoxicity, case of making morphological changes at nanoscale, low cost, high conductivity, and good optical properties (Greene et al. 2003; Tian et al. 2003; Yang et al. 2004; Yang et al. 2009; Sun et al. 2012) has been extensively investigated as an excellent photocatalyst and photoanode candidate for solar energy conversion (Wang et al. 2010; Schölin et al. 2011; Zhang et al. 2014). However, ZnO can only respond to UV light (4% in the solar spectrum) due to the wide band gap of ~3.37 eV. This limits the performance of a PEC cell for solar water-splitting performances over ZnO photoanodes. Therefore, in order to acquire high PEC efficiency, ZnO-based photoanodes need to be sensitized to visible light. Effective and controlled doping with metal/nonmetal elements and the addition of sensitizers are the common means to modify wide band gap semiconductors (Yang et al. 2009; Duan et al. 2012).

7.4.3.1   Doping

Doping is a technique to introduce dopants (>10% of the main semiconductor) into crystal lattice of the main semiconductor during crystal growth via different chemical and physical methods such as a sol–gel or hydrothermal process (Zhang and Que 2010; Phadke et al. 2011). Usually, the doped impurities tend to aggregate on the host's surface during the crystal growth but on increasing dopant concentration they start entering into the host crystal lattice (Ni et al. 2007). Heavy doping of metals frequently leads to the destruction of the crystal lattice to some extent and subsequently the functional nanostructures of host materials (Ahn et al. 2009; Kronawitter et al. 2012). Recently, Wang et al. reported an interesting example of controlled N doping on lattice of ZnO nanorod arrays (NRAs), fabricated using an advanced ion implantation method and elaborated by Figure 7.17a–e. The controlled distribution of the N dopants to ZnO NRAs not only extended the optical absorption edges limited to the visible light region, but also introduced a beautiful terraced band structure. As a consequence, N gradient–doped ZnO NRAs can not only utilize the visible light irradiation but also efficiently drive photoinduced electron and hole transfer via the terraced band structure. Subsequently, photoanodes for PEC water splitting under visible light irradiation that displayed a remarkably enhanced visible light–driven PEC photocurrent density of ~160 $\mu A/cm^2$ at 1.1 V versus SCE were utilized by the assembly, which is about double that of the pristine ZnO NRAs (Wang et al. 2015a).

In postgrowth doping strategy, NH_3 treatment is effective to incorporate N dopants into wide band gap materials (e.g., ZnO NRs) without disturbing the original morphology.

7.4.3.2   Sensitization

Dang et al. employed ITO/ZnO/CdS bilayered film in 2010 as a working electrode in a PEC cell and found better PEC performance of bilayered film, with the V_{OC} ~465 mV and J_{sc} ~412 mA and thickness of CdS (70–180 nm) than ITO/ZnO film. Here, CdS acted as a visible sensitizer while ZnO, being a wide band semiconductor, apparently responsible for charge separation, thereby suppressed the recombination process. Hotchandani et al. (1992) developed the coupled semiconductor films, ZnO coupled with CdSe, for PEC water splitting and elucidated the charge transfer processes using picosecond laser flash photolysis experiments. This confirmed that the long lifetime of charge carriers in CdSeZnO retards the recombination of trapped charge carriers and enhances the efficiency of the interfacial charge transfer to the adsorbed substrate.

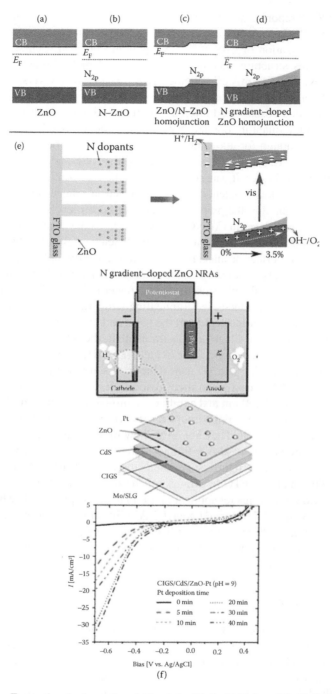

FIGURE 7.17 Energy band schematics of (a) pristine ZnO, (b) N-doped ZnO, (c) ZnO/N-ZnO homojunction, and (d) N gradient–doped ZnO homojunction. (e) Schematics of N gradient–doped ZnO nanorod arrays (NRAs) and terraced band structure promoting charge separation. (f) tandem cell of structure CIGS/CdS/ZnO/Pt that exhibited maximum -32.5 mA/cm² photocurrent density at pH=9, against the Ag/AgCl reference electrode. (From Wang, M. et al., *Sci. Rep.*, 5, 12925, 2015 and Mali M. G., et al. 2015. *ACS Appl. Mater. Interfaces* 7 (38): 21619–21625.)

Wang et al. (2010) reported quantum dot CdS and CdSe cosensitized ZnO nanowired photoanodes for PEC water splitting. ZnO NW arrays were deposited on ITO substrate followed by sensitization of CdS and CdSe quantum dots on each side. (Bawendi et al. (1990)) This structure was analogous to a tandem CdSeZnO/ZnO/CdS cell structure and exhibited 12 mA/cm^2 photocurrent density at 0.4 V versus an Ag/AgCl reference cell (Figure 7.17f). The photocurrent efficiency and IPCE are better than those of single-sided cosensitized layered structures, which was attributed to direct contact between the quantum dot and NW. The conduction band edges of CdS and CdSe were close enough for allowing electrons to delocalize between their respective conduction bands. And the electrons created in CdSe had been transferred to ZnO through the CdS layer. It was concluded that the electron transfer in the CdSe/CdS/ZnO device was less efficient compared to electron transfer in CdSeZnO/ZnO/CdS, which is attributed to the proper alignment of Fermi levels of CdS, CdSe, and ZnO. The intermediate layer in CdSe/CdS/ZnO increases the chance of electron–hole recombination and limits its electron collection efficiency. Besides the engineering of the band gap, the effect of annealing temperature on the optical properties and band gap energy was also studied by Ajuba et al. (2010) for ZnO/NiO multilayer TFs, fabricated by the chemical bath deposition (CBD) method. They found that the band gap energy decreases with increase in annealing temperature, which may be either due to evaporation of water molecules off the films and/or reorganization of the films.

Furthermore, Hsu and Chen demonstrated that the ITO/FTO substrate-free, hydrogen-treated, aluminum-doped zinc oxide (AZO) NRA TFs can act as the transparent conducting oxide TF and 1D nanostructured semiconductor simultaneously. Moreover, on CdS NP sensitization their absorption range extended up to the visible light region around 460 nm and exhibited significantly improved PEC property. After further heat treatment, a maximum short current density of 5.03 mA/cm^2 was obtained under illumination. They not only show higher efficiency than those have no CdS NPs sensitization but also with those having no Al doping and/or hydrogen treatment and even show slightly superior to the state of art CdS-sensitized zinc oxide NRA TFs on ITO or FTO substrates. Furthermore, Mali et al. used Pt-modified Cu(InGa)Se$_2$/CdS/ZnO films for application to solar water splitting. The highest photocurrent density of −32.5 mA/cm^2 under 1.5 AM illumination was achieved with an electroplating time of 30 minutes at a pH of 9. This photocurrent density is higher than that reported in literature due to the efficient charge separation at the interface junction facilitated (Mali et al. 2015).

7.4.4   n-SrTiO$_3$ Thin Films

Chemically and mechanically rough and tough material SrTiO$_3$ (STO; bluish-black, indirect E_g = 3.25 eV; direct E_g = 3.75) is a perovskite structured HER used for water splitting. STO absorbs light considerably better than the other member of the perovskite family, that is, germanates, tantalates, and niobates. Another desirable feature of the STO system is that no rare elements are required for its functioning. Wrighton et al. (1976b) first recognized the single crystal n-SrTiO$_3$ for a photoanode and Pt as cathode for overall splitting of water without external bias in an electrochemical cell. Assembly generates a reasonably good amount of photocurrent that could initiate stoichiometric water splitting without bias and with a slight increase in external bias the photocurrent increases more rapidly than TiO_2 or SnO_2. The PEC properties of $SrTiO_3$ electrodes are similar to those of TiO_2, although the onset potential of photocurrent shifts negative by about 0.3–0.35 V. Therefore, the maximum external quantum efficiency is recorded higher up to 10%, which is about an order of magnitude greater than the highest value obtained for PE in cells with crystalline TiO_2 electrodes in the absence of a bias voltage. This increase in quantum efficiency is due to the increased band bending of $SrTiO_3$, which is about 0.2 eV for a cell without a bias voltage and for increasing electron–hole pair separation efficiency, a bias voltage is applied to make the anode Fermi level more positive than the cathode Fermi level, which is equivalent to E_{red} (H^+/H_2) in accordance with increasing amount of band bending (Mavroides et al. 1976). To extend the photoresponse boundary of $SrTiO_3$ electrodes up to the visible region, single crystals of $SrTiO_3$ photoelectrodes have been doped with various transition metal ions (M^{n+}) or metal oxides. The response to visible light for water decomposition has been found to decrease in the following order: $Cr^{3+} > Co^{2+} > Ni^{2+} > Mg^{2+} > Rh^{3+}$ (De Haart et al. 1981). The absorption spectrum of chromium-doped $SrTiO_3$ single crystals extends the fundamental band edge from 390 nm to about 600 nm, which is attributed to $Cr^{3+} \rightarrow Ti^{4+}$ charge transfer. Townsend et al. (2012) find that 6.5-nm-sized STO liberates 3.0 μmol $H_2/h/g$, 30-nm-sized NiO–STO composites show lower water-splitting efficiency19.4 μmol $H_2/h/g$ than that of bulk NiO–$SrTiO_3$ (28 μmol $H_2/h/g$). Yin et al. (2004) demonstrated that single crystal photoanodes made of niobium-doped strontium titanate $SrTiO_3$:Nb (Nb = 0.07, 0.69 mol%) showed relatively efficient water PE under irradiation from a 500 W Xe lamp in the presence of 0.1 M Na_2SO_4 (pH = 5.92). Under monochromatic light irradiation of 298.2 and 448.2 nm, the corresponding IPCE values are 15.67% and 0.26% (for 0.07 mol% Nb-doped $SrTiO_3$), 4.32% and 0.19% (for 0.69 mol% Nb-doped $SrTiO_3$), with respect to the IPCE value for a pure $SrTiO_3$ anode (0.92% at 298.2 nm).

A silicon-based photocathode was fabricated by capping of a thin epitaxial layer of STO grown directly on Si(001). A metal–insulator–semiconductor photocathode interface facilitates photogenerated electron transport because of the conduction band alignment and lattice match between single-crystalline $SrTiO_3$ and silicon. The device exhibited maximum photocurrent density of 35 mA/cm^2 and an open circuit potential of 450 mV under a broad spectrum illumination at 100 mW/cm^2 (Figure 7.18a). There was no observable decrease in performance after 35 hours of operation in 0.5 M H_2SO_4. (Figure 7.18b). Furthermore, mesh-like Ti/Pt nanostructures were loaded on STO using a nanosphere lithography lift-off process that can achieve an applied-bias photon-to-current efficiency of 4.9% (Ji et al. 2015).

Heterostructures of $SrTiO_3$ TFs performed well for water splitting compared to those of $SrTiO_3$. Choudhary et al. fabricated CuO/$SrTiO_3$ bilayered TFs deposited on indium tin oxide substrate with varying thickness of CuO using a sol–gel spin coating technique and utilized them as photoelectrodes in the PEC cell for a water-splitting reaction. A 590-nm-thick CuO/$SrTiO_3$ bilayered photoelectrode exhibits maximum photocurrent density of 1.85 mA/cm^2 at −0.9 V versus a saturated calomel electrode, which is approximately eight times higher than that of pristine CuO and 30 times higher than that of pristine $SrTiO_3$. Ameliorated separation of the photogenerated carriers at the CuO/$SrTiO_3$ interface and higher value of flat-band potential were attributed to the

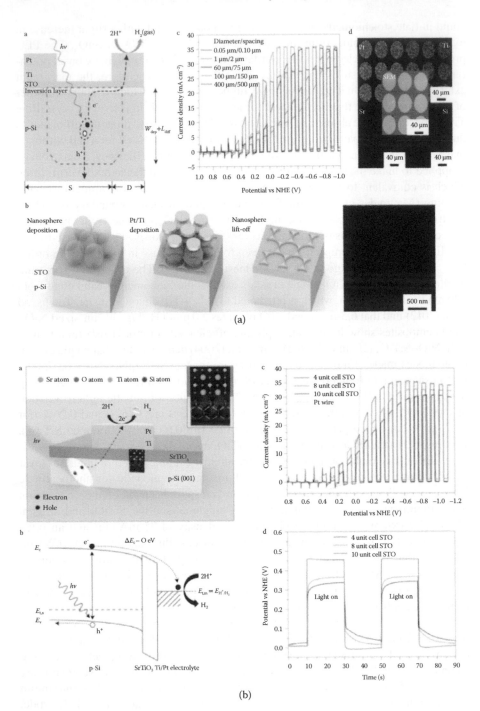

FIGURE 7.18   (a) Mesh-like Ti/Pt nanostructures were loaded on strontium titanate (STO), and used as an electrode for hydrogen generation. (b) Metal–insulator–semiconductor photocathode interface, facilitate the photogenerated electrons to transport because of the conduction-band alignment and lattice match between single-crystalline SrTiO₃ and silicon. (From Ji, L. et al. *Nat. Nanotechnol.*, 10, 84–90, 2015.)

increased photocurrent density and enhanced photoconversion efficiency (Choudhary et al. 2013). $SrTiO_3$–TiO_2 nanotube array composite TFs, prepared by a hydrothermal method, show better PV efficiency in photocatalytic splitting of water than TiO_2 (4.9 times better) and $SrTiO_3$ (2.1 times better) arrays alone (Ng et al. 2010).

7.4.5 OTHER THIN FILMS

$BiVO_4$ (2.4–2.5 eV) exists in three polymorphic forms, that is, orthorhombic, tetragonal, and monoclinic. All exhibited n- and p-type semiconducting properties (Vinke et al. 1992), with reasonably good band edge alignment with respect to the water redox potentials (Long et al. 2008). In addition, it has high photon-to-current conversion efficiencies (>40%). However, the thermodynamically stable monoclinic phase exhibits good PEC properties. The $BiVO_4$ TF electrode prepared at 673–723 K using 200–300 mmol/L concentration of the precursor solution ($Bi(NO_3)_3$ and NH_4VO_3) gave an excellent anodic photocurrent with 73% IPCE at 420 nm at 1 V versus Ag/AgCl. PEC water splitting using the $BiVO_4$ TF as photoanode and Pt electrodes works for solar energy conversion of 0.005% on applied external bias at 1 V versus CE. Furthermore, the CoO-modified $BiVO_4$ TF achieved 0.04% solar energy conversion efficiency even at a smaller external bias (0.6 V vs. CE) than a theoretical voltage (1.23 V) required for electrolysis of water (Jiaa et al. 2012).

TFs of n-type semiconductor tin dioxide (SnO_2 E_g = 3.5 eV) have wide applications in transparent conductive films, gas sensors, lithium-ion batteries, water splitting, and solar cells. Compared to TiO_2, SnO_2 requires a slightly larger potential to achieve onset photocurrent. Wrighton et al. (1976a) studied the single crystal TF of Sb-doped SnO_2 that can serve as photoanodes to electrolyze water to H_2 and O_2. The device substantially responded with quite good photocurrent (3 mA) at an external bias voltage of +0.5 V upon illumination from a 200 W high pressure Hg lamp. Recently, Wang and Rogach (2014) explored the wide variety of porous and hollow hierarchical structures of 3D complex of low-dimensional nanosized tin dioxide (SnO_2) with tunable and improved physicochemical properties, which can be modified by doping and loading with other elements and can be used in optoelectronic devices (Figure 7.14b).

Butler et al. (1976) and Butler (1977) introduced tungsten trioxide, WO_3 (E_g = 2.7 eV), as a potential photoanode material for PE of water in proper electrolyte, for example, in a 1.0 M sodium acetate solution. It has the potential to utilize nearly 12% of the AM 1 solar spectrum as compared to 4% for TiO_2 with a diffusion length comparable to its optical absorption depth. The latest approaches to improve efficiency are nanostructuring, increasing visible absorption via intentional doping, improving carrier separation with heterojunctions, tailoring photocatalytic selectivity toward water oxidation with electrocatalysts, and enhancing WO_3 stability in neutral solution using surface coating. In recent times, Frank E. Osterloh (2015) fabricated a 30 nm WO_3 TF electrode to record a water oxidation photocurrent of 3.8 mA/cm^2, under +1.36 V (vs. reversible hydrogen electrode [RHE]) applied bias in 0.1 M Na_2SO_4 solution at pH = 3.5, which is achieved with 50 mW/cm^2 unfiltered Xe illumination (Zhao et al. 2015). Wang et al. (2016) synthesized nanostructured $BaTaO_2N$ TFs on metallic Ta substrates and applied them as photoanodes for solar-driven PEC water oxidation (Figure 7.19a). Photooxidative corrosion of the $BaTaO_2N$ electrode could be assuaged by the deposition of a cobalt phosphate layer on its surface. Under the 1.5G simulated light, the modified electrode maintained a photoanodic current of 0.75 mA/cm^2 at 1.23 V versus the RHE with unit Faradaic efficiency. Oxidized TiN TFs were prepared via radio frequency (RF) magnetron reactive sputtering by controlling deposition pressure and subsequent annealing temperature of the films in a closed furnace. This forms different phases of TiO$_2$ on TiN to optimize visible light absorption and PEC activity. The resulting TiN/TiO$_2$ TFs showed drastic changes in their crystal structure, optical properties, and PEC performance. These films can also be utilized for many solar-driven optoelectronic devices (Figure 7.19b). Semiconductor CIGS, CuIn$_x$ Ga$_{1-x}$ Se$_2$, with band gap energy of 1.3–1.5 eV, is a highly interesting compound used to drive overall solar water splitting. Now, STH conversion stands at a point where the efficient PEC cells are highly expensive and cheaper cells are not as efficient. Licht et al. (2000, 2001) studied the concept of a multiple band gap tandem for PEC water splitting (Licht 2001; Licht et al. 2000). To fill up this gap polycrystalline TF solar cells of CIGS$_2$ were used in PEC water splitting. CIGS$_2$ TF PEC cells have a few advantages over III–V based and a-Si:H solar cells, such as low-cost CIGS$_2$ TF due to the use of cheaper and more robust soda-lime glass substrate and 100–200 times thinner semiconductor than the (GaIn)P/GaAs and GaAs/AlGaAs III–V cells. Large area depositions, more easily achievable integral interconnect, are free from intrinsic degradation mechanisms such as the Staebler–Wronski effect in an Si:H cells. Furthermore, two TFs of CuIn$_{1-x}$ Ga$_x$ S$_2$/CdS (CIGS$_2$) (~0.43 cm^2 sized) connected in series and having transparent and conducting back contacts in PV cells show high PEC efficiency of 5.95%, where the same bonded to RuS$_2$ photoanode and platinum cathode exhibited 2.99% at (AM1.5) for oxygen and hydrogen generation by water splitting (Figure 7.19c). Corresponding efficiencies of both of the above cases with PV cells having opaque Mo-back contact are 11.99% and 8.78% (AM1.5) (Jahagirdar and Dhere 2007). Similarly, a monolithic device of three PV cells of CIGS$_2$ with Pt cathode, connected into a series, provided a sufficient driving force to complete the water-splitting reaction with 10% STH efficiency (Figure 7.19c) (Jacobsson et al. 2013).

7.5   MESOPOROUS MATERIALS

The class of mesoporous materials (2 nm < pore diameters >50 nm) lies between the microporous materials (pore diameters < 2 nm) and macroporous materials (pore diameters > 50 nm) (Kresge et al. 1992) and are also known as mesoporous molecular sieves with well-defined pore size and surface areas up to 1000 m^2/g (Figure 7.20a and b). They also occupy a unique place between crystalline zeolites and 3D-structured materials (3DSM) and direct-write materials with characteristic ordered features of size >100 nm. Mesoporous materials can be prepared by a self-assembly process by including the combined solutions of sol–gel precursors (e.g., metal alkoxides) and structure-directing amphiphiles, usually block copolymers or surfactants (Figure 7.20c) (Beck et al. 1992; Wan and Zhao 2007). Sometimes, flexible, "one-pot synthesis" employed self-assembling templates to stimulate the controlled size and 3D mesoporous pores. Furthermore, surface functionality of the pores can be modified by decorating them with organically modified precursors, for example, organosiloxanes RSi(OR')$_3$ or bis (organosiloxanes) (R'O)$_3$Si-R-Si(OR')$_3$. It is

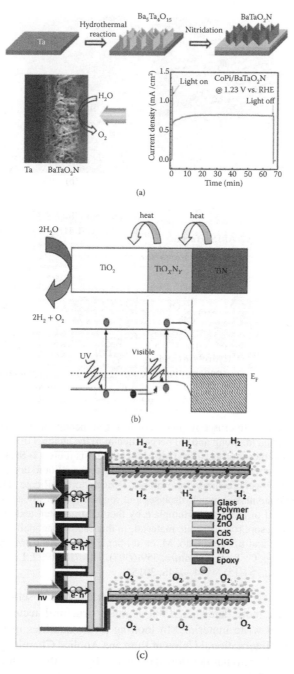

FIGURE 7.19 Nanostructured TFs for water splitting: (a) BaTaO$_2$N, (b)TiO$_2$/TiO$_x$ N$_y$ /TiN, (c) monolith of three cells of CuIn$_x$ Ga$_{1-x}$ Se$_2$ connected into a series. (From Wang, C. et al., *J. Phys. Chem. C*, 120(29), 15758–764; Smith, W. et al., *J. Phys. Chem. C*, 116(30), 15855–15866, 2012; Jahagirdar, A.H. and Dhere, N.G., *Sol. Energy Mater. Sol. Cells*, 91(15–16), 1488–1491, 2007.)

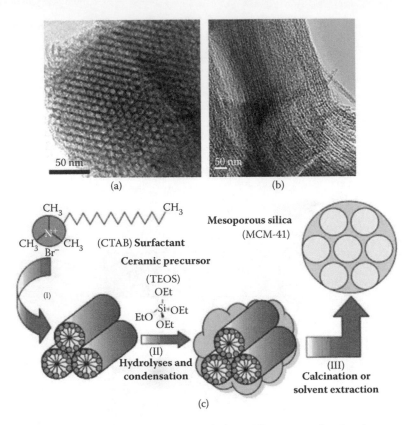

FIGURE 7.20 Electron microscopy images of nitrogen-incooperated ordered mesoporous carbon (N-OMC) taken (a) along and (b) perpendicular to the channel direction and (c) schematic of the classical mesoporous silica (MCM-41) synthesis route. (**1**) Surfactant, for example, cetyltrimethylammonium bromide (CTAB), is used to form liquid crystalline micelles in water. (**2**) Ceramic sol–gel precursor, for example, tetraethylorthosilicate (TEOS), is added to this micellar solution to make, upon hydrolyses and condensation, a silica network around the micelles. (**3**) Removal of the organic template by thermal treatment (calcination) or solvent extraction yields a mesoporous ceramic material, in this case hexagonally ordered MCM-41 silica framework. [(a and b) From Guo, M. et al., *Sci. Technol. Adv. Mater.*, 15(3), 035005, 2014. (c) From Kresge, C. T. et al., *Nature*, 359(6397), 710–712, 1992; Lunkenbein, T. et al., *Angew. Chem. Int. Ed.*, 54(15), 4544–4548, 2015.]

relatively difficult to control long range order and orientation of self-assembled structures and they have more defects and not as much structural accuracy as compared to 3DOM or direct write materials. In looking to the qualities of the mesoporous materials, interest (International Union of Pure and Applied Chemistry [IUPAC] definition: pore size 2–50 nm) has developed dramatically over the last few years because the pore structure of these materials provides an extremely large surface area with respect to the small volume of the material. Thus, the materials can be used in the field of catalysis, chemical sensors, and molecular separation. Structural strains and defects are the key factors in solid catalysis. Ordered mesoporous materials templated by "soft" amphiphilic templates overcome pore size constraints of zeolites to allow more facile diffusion of bulky molecules. This lends them to applications in catalysis and absorption technologies where requirements for long range material order can be less important. For example, acidic aluminosilicates are investigated for uses in fluid catalytic cracking and condensed-media chemical conversion processes. Surface functionalized mesoporous sieves can be used in active elements of sensors. Large, optically active molecules, such as dyes (e.g., rhodamine 6G) and conjugated polymers (poly [2-methoxy-5-(2′-ethylhexyloxy)-1, 4-phenylene vinylene] [MEH-PPV]), can be incorporated into mesoscale pores to make hybrid materials with unique optoelectronic properties. As regular materials do not perform very well in the solid-state heterogeneous catalysis, the materials with well-organized designer mesoscale effects certainly proved to be a game changer (Lunkenbein et al. 2015). But catalytic performance without the Rietveld refined active phase of the catalyst is not able to resolve the scientific core of the problem. Thermodynamically, one can stop growth in such structures by adding an energy bias after attaining a distinct size. Mechanical stresses or torsion buildup between elementary constituting units (say metal oxide octahedra) might be the energy biases at mesoscale lengths (Mueller et al. 1995). Additionally, the elementary dipole moments can also control the well-defined equilibrium size of biominerals and mesocrystals to artify the desired objects with pores, by restricting the practical size of those units (Colfen and Antonietti. 2005). Therefore, one can say that the mesophases have opened up new perspectives in materials science toward the design and engineering of materials in new avatars such as self-organized supramolecular materials. The liquid crystalline medium offers an ideal vehicle to explore and control the organization of matter on the nanoscale to microscale. Monte Carlo simulations predict that the dynamic disorder is crucial in defining mesophase behavior, and the apparent kinetic barrier for the liquid to mesophase (rotational) transition is much lower for liquid crystals (orientational order) than for plastic solids (translational order) in polyhedral structures namely, truncated octahedrons, rhombic dodecahedrons, hexagonal prisms, cubes, gyrobifastigiums and triangular prisms (Agarwal and Escobedo 2011). But the cost is still a prime aspect of the study, which is concerned with the synthesis and processing cost of the mesoporous materials.

Representative mesoporous structures a used for water splitting are discussed in this section. Ma et al. synthesized mesoporous nanostructure of $LiNi_{1-x}Fe_xPO_4@C(0 \leq x \leq 1)$ through a spray dry method as a highly effective catalyst for electrochemical oxygen evolution reaction (OER). In particular, $LiNi_{1-x}Fe_xPO_4@C$ (3:1) shows superior activity as a highly effective catalyst for OER to those state-of-the-art noble metal catalysts (e.g., RuO_2 and IrO_2); it only needs an overpotential of 311 mV for a current density of 10 mA/cm² and maintains its high catalytic activity after 1000 cycles (Figure 7.21a) (Ma et al. 2015).

A conventional TiO_2 photoelectrode was integrated with ordered mesoporous carbon material (CMK-3) and Au metal NPs to improve the photocatalytic efficiency of water splitting under visible light irradiation. Compared to TiO_2, Au/TiO_2-CMK-3 photoelectrode demonstrated over two orders of magnitude enhancement of photocurrent due to the generation of hot electrons, near field from Au NPs, improvement of free carrier transport and additional long-wavelength absorption under 532-nm laser irradiation (Hung et al. 2015). A template-directing self-assembling method was used to prepare mesoporous photocatalyst $InVO_4$. Compared with the

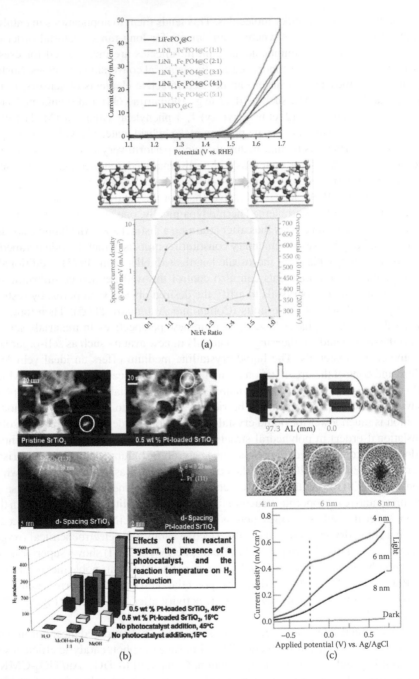

FIGURE 7.21 Splitting water using mesoporous assemblies such as (a) $LiNi_{1-x}Fe_xPO_4@C$ ($0 \leq x \leq 1$), (b) 0.5 wt.% Pt-loaded, mesoporous $SrTiO_3$, and (c) TiO_2 mesocrystal for water splitting. [(a) From Ma, S. et al., *Chem. Commun.*, 51, 15815–15818, 2015; Puangpetch, T. et al., *J. Mol. Catal. A Chem.*, 312(1–2), 97–100, 2009; Srivastava, S. et al., *ACS Nano*, 8(11), 11891–11898, 2014.]

anatase TiO_2 and conventional $InVO_4$, the mesoporous $InVO_4$ was more responsive toward visible light (H_2 evolution rate 1836 μmol/g/h for water) (Xu et al. 2006). Grewe and Tüysüz prepared (Ca, Ba, Sr, Na, K) metal incorporated ordered mesoporous composite materials, out of which a sodium-based composite showed a large surface area (108–120 m^2/g) by variation of the Ta/Na ratios through a soft templating route. The efficiency of this sample was further improved through decorated NiO_x as cocatalyst. A 2.5 wt.% NiO_x loading was found to be the optimal loading amount, generating 64 μmol/h H_2 and 31 μmol/h O_2 when tested for overall water splitting (Grewe and Tüysüz 2016). A typical 0.5 wt.% Pt loaded, mesoporous, $SrTiO_3$ assembly in the presence of a hole scavenger, that is, 50 vol.% aqueous methanol solution, provides the highest photocatalytic activity, with hydrogen production rates of 276 and 188 μmol/h/g_{cat} and quantum efficiencies of 1.9% and 0.9% under UV and visible light irradiation, respectively. The pristine mesoporous assembly of $SrTiO_3$ photocatalysts exhibited much higher photocatalytic activity in hydrogen production via the photocatalytic water splitting using methanol as the hole scavenger than both nonmesoporous, commercial $SrTiO_3$ and commercial TiO_2 (Degussa P25), albeit with lower specific surface areas compared to both commercial photocatalysts. These results point out that the mesoporous assembly of NCs with high pore uniformity plays a significant role in affecting the photocatalytic hydrogen production activity of the $SrTiO_3$ (Figure 7.21b) (Puangpetch et al. 2009). Size-selected TiO_2 nanoclusters (NCs) gain immense interest as they offer the link between the distinct behavior of atoms and NPs. Srivastava et al. precisely deposited the size-selected TiO_2 NCs (from 2 to 15 nm) on H-terminated Si (100) substrate by special magnetron sputtering. The sample was tested for PEC catalytic performance, and significant enhancement in photocurrent density (0.8 mA/cm^2) with decreasing NC size was observed with a low saturation voltage of -0.22 V versus Ag/AgCl (0.78 V vs. RHE), as shown in Figure 7.21c (Srivastava et al. 2014).

7.6 ADVANCED NANOSTRUCTURES FOR WATER SPLITTING

7.6.1 BIOINSPIRED DESIGN OF REDOX REACTION–ACTIVE LIGANDS FOR MULTIELECTRON CATALYSIS

Biological reactions are the great inspiration for synthetic products in laboratories. To explore new designer material for the two-electron reduction and four-electron oxidation of water to hydrogen, investigators are attracted toward the integral role and ubiquity of redox active ligands in numerous biological systems. And in this search the metalloenzymes are known for routine performance of multielectron reactions with near thermodynamically allowed potentials under physiological conditions by accumulating multiple redox equivalents over proximal sites involving ligated or adjacent redox active cofactors (Denisov et al. 2005; Groves 2006; Hiromoto et al. 2009; Pratt et al. 2012; Lyons and Stack 2013). Such redox reaction active moieties with finely tuned potentials are positioned at apt positions within metalloenzyme active sites to promote synergistic redox chemistry at an optimal rate. For fulfilling that particular interest, metalloenzymatic systems composed of a single metal active site function in harmony with redox active organic accessories to execute multielectron transformations. (Stubbe and van der Donk 1998) These enzymatic classes include the following:

1. Galactose oxidase (GO) which catalyzes the two-electron active oxidation of primary alcohols to aldehydes via cooperative Cu(II)-centered coordinated phenoxyl radical (Whittaker 2003; Verma et al. 2011).
2. Copper amine oxidase (CAO) which utilizes an o-quinone moiety (2,7, 9-tricarboxy-1H-pyrrolo[2,3-f]quinoline [TPQ]) to catalyze two half reactions en route to transforming primary amines to aldehydes (Dawkes and Phillips 2001; Mure et al. 2004; Du Bois and Klinman 2005).
3. Mononuclear iron hydrogenase comprising a tautomeric 2-hydroxypyridine/pyridone ligand and a closely spaced, redoxactive pterin cofactor that together enable efficient hydrogen process.
4. Family of cobalt complexes of the parent pentadentate/PY_5Me_2 ligand and derivatives containing peripheral substitutions on the pyridine ring by varying the location of the redox active pyrazine reservoir(s) (Khnayzer et al. 2014).

Sun et al. (2011, 2013) and King et al. (2013) showed the redox noninnocent nature of the pyrazine donors that perform the overall water splitting. Figure 7.22a represents the mechanism of electrochemical proton reduction catalyzed by a molecular organic framework $[(PY5Me_2)Co]^{2+}$. Another class of Co(II) complexes supported by tetradentate polypyridine ligands ought to be a more active photocatalytic composition than similar catalysts with pentadentate ligands and also ought to retain high catalyst stability. Usually, the metal complexes with tetradentate ligands that promote cis-open coordination sites appear to be more active for hydrogen evolution than catalysts with $trans$-open sites (Tong et al. 2014). Mononuclear metalloenzymes operated in concert with precisely positioned redox active cofactors to organize the multielectron reactions, under protein-compatible conditions in water at neutral, physiological pH and temperature (36.7°C). Isostructural analogs of metalloenzymes, coordinated by redox inactive zinc(II) or Co(II), were prepared and characterized to disentangle the contributions of ligand-based and metal-based redox chemistry of these systems. This synthetic system has conceptual parallels to mononuclear metalloenzymes, such as galactose oxidase, copper amine oxidase, and [Fe]-hydrogenase, that combine a single metal center and pendant redox active organic cofactor with strict conformational demands and reveals the one electronic balance between the metal center and ligand in dictating reactivity and function. These results highlight the importance of electronic structure considerations regarding the placement of redox noninnocent ligands in catalyst design, which has broader implications for the use of electron–hole reservoirs for multielectron chemical transformations like, water splitting as shown in Figure 7.22b. The precious metal-free CdSe NCs tightly coupled with tripodal S donor capping agents $(Co(bdt)_2)^-$ (bdt, benzene-1,2-dithiolate) through the carboxylate are also found to be excellent catalysts for H_2 generation from water that exhibits outstanding activity with a quantum

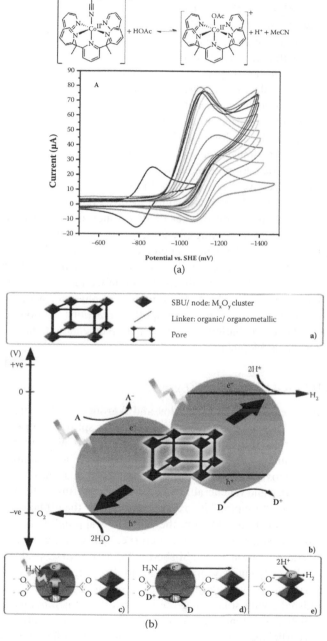

FIGURE 7.22 (a) Metal-organic frameworks for water splitting mechanisms of electro-chemical proton reduction catalyzed by [(PY5Me₂)Co]²⁺. (b) The inorganic node (secondary building unit [SBU]), organic linker, and pore are the basic components of a metal-organic framework (MOF) used for overall water splitting. (From Meyer, K. et al., *Energy Environ. Sci.*, 8, 1923–1937; Sun, Y. et al., *Chem. Sci.*, 4, 118–124, 2013.)

yield for H_2 formation of 24% at 520 nm light and durability with >300,000 turn-overs relative to catalyst in 60 hours (Das et al. 2013).

Recently, Mayer et al. (2015) introduced metal-organic frameworks (MOFs) as a new arrival in the world of photocatalytic material for water splitting. They debuted as a solid support with high porosity and structural versatility that can offer a tanta-lizing merger of ions of the components needed for solar light harvesting and water splitting. They employ electrocatalysis, chemically introduced redox partners, and photocatalysts to generate dioxygen and dihydrogen from water. The beauty of these systems lie in the ease of variation and modification in structures by engineering different pore sizes, connectivity, and chemical functionality in the systems that render them as ideal candidates for water splitting. The conceptual schematics of photocatalyzed water oxidation or reduction using an MOF in the presence of an acceptor or donor, respectively, are illustrated in Figure 7.22b. On the right side, the photogenerated electrons reduce H^+ and simultaneously produce holes to oxidize the donors. On the left, photogenerated holes oxidize water and synchronize electrons to reduce acceptors. Light-harvesting phenomena was accomplished by an organic linker. Under light exposure the photoexcited electron transferred to the metal oxide node for subsequent proton reduction. The role of semiconducting MOFs in these systems is to control the band gap by linker functionalization and doping. MOFs are considered a holistic material, which offers impressive physical, spatial, and chemi-cal versatility to support and sustain water-splitting reactions. Although MOFs cur-rently suffer with the major challenge of practical implementation, they have copious opportunities for the development.

7.6.2 HYDROSOL: MONOLITH REACTORS

The HYDROSOL team (Konstandopoulos and Agrofiotis 2006) has constructed an innovative solar thermochemical reactor for water splitting to produce hydro-gen and oxygen from thin-walled multichanneled (honeycomb) monoliths of ceramic refractory material that absorbs solar radiation. These monolith channels were coated with active water-splitting materials (NP materials with very high water-splitting activity and regenerability) capable of splitting water vapor passing through the reactor by "trapping" its oxygen and leaving the effluent gas stream as pure hydrogen product. In a next step, the oxygen "trapping" material (based on doped iron oxides) is solar-aided regenerated (i.e., releases the oxygen absorbed) and hence a cyclic operation is established. This is the first kind of pilot-scale reactor that can be continuously operated up to 40 cycles (2-day continuous H_2 production). Furthermore, for scaling up the hydrogen production this technology is going to handshake with solar concentration systems. Such plants can prove to be a boon for the solar potential regions of the world. Reactor efficiencies up to 28% and process efficiencies of up to 9% were observed. In fact the HYDROSOL technology can make available masses of hydrogen and oxygen gases at the cost of 24 Eurocent/kWh in the medium term and 10 Eurocent/kWh in the long term (Figure 7.23a–c).

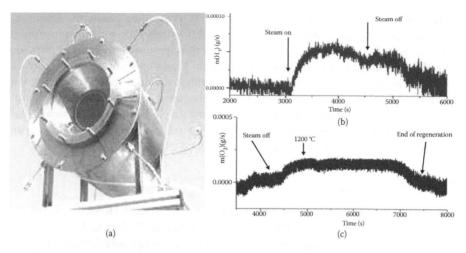

FIGURE 7.23 (a) Front view of the solar water-splitting receiver reactor. (b) Mass flow of hydrogen during splitting of water. (c) Mass flow of oxygen during regeneration. (From Konstandopoulos, A.G. and Agrofiotis, C., *Rev. Energies Renouv.*, 9(N3), 121–126, 2006.)

7.6.3 PLASMON-RESONANT NANOSTRUCTURES

The energy potential of solar radiation is used in the effective conversion of water to chemical fuels such as hydrogen and oxygen, which can be stored and used when desired. This conversion can be pooled through photocatalytic material. The material should be inexpensive, stable, and highly active. Unfortunately, until now no single material combines all of these challenging requirements. Even excellent materials suffer from very short carrier diffusion lengths (~10 nm) and prevent efficient collection of photogenerated carriers at about absorption depth (~0.1–10 μm) away from the catalyst/liquid interface. Numerous actions were taken to overcome these issues by either shortening the distance needed to be traveled by carriers or enhancing the light absorption in the near-surface region. The first includes the use of nanostructures, 3–9 nm with a large surface to volume ratio to facilitate the carriers to reach the surface, but they come with undesirable optical losses. The second approach includes the use of plasmon-resonant materials to produce energetic, hot electrons capable of driving water-splitting reactions (Lee et al. 2012b; Dotan et al. 2013). One way to get around this issue is to take advantage of both approaches, which means using light impregnated metallic nanostructures (plasmonic) to produce energetic, hot electrons capable of driving water-splitting reactions (Warren and Thimsen 2012). Kim et al. (2014) proposed an alternative pathway that thwarts such losses while maintaining effective light concentration. This is accomplished by nanofashioning high-index photocatalyst materials into nanobeam arrays that support strong optical resonances.

Many of the high-performance, inexpensive photocatalysts including α-Fe$_2$O$_3$ (hematite) suffer from very poor charge transport with minority carrier diffusion lengths that are significantly shorter (nanometer scale) than the absorption depth of light (micrometer scale near the band edge). As a result, most of the photoexcited carriers recombine rather than participate in water-splitting reactions. Plasmon-resonant nanostructures have been employed to effectively enhance light absorption in the near-surface region of photocatalysts, but this approach suffers from intrinsic optical losses in the metal. To avoid these optical losses optical resonances may be cultivated in the active nanopatterned photocatalyst material that supports optical Mie and guided resonances for substantial enhancement in the photocarrier generation rate within 10–20 nm from the water/photocatalyst interface. Graphical comparison of the enhancement factor in different particle-sized nanostructures with bulk material efficiency is demonstrated in Figure 7.24a.

7.6.4 META MATERIALS

Metamaterials were first discovered by the physicist W. E. Kock (Kock 1946), but the term "metamaterial" (Walser 2001) was coined by Rodger Walser to describe meta-materials as materials that can divert light or sound in the opposite direction in comparison to the conventional refractive medium (negative indices of refraction) (Figure 7.24a and b) and offer exciting new prospects for manipulating light (Kock 1946). A light beam passing from air or a vacuum into a common refractive medium such as glass, water, or quartz is bent at the surface boundary so its path inside the material is more nearly perpendicular to the surface than its path outside the material (Figure 7.24a). The extent of the bending depends on the angle at which the ray strikes the boundary and also on the index of refraction of the medium. All common transparent materials have positive indices of refraction. A metamaterial bends an incident light ray or sound wave so its internal direction is reversed (Figures 7.24b and 7.25a–d). Possible applications of transparent metamaterials with negative indices of refraction include red and infrared (IR) lasers, optical communications systems, optoelectronics, spectrometry, monitoring systems to detect trace gases in the atmosphere, medical diagnostic equipment, and optical cloaking devices. (Pendry et al. 2006) Certain meta-materials with unique surface structures can bend visible light (sometimes IR light) rays in the opposite sense from traditional refractive media as depicted in Figure 7.24a and b. Therefore, scientists have begun to look to these uniquely structured materials with great hope. Metamaterials are basically electromagnetic materials that possess engineered subwavelength structures, which can display strong coherence with the magnetic and/or electric component of an incident electromagnetic light wave. This phenomenon leads to exclusive properties such as abnormal reflection/refraction; perfect absorption; tunability of the transmission, reflection, and absorption phenomenon of the incident waves; and subwavelength focusing. Tunability in metamaterials can be achieved by manipulating and controlling the interaction between the metamaterials and the incident waves lithographic features of the metamaterials can be imprint on flexible materials. Tunability in metamaterials can be achieved by manipulating and controlling the interaction between the metamaterials and the incident waves through structural features. Moreover the beauty of metamaterials is that the lithographic features of the metamaterials can be imprinted at micro or nanoscale on flexible materials that make them able to bend, stretch, and roll. Reconfiguration of the metamaterials is possible by electromechanical displacements, lattice displacement, thermal annealing, and change in superfluid density and mechanical tuning. Although, over several

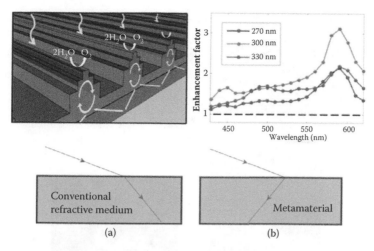

FIGURE 7.24 (a) Plasmon-resonant nanostructures and (b) metamaterials.

decades, this terahertz technology has been applied for cloaking (Li et al. 2013), sensing (Kim et al. 2012), superlensing (an optical lens with resolution below diffraction limit) (Shalaev 2007; Huang et al. 2013), on-chip photonic and optoelectronic devices (Luo et al. 2003; Shalaev 2007), perfect absorbers (Jiang et al. 2011; Watts et al. 2012), and energy harvesting (Ramahi et al. 2012; Hawkes et al. 2013), it has never been used in PEC devices and has tremendous possibilities in this field. Metamaterials designed on flexible, low surface energy elastomeric substrates also integrate with nonplanar surfaces. Flexible substrates utilized for metamaterial devices include metaflex (Xu et al. 2011), polyethylene naphthalene (PEN) (Han et al. 2011), polyethylene terephthalate (PET) (Miyamaru et al. 2009; Tao et al. 2009; Tumkur et al. 2011), polymethylmethacrylate (PMMA) (Tumkur et al. 2011), polystyrene (Gibbons et al. 2009), and polydimethylsiloxane (PDMS). Substrates properties like dielectric permittivity, loss tangent, refractive index, and Young's modulus play a critical role in the operation of metamaterials. Some are multilayered fishnet structures of metamaterials that are well illustrated by Figure 7.25.

Robatjazi et al. made a metamaterialistic setup that features three layers of materials. The bottom layer is a thin sheet of shiny aluminum. This layer is covered with a thin coating of transparent nickel oxide, which is scattered atop with plasmonic gold NPs and forms a puck-shaped disc about 10–30 nm in diameter (Figure 7.26a) (Robatjazi et al. 2015). When sunlight hits the discs, either directly or as a reflection from the aluminum, the discs convert the light energy into hot electrons. The aluminum attracts the resulting electron–holes and the nickel oxide/polymer allows these to pass while also acting as an impermeable barrier to the hot electrons, which stay on the gold. The sheet of flat material dipped into the water under light irradiation, where the gold NPs act as catalysts for water splitting. The efficiency is reported in terms of photocurrent available for water splitting rather than directly measuring the evolved hydrogen and oxygen gases produced by splitting. The next example in this context is uniform and scalable nanocomposite film of Fe_2O_3 NRs and NiO_x NPs, which have been applied to enhance solar water reduction in neutral pH water on the surface

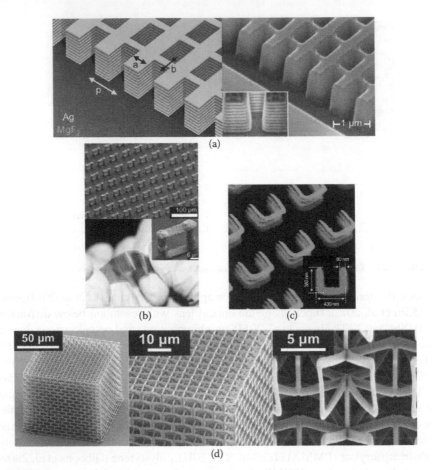

FIGURE 7.25 (a) Multilayered fishnet structure defined by focused ion beam milling of 21 layers—11 of Ag and 10 of magnesium fluoride (MgF). In this design, $p = 860$ nm, $a = 565$ nm, and $b = 265$ nm. (b) Scanning electron micrograph showing vertical, three-dimensional (3D) magnetic terahertz metamaterials defined on polyimide substrates. (b) Photograph of the flexible device with an inset showing an individual 3D structure. (c) A multilayered metamaterial comprising four layers of gold split-ring resonators (SRRs) planarized with a PC403 spacer layer defined by electron beam lithography. (d) Scanning electron micrographs of a 3D mechanical metamaterial showing three different magnifications of the complex structure defined using direct laser writing. [(a) From Valentine, J., *Nature*, 455, 376–379, 2008. (b) From Fan, K. B. et al., *Opt. Exp.*, 19, 12619–12627, 2011. (c) From Liu, N. et al., *Nat. Mater.*, 7, 31–37, 2008. With permission. (d) From Buckmann, N. et al., *Adv. Mater.*, 24, 2710–2714, 2012.]

of p-Si photocathodes (Kargar et al. 2015). Finally, a NiO_x –Fe_2O_3–SnO_2–coated p-Si photocathode was fabricated for a net photocathodic current ~ 0.25 mA/cm^2 at 0 V versus RHE and a cathodic onset potential of 0.25 V versus RHE, generated under the light current–dark current situation.

The light reflected from a metallic mirror used as an electric contact and optical mirrors produces a standing wave with reduced intensity near the reflective surface. This effect is highly undesirable in optoelectronic devices because it dictates a

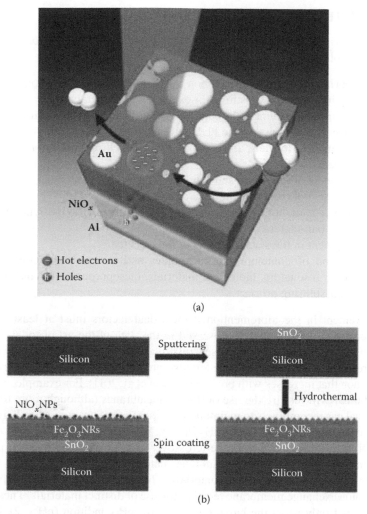

FIGURE 7.26   Three-layered metamaterialistic materials setup for water splitting: (a) puck-shaped discs about 10–30 nm in diameter with Al substrate, NiO_x, and Au NPs; (b) thin sheet of Si, SnO_2, and NiO_x decorated Fe_2O_3 nanorods (NRs).

minimum spacing between the metal and the underlying active semiconductor layers, therefore posing a fundamental limit to the overall thickness of the device. These challenges were conquered by Esfandyarpour and his team, who uses a metamaterial mirror whose reflection phase is tunable from that of a perfect electric mirror ($\varphi = \pi$) to that of a perfect magnetic mirror ($\varphi = 0$). This tunability in the reflection phase can also be exploited to optimize the standing wave profile in planar devices to maximize light–matter interaction. Specifically, we show that light absorption and photocurrent generation in a sub-100-nm active semiconductor layer of a model solar cell can be enhanced by ~20% over a broad spectral band.

This brand new field needs to be explored a lot but in the future metamaterials could be potential candidates for water splitting.

7.7 CHALLENGES AND PERSPECTIVES

The current prime challenges that are associated with the PEM electrolysis of water are given as follows:

1. Appropriate materials for reduction (metal loading) and oxidation of water are required that can substitute for expensive noble catalysts.
2. Decreasing the amount of the catalyst to be utilized in overall water spitting.
3. Enhance the efficiency of the PEC cell more than 10%.
4. Development of low cost and corrosion-resistant current collectors and separator plates.
5. Improvement of long-term stability/durability of all components.
6. Improvement of overall membrane characteristics.
7. Development of both empirically and physically predictive relations for operating parameters.
8. Development in the stacks concepts (Tadem cells).
9. Application of contemporary techniques and materials (metamaterials, plasmonic resonators, biological materials, mesoporous materials, etc.) in the water-splitting process.

Improvement in the aforementioned individual factors must at least maintain the same overall performance as observed in the state-of-the-art of today, because improving on one area may degrade several others. Problems related to higher operational pressures in PEM electrolysis are also present, such as the cross-permeation phenomenon that increases with pressure (Millet et al. 2011). For example, pressures above 100 bar will require the use of thicker membranes (although more resistant), and internal gas recombiners to maintain the critical concentrations (mostly H_2 in O_2) under the safety threshold (4 vol.% H_2 in O_2) (Millet et al. 2010, 2011). Lower gas permeability through the membrane (crossover) can be obtained by incorporating miscellaneous fillers inside the membrane material (Ornelas et al. 2009) but this normally leads to less conducting materials. The corrosive acidic regime provided by the proton exchange membrane requires the use of distinct materials. These materials must not only resist the harsh corrosive low pH condition (pH < 2), but also sustain the high applied overvoltage (<2 V), especially at high current densities. Corrosion resistance applies not only for the catalysts used, but also current collectors and separator plates. Only a few materials can be selected that would perform in this harsh environment. This will demand the use of scarce, expensive materials and components such as noble catalysts (platinum group metals [PGM], e.g., Pt, Ir, and Ru), titanium-based current collectors, and separator plates.

In addition to the plentiful advantages of nanomaterials, they have some disadvantages over conventional ones, that is, synthesis at various size scales, space charge layers that are not effective for separating electron–hole pairs, recombination that is enhanced in the absence of a bias, and a larger surface area that promotes defects, which induces recombination of carriers and makes electron–hole separation more difficult to achieve, reduces the photovoltage drawn from the absorber, and so on. Moreover, their long-term instability and high cost remain the key concerns for commercialization. A number of recombination phenomena including internal conversion, interface recombination, prominent thermal losses, and exciton binding losses are considered. Furthermore, the effects of new NPs and nanomaterials on the behavior of biological cells and health of living beings are not clear to date. Single-junction PV cells are typically limited by four fundamental loss mechanisms (Hanna and Nozik 2006). Nanostructured water-splitting devices have a few drawbacks such as incomplete solar spectrum absorption, thermalization of hot carriers (or excitons) in the form of excess heat, chemical potential (thermodynamic) losses dictating that the photovoltage must be less than the band gap (E_g) for relaxed carriers, and radiative recombination. The presence of other nonradiative losses, for example, Auger recombination (Green 1984), internal or phonon conversion, surface recombination (Tiedje et al. 1984), and so on, can further reduce cell performance (Green et al. 2011).

7.8 SUMMARY

This chapter deals with the various practical and theoretical aspects of the prominent oxide nanomaterial structures and their energetic transport dynamics along with the designing of the photocatalytic material. A small introduction and classification of optical devices is given on the basis of energetic transport dynamics. Limitation of the macrocrystalline and nanomaterials is also described. In looking to these limitations various strategies are planned to enhance the photoresponses of the nanomolecular device such as surface passivation, and development of new oxide or nonoxide or semioxide materials. Roles of different components of the water-splitting devices in the making of an efficient molecular device were discussed in light of interfacial electron-transfer reactions. This chapter highlights the importance of nanostructured materials by specifying their unique properties. The application of TFs, nanocrystalline, and mesoporous substances for water splitting is also discussed in detail. Nanomaterials can be employed as a photoelectrode material to improve PEC water-splitting performance to some extent, by increasing electrolyte accessible area and shortening minority carrier diffusion distance, but nanostructure engineering cannot change the intrinsic electronic properties of the material. Therefore, recent advancement in chemical, mechanical, and optical modification of nanostructured electrodes, including surface modification with catalyst and plasmonic metallic structures, metastructuring, element doping, and incorporation of the functional heterojunctions, led to significant enhancements in efficiencies by improvement in charge separation, transport, collection, solar energy harvesting, and suppression of recombination of charge carriers. Finally, the current challenges, modified research strategies, and future directions for water splitting are discussed with recommendations to facilitate the further exploration of new photoelectrode materials and their associated technologies by in-depth understanding of photocatalysts and photoactive material for solar-driven water splitting.

REFERENCES

Agarwal, U. and F. A. Escobedo. 2011. Mesophase behaviour of polyhedral particles. *Nat. Mater.* 10: 230–35.
Aharon-Shalom, E. and A. Heller. 1982. Efficient p-InP (Rh-H alloy) and p-InP (Re-H alloy) hydrogen evolving photocathodes. *J. Electrochem. Soc.* 129: 2865–66.

Ahn, C. H., W. S. Han, B. H. Kong, and H. K. Cho. 2009. Ga-doped ZnO nanorod arrays grown by thermal evaporation and their electrical behavior. *Nanotechnology* 20: 015601.

Ajuba, A. E., S. C. Ezugwu, P. U. Asogwa, and F. I. Ezema. 2010. Composition and optical characterization of ZnO/NiO multilayer thin film: Effect of annealing temperature. *Chalcogenide Lett.* 7: 573–79.

Al'tshuler, B. L. and P. A. Lee. 1988. Disordered electronic systems. *Phys. Today* 41(12): 36–44.

Atwater, H. A. and A. Polman. 2010. Plasmonics for improved photovoltaic devices. *Nat. Mater.* 9(3): 205–13.

Bard, A. J., 1979. Photoelectrochemistry and heterogeneous photocatalysis at semiconductors. *J. Photochem.* 10(1): 59–75.

Bard, A. J. and L. R. Faulkner. 2001. *Electrochemical Methods: Fundamentals and Applications*, 2nd edition. New York, NY: Wiley, p. xxi, 833.

Bawendi, M. G., M. L. Steigerwald, and L. E. Brus. 1990. The quantum mechanics of larger semiconductor clusters (quantum dots). *Annu. Rev. Phys. Chem.* 41: 477–96.

Beck, J. S., J. C. Vartuli, W. J. Roth, et al. 1992. A new family of mesoporous molecular sieves prepared with liquid crystal templates. *J. Am. Chem. Soc.* 114: 10834–43.

Becquerel AE. 1839. Recherches sur les effets de la radiation chimique de la lumiere solaire au moyen des courants electriques. *Comptes Rendus de L'Academie des Sciences* 9:145–149.

Bolton, J. R., S. J. Strickler, and J. S. Connolly. 1985. Limiting and realizable efficiencies of solar photolysis of water. *Nature* 316(6028): 495–500.

Bolts, J. M. and M. S. Wrighton. 1976. Correlation of photocurrent voltage curves with flat-band potential for stable photoelectrodes for the electrolysis of water. *J Phys. Chem.* 80: 2641–46.

Brabec, C. J., N. S. Sariciftci, and J. C. Hummelen. 2001. Plastic solar cells. *Adv. Funct. Mater.* 11: 15–26.

Butler, M. A. 1977. Photoelectrolysis and physical properties of semiconducting anode. *J. Appl. Phys.* 48: 1914–20.

Butler, M. A., R. D. Nasby, and R. K. Quinn. 1976. Tungsten trioxide as an electrode for photoelectrolysis of water. *Sol. State Commun.* 19: 1011–14.

Bux, S. K., R. G. Blair, P. K. Gogna, et al. 2009. Nanostructured bulk silicon as an effective thermoelectric material. *Adv. Funct. Mater.* 19: 2445–52.

Chaudhuri, R. G. and S. Paria. 2014. Visible light induced photocatalytic activity of sulfur doped hollow TiO_2 nanoparticles, synthesized via a novel route. *Dalton Trans.* 43: 5526–34.

Choudhary, S., A. Solanki, S. Upadhyay, et al. 2013. Nanostructured $CuO/SrTiO_3$ bilayered thin films for photoelectrochemical water splitting. *J. Sol. Stat. Electrochem.* 17(9): 2531–38.

Colfen, H. and M. Antonietti. 2005. Mesocrystals: Inorganic superstructures made by highly parallel crystallization and controlled alignment. *Angew. Chem. Int. Ed.* 44 : 5576–91.

Dai, P., W. Li, J. Xie, et al. 2014. Forming buried junctions to enhance the photovoltage generated by cuprous oxide in aqueous solutions. *Angew. Chem. Int. Ed.* 53: 13493–97.

Dang, T. C., D. L. Pham, H. L. Nguyen, and V. H. Pham. 2010. CdS sensitized ZnO electrodes in photoelectrochemical cells. *Adv. Nat. Sci. Nanosci. Nanotechnol.* 1(035010): 6.

Dao, M., L. Lu, R. J. Asaro, J. T. M. De Hosson, and E. Ma. 2007. Toward a quantitative understanding of mechanical behavior of nanocrystalline metals. *Acta. Mater.* 55: 4041–65.

Das, A., Z. Han, M. G. Haghighi, and R. Eisenberg. 2013. Photogeneration of hydrogen from water using CdSe nanocrystals demonstrating the importance of surface exchange. *Proc. Natl. Acad. Sci. USA* 110(42): 16716–23.

Dawkes, H. C. and S. E. V. Phillips. 2001. Copper amine oxidase: Cunning cofactor and controversial copper. *Curr. Opin. Struct. Biol.* 11: 666–73.

De Haart, L. G. J., A. W. Wiersma, G. Blasse, A. H. A. Tinnemans, and A. Mackor. 1981. The sensitization of SrTiO3 photoanodes by doping with various transition metal ions. *Mater. Res. Bull.* 16: 1593–600.

De Jongh, P. E., D. Vanmaekelbergh, and J. J. Kelly. 1999. Cu_2O: A catalyst for the photo-chemical decomposition of water? *Chem. Commun.* 12: 1069–70.

Debe, M. K., A. J. Steinbach, and K. Noda 2006. Stop-start and high-current durability testing of nanostructured thin film catalysts for PEM fuel cells. *ECS Trans.* 3(1): 835–53.

Debe, M. K., S. M. Hendricks, G. D. Vernstrom, et al. 2012. Initial performance and durability of ultra-low loaded NSTF electrodes for PEM electrolyzers. *J. Electrochem. Soc.* 159(6): K165–76.

Denisov, I. G., T. M. Makris, S. G. Sligar, and I. Schlichting. 2005. Structure and chemistry of cytochrome P450. *Chem. Rev.* 105: 2253–78.

Dong, F., S. Guo, H. Li, X. Wang, and Z. Wu. 2011. Enhancement of the visible light photocatalytic activity of C-doped TiO_2 nanomaterials prepared by a green synthetic approach. *J. Phys. Chem. C* 115: 13285–92.

Dotan, H., O. Kfir, E. Sharlin, et al. 2013. Resonant light trapping in ultrathin films for water splitting. *Nat. Mater.* 12: 158–64.

Du Bois, J. L. and J. P. Klinman. 2005. Mechanism of post-translational quinone formation in copper amine oxidases and its relationship to the catalytic turnover. *Arch. Biochem. Biophys.* 433: 255–65.

Duan, Y., N. Fu, Q. Liu, et al. 2012. Sn-doped TiO_2 photoanode for dye-sensitized solar cells. *J. Phys. Chem. C* 116(16): 8888–93.

Esfandyarpour, M., E. C. Garnett, Y. Cui, M. D. McGehee, and M. L. Brongersma. 2014. Metamaterial mirrors in optoelectronic devices *Nat. Nanotechnol.* 9: 542–7.

Fominykh, K., J. M. Feckl, J. Sicklinger, et al. 2014. Ultrasmall dispersible crystalline nickel oxide nanoparticles as high-performance catalysts for electrochemical water splitting. *Adv. Funct. Mater.* 24(21): 3123–29.

Fominykh, K., P. Chernev, I. Zaharieva, et al. 2015. Iron-doped nickel oxide nanocrystals as highly efficient electrocatalysts for alkaline water splitting. *ACS Nano* 9(5): 5180–88.

Forrest, S. R. 2005. The limits to organic photovoltaic cell efficiency. *MRS Bull.* 30: 28–32.

Fujishima, A. and K. Honda. 1971. Electrochemical evidence for the mechanism of the primary stage of photosynthesis. *Bull. Chem. Soc. Jpn.* 44: 1148–50.

Fujishima, A. and K. Honda. 1972. Electrochemical photolysis of water at a semiconductor electrode. *Nature* 238: 37–8.

Gibbons, N., J. J. Baumberg, C. L. Bower, M. Kolle, and U. Steiner. 2009. Scalable cylindrical metallodielectric metamaterials. *Adv. Mater.* 21: 3933–36.

Gratzel, M. 2001. Photoelectrochemical cells. *Nature* 414: 338–44.

Green, M. A. 1984. Limits on the open-circuit voltage and efficiency of silicon solar cells imposed by intrinsic Auger processes. *IEEE Trans. Electron Dev.* 31: 671–78.

Green, M. A., K. Emery, Y. Hishikawa, and W. Warta. 2011. Solar cell efficiency tables (version 37). *Prog. Photovoltaics Res. Appl.* 19(1): 84–92.

Greene, L. E., M. Law, J. Goldberger, et al. 2003. Low-temperature wafer-scale production of ZnO nanowire arrays. *Angew. Chem. Int. Ed.* 42: 3031–34.

Grewe, T. and H. Tüysüz. 2016. Alkali metals incorporated ordered mesoporous tantalum oxide with enhanced photocatalytic activity for water splitting. *J. Mater. Chem. A* 4: 3007–17.

Groves, J. T. 2006. High-valent iron in chemical and biological oxidations. *J. Inorg. Biochem.* 100: 434–47.

Han, N. R., Z. C. Chen, C. S. Lim, B. Ng, and M. H. Hong. 2011. Broadband multi-layer terahertz metamaterials fabrication and characterization on flexible substrates. *Opt. Express* 19: 6990–98.

Han, S., Y.-C. Pu, L. Zheng, L. Hu, J. Z. Zhang, and X. Fang. 2016. Uniform carbon-coated CdS core–shell nanostructures: Synthesis, ultrafast charge carrier dynamics, and photo-electrochemical water splitting. *J. Mat. Chem. A* 4: 1078–86.

Hanna, M. C. and A. J. Nozik. 2006. Solar conversion efficiency of photovoltaic and photo-electrolysis cells with carrier multiplication absorbers. *J. Appl. Phys.* 100: 074510.

Hara, M., T. Kondo, M. Komoda, et al. 1998. Cu_2O as a photocatalyst for overall water splitting under visible light irradiation. *Chem. Commun.* 3: 357–58.

Hawkes, A. M., A. R. Katko, and S. A. Cummer. 2013. A microwave metamaterial with integrated power harvesting functionality. *Appl. Phys. Lett.* 103: 163901.

Henderson, M. A. 2011. A surface science perspective on TiO_2 photocatalysis. *Surf. Sci. Rep.* 66: 185–297.

Hey, T. and P. Walters. 2003. *The New Quantum Universe*. Cambridge, UK: Cambridge University Press, p. 185.

Hiromoto, T., E. Warkentin, J. Moll, U. Ermler, and S. Shima. 2009. The crystal structure of an [Fe]-hydrogenase–substrate complex reveals the framework for H_2 activation. *Angew. Chem. Int. Ed.* 48: 6457–60.

Holmes, M. A., T. K. Townsend, and F. E. Osterloh. 2012. Quantum confinement controlled photocatalytic water splitting by suspended CdSe nanocrystals. *Chem. Commun.* 48: 371–73.

Hotchandani, S. and P.V. Kamat. 1992. Charge-transfer processes in coupled semiconductor systems. Photochemistry and photoelectrochemistry of the colloidal CdSeZnO system. *J. Phys. Chem.* 96: 6834–9.

Hsu, C.-H. and D.-H. Chen. 2012. CdS nanoparticles sensitization of Al-doped ZnO nanorod array thin film with hydrogen treatment as an ITO/FTO-free photoanode for solar water splitting. *Nano Exp. Nanoscale Res. Lett.* 7: 593.

Hu, Y.-S., A. Kleiman-Shwarsctein, A. J. Forman, D. Hazen, J.-N. Park, and E. W. McFarland. 2008. Pt-Doped α-Fe_2O_3 thin films active for photoelectrochemical water splitting. *Chem. Mater.* 20(12): 3803–805.

Huang, L., X. Chen, H. Muehlenbernd, et al. 2013. Three-dimensional optical holography using a plasmonic metasurface. *Nat. Commun.* 4: 2808.

Huang, Y., R. J. Nielsen, W. A. Goddard, and M. P. Soriaga. 2015. The reaction mechanism with free energy barriers for electrochemical dihydrogen evolution on MoS_2. *J. Am. Chem. Soc.* 137(20): 6692–98.

Hung, W. H., S. N. Lai, C. Y. Su, et al. 2015. Combined Au-plasmonic nanoparticles with mesoporous carbon material (CMK-3) for photocatalytic water splitting. *Phys. Lett.* 107: 073904.

Ibadurrohman, M. and K. Hellgardt. 2015. Morphological modification of TiO_2 thin films as highly efficient photoanodes for photoelectrochemical water splitting. *ACS Appl. Mater. Interfaces* 7(17): 9088–97.

Indrea, E., S. Dreve, T. D. Silipas, et al. 2009. Nanocrystalline semiconductor materials for solar water-splitting. *J. Alloy. Compd.* 483(1–2): 445–49.

Jacobsson, T. J., V. Fjällström, M. Sahlberg, M. Edoff, and T. Edvinsson. 2013. A monolithic device for solar water splitting based on series interconnected thin film absorbers reaching over 10% solar-to-hydrogen efficiency. *Energy Environ. Sci.* 6: 3676–83.

Jahagirdar, A. H. and N. G. Dhere. 2007. Photoelectrochemical water splitting using $CuIn_{1-x}Ga_xS_2$/CdS thin-film solar cells for hydrogen generation. *Sol. Energy Mater. Sol. Cells* 91(15–16): 1488–91.

Jaramillo, T. F., S. H. Baeck, A. Kleiman-Shwarsctein, K. S. Choi, G. D. Stucky, and E. W. McFarland. 2005. Automated electrochemical synthesis and photoelectrochemical characterization of $Zn_{1-x}Co_xO$ thin films for solar hydrogen production. *J. Comb. Chem.* 7(2): 264–71.

Ji, L., M. D. McDaniel, S. Wang, et al. 2015. A silicon-based photocathode for water reduction with an epitaxial $SrTiO_3$ protection layer and a nanostructured catalyst. *Nat. Nanotechnol.* 10: 84–90.

Jiaa, Q., K. Iwashinaa, and A. Kudo. 2012. Facile fabrication of an efficient $BiVO_4$ thin film electrode for water splitting under visible light irradiation. *Proc. Natl. Acad. Sci. USA* 109(29): 11564–69.

Jiang, Z. H., S. Yun, F. Toor, et al. 2011. Conformal dual-band near-perfectly absorbing mid-infrared metamaterial coating. *ACS Nano* 5(6): 4641–47.

Jurss, J. W., R. S. Khnayzer, J. A. Panetier, et al. 2015. Bioinspired design of redox-active ligands for multielectron catalysis: Effects of positioning pyrazine reservoirs on cobalt for electro- and photocatalytic generation of hydrogen from water. *Chem. Sci.* 6: 4954–72.

Kargar, A., J. S. Cheung, and C.-H. Liu. et al. 2015. NiO_x-Fe_2O_3-coated p-Si photocathodes for enhanced solar water splitting in neutral pH water. *Nanoscale* 7: 4900–05.

Kato, H., K. Asakura, and A. Kudo. 2003. Highly efficient water splitting into H_2 and O_2 over lanthanum-doped $NaTaO_3$ photocatalysts with high crystallinity and surface nanostructure. *J. Am. Chem. Soc.* 125(10): 3082–89.

Kato, H., Y. Sasaki, N. Shirakura, and A. Kudo. 2013. Synthesis of highly active rhodium-doped $SrTiO_3$ powders in Z-scheme systems for visible-light-driven photocatalytic overall water splitting. *J. Mater. Chem. A* 1(39): 12327–333.

Katz, J. E., T. R. Gingrich, E. A. Santori, and N. S. Lewis. 2009. Combinatorial synthesis and highthroughput photopotential and photocurrent screening of mixed-metal oxides for photoelectrochemical water splitting. *Energy Environ. Sci.* 2(1): 103–12.

Kay, A., I. Cesar, and M. Grätzel. 2006. New benchmark for water photooxidation by nanostructured α-Fe_2O_3 films. *J. Am. Chem. Soc.* 128(49): 15714–21.

Khan, S. U. M. and J. Akikusa. 1999. Photoelectrochemical splitting of water at nanocrystalline n-Fe_2O_3 thin-film electrodes. *J. Phys. Chem. B* 103(34): 7184–89.

Khnayzer, R. S., V. S. Thoi, M. Nippe, et al. 2014. Towards a comprehensive understanding of visible-light photogeneration of hydrogen from water using cobalt (ii) polypyridyl catalysts. *Energy Environ. Sci.* 7: 1477–88.

Kibria, M. G. and Z. Mi. 2016. Artificial photosynthesis using metal/nonmetal-nitride semiconductors: Current status, prospects, and challenges. *J. Mater. Chem. A* 4: 2801–20.

Kim, C. W., S. J. Yeoh, H.-M Cheng, and Y. S. Kang. 2015. A selectively exposed crystal facet-engineered TiO_2 thin film photoanode for the higher performance of the photoelectrochemical water splitting reaction. *Energy Environ. Sci.* 8: 3646–53.

Kim, J., H. Son, D. J. Cho, et al. 2012. Electrical control of optical plasmon resonance with graphene. *Nano Lett.* 12: 5598–602.

Kim S. J. , I. Thomann, J. Park, J. H. Kang , A. P. Vasudev, M. L. Brongersma. 2014. Light trapping for solar fuel generation with Mie resonances. *Nano Lett* 14: 1446–52.

Kim, W. D., J.-H. Kim, S. Lee, et al. 2016. Role of surface states in photocatalysis: Study of chlorine-passivated CdSe nanocrystals for photocatalytic hydrogen generation. *Chem. Mater.* 28(3): 962–68.

King, A. E., Y. Surendranath, N. A. Piro, J. P. Bigi, J. R. Long, and C. J. Chang. 2013. A mechanistic study of proton reduction catalyzed by a pentapyridine cobalt complex: Evidence for involvement of an anation-based pathway. *Chem. Sci.* 4: 1578–87.

Kock, W. E. 1946. Metal-lens antennas. *IRE Proc.* 34: 828–36R.

Konstandopoulos, A. G. and C. Agrofiotis. 2006. Hydrosol: Advanced monolithic reactors for hydrogen generation from solar water splitting. *Rev. Energies Renouv.* 9(N 3): 121–26.

Kresge, C. T., M. E. Leonowicz, W. J. Roth, J. C. Vartuli, and J. S. Beck. 1992. Ordered mesoporous molecular sieves synthesized by a liquid-crystal template mechanism. *Nature* 359(6397): 710–12.

Krol, R. 2012. Principles of photoelectrochemical cells. In: *Photoelectrochemical Hydrogen Production* (eds. R. van de Krol and M. Grätzel), Vol. 102. New York, NY: Springer, pp. 13–67.

Kronawitter, C. X., Z. Ma, D. Liu, S. S. Mao, and B. R. Antoun. 2012. Engineering impurity distributions in photoelectrodes for solar water oxidation. *Adv. Energy Mater.* 2: 52–7.

Kudo, A. and Y. Miseki. 2009. Heterogeneous photocatalyst materials for water splitting. *Chem. Soc. Rev.* 38: 253–78.

Kumar, P., P. Sharma, R. Shrivastav, S. Dass, and V. R. Satsangi. 2011. Electrodeposited zirconium-doped α-Fe_2O_3 thin film for photoelectrochemical water splitting. *Int. J. Hydrogen Energy* 36(4): 2777–784.

Lee, J., S. Mubeen, X. Ji, G. D. Stucky, and M. Moskovits. 2012a. Plasmonic photoanodes for solar water splitting with visible light. *Nano Lett.* 12: 5014–19.

Lee, M. H., K. Takei, J. Zhang, R. Kapadia, et al. 2012b. p-Type InP nanopillar photocathodes for efficient solar-driven hydrogen production. *Angew. Chem. Int. Ed.* 51: 10760–64.

Lewis, N. S., G. Crabtree, A. J. Nozik, M. R. Wasielewski, and A. P. Alivisatos. 2005. Basic research needs for solar energy utilization. Department of Energy. Available at www.science.energy.gov/bes/news-and-resources/reports/.

Li, J., C. M. Shah, W. Withayachumnankul, et al. 2013. Flexible terahertz metamaterials for dual-axis strain sensing. *Opt. Lett.* 38(12): 2104–06.

Liao, L., Q. Zhang, Z. Su, et al. 2014. Efficient solar water-splitting using a nanocrystalline CoO photocatalyst. *Nat. Nano* 9(1): 69–73.

Licht, S. 2001. Multiple band gap semiconductor/electrolyte solar energy conversion. *J. Phys. Chem. B* 105: 6281–94.

Licht, S., B. Wang, S. Mukerji, T. Soga, M. Umeno, and H. Tributsch. 2000. Efficient solar water splitting, exemplified by RuO_2-catalyzed AlGaAs/Si photoelectrolysis. *J. Phys. Chem. B* 104: 8920.

Lin Y., R. Kapadia, J. Yang, et al. 2015. Role of TiO_2 surface passivation on improving the performance of p-InP photocathodes. *J. Phys. Chem. C* 119(5): 2308–313.

Lin, Y., S. Zhou, S. W. Sheehan, and D. Wang. 2011. Nanonet-based hematite heteronanostructures for efficient solar water splitting. *J. Am. Chem. Soc.* 133: 2398–401.

Liu, C., J. Sun, J. Tang, and P. Yang. 2012. Zn-doped p-Type gallium phosphide nanowire photocathodes from a surfactant-free solution synthesis. *Nano Lett.* 12: 5407–11.

Liu, J., Y. Liu, N. Liu, et al. 2015. Metal-free efficient photocatalyst for stable visible water splitting via a two-electron pathway. *Sci.* 347(6225): 970–974.

Loiudice, A., J. K. Cooper, L. H. Hess, T. M. Mattox, I. D. Sharp, and R. Buonsanti. 2015. Assembly and photocarrier dynamics of heterostructured nanocomposite photoanodes from multicomponent colloidal nanocrystals. *Nano Letters* 15(11): 7347–54.

Long, M. C., W. M. Cai, and H. Kisch. 2008. Visible light induced photoelectrochemical properties of n-$BiVO_4$ and n-$BiVO_4$/p-Co_3O_4. *J. Phys. Chem. C* 112: 548–54.

Lucht, K. P. and J. L. Mendoza-Cortes. 2015. Birnessite: A layered manganese oxide to capture sunlight for water-splitting catalysis. *J. Phys. Chem. C* 119(40): 22838–846.

Lunkenbein, T., J. Schumann, M. Behrens, R. Schlögl, and M. G. Willinger. 2015. Formation of a ZnO overlayer in industrial Cu/ZnO/Al_2O_3 catalysts induced by strong metal–support interactions. *Angew. Chem. Int. Ed.* 54(15): 4544–48.

Luo, C. Y., S. G. Johnson, J. D. Joannopoulos, and J. B. Pendry. 2003. Subwavelength imaging in photonic crystals. *Phys. Rev. B* 68: 045115.

Luo, J., J.-H. Im, M. T. Mayer, et al. 2014. Water photolysis at 12.3% efficiency via perovskite photovoltaics and earth-abundant catalysts. *Science* 345: 1593–96.

Lyons, C. T. and T. D. P. Stack. 2013. Recent advances in phenoxyl radical complexes of salen-type ligands as mixed-valent galactose oxidase models. *Coord. Chem. Rev.* 257: 528–40.

Ma, S., Q. Zhu, Z. Zheng, W. Wang, and D. Chen. 2015. Nanosized $LiNi_{1-x}Fe_xPO_4$ embedded in a mesoporous carbon matrix for high-performance electrochemical water splitting. *Chem. Commun.* 51: 15815–18.

Maeda, K. 2013. Z-Scheme water splitting using two different semiconductor photocatalysts. *ACS Catal.* 3(7): 1486–503.

Maeda, K., K. Teramura, D. L. Lu, et al. 2006. Photocatalyst releasing hydrogen from water-enhancing catalytic performance holds promise for hydrogen production by water splitting in sunlight. *Nature* 440 (7082): 295.

Maeda, K., K. Teramura, and K. Domen. 2008. $Ga_{1-x}Zn_x)(N_{1-x}O_x)$ solid solution for overall water splitting under visible light. *J. Catal.* 254: 198–204.

Mali, M. G., H. Yoon, B. N. Joshi, et al. 2015. Enhanced photoelectrochemical solar water splitting using a platinum-decorated CIGS/CdS/ZnO Photocathode. *ACS Appl. Mater. Interfaces* 7(38): 21619–25.

Malingowski, A. C., P. W. Stephens, A. Huq, Q. Z. Huang, S. Khalid, and P. G. Khalifah. 2012. Substitutional mechanism of Ni into the wide-band-gap semiconductor in TaO_4 and its implications for water splitting activity in the wolframite structure type. *Inorg. Chem.* 51(11): 6096–103.

Malviya, K. D., H. Dotan, D. Shlenkevich, A. Tsyganok, H. Mor, and A. Rothschild. 2016. Systematic comparison of different dopants in thin film hematite (α-Fe_2O_3) photoanodes for solar water splitting. *J. Mater. Chem.* A 4: 3091–99.

Matoba, T., K. Maeda, and K. Domen. 2011. Activation of $BaTaO_2N$ photocatalyst for enhanced non-sacrificial hydrogen evolution from water under visible light by forming a solid solution with $BaZrO_3$. *Chem. A Eur. J.* 17(52): 14731–35.

Mavroides, D., I. Tchernev, J. A. Kafalas, and D. F. Kolesar. 1975. Photoelectrolysis of water in cells with TiO_2 anodes. *Mater. Res. Bull.* 10: 1023–30.

Mavroides, J. G., J. A. Kafalas, and D. F. Kolesar. 1976. Photoelectrolysis of water in cells with SrTiO3 anodes. *Appl. Phys. Lett.* 28: 241–43.

McKone, J. R., A. P. Pieterick, H. B. Gray, and N. S. Lewis. 2012. Hydrogen evolution from Pt/Ru-coated p-type WSe_2 photocathodes. *J. Am. Chem. Soc.* 135: 223–31.

Meyer, K., M. Ranocchiari, and J. A. Van Bokhoven. 2015. Metal organic frameworks for photo-catalytic water splitting. *Energy Environ. Sci.* 8: 1923–37.

Millet, P., R. Ngameni, S., A. Grigoriev, et al. 2010. PEM water electrolyzers: From electrocatalysis to stack development. *Int. J. Hydrogen Energy* 35(10): 5043–52.

Millet, P., R. Ngameni, S., A. Grigoriev. and V. N. Fateev. 2011. Scientific and engineering issues related to PEM technology: Water electrolysers, fuel cells and unitized regenerative systems. *Int. J. Hydrogen Energy* 36(6): 4156–63.

Miyamaru, F., M. W. Takeda, and K. Taima. 2009. Characterization of terahertz metamaterials fabricated on flexible plastic films: Toward fabrication of bulk metamaterials in terahertz region. *Appl. Phys. Exp.* 2: 042001.

Moriya, M., T. Minegishi, H. Kumagai, M. Katayama, J. Kubota, and K. Domen. 2013. Stable hydrogen evolution from CdS-modified $CuGaSe_2$ photoelectrode under visible-light irradiation. *J. Am. Chem. Soc.* 135: 3733–35.

Mueller, A., E. Krickmeyer, and J. Meyer. 1995. $Mo_{154}(NO)_{14}O_{420}(OH)_{28}(H_2O)_{70}]^{(25 \pm 5)-}$: A water-soluble big wheel with more than 700 atoms and a relative molecular mass of about 24000. *Angew. Chem. Int. Ed.* 34: 2122–24.

Muñoz, A. G., C. Heine, M. Lublow, et al. 2013. Photoelectrochemical conditioning of MOVPE p-InP films for light-induced hydrogen evolution: Chemical, electronic and optical properties. *ECS J. Solid State Sci. Technol.* 2: Q51–58.

Mure, M. 2004. Tyrosine-derived quinone cofactors. *Acc. Chem. Res.* 37: 131–39.

Nair, M. P., K. V. C. Rao, and C. G. R. Nair. 1991. Investigation of the mixed oxide materials—TiO_2- SiO_2, TiO_2- SiO_2- Al_2O_3, TiO_2- SiO_2- In_2O_3 and TiO_2-SiO_2-RuO_2—In regard to the photoelectrolysis of water. *Int. J. Hydrogen Energy* 16(7): 449–59.

Ng, J., S. Xu, X. Zhang, H. Y. Yang, and D. D. Sun. 2010. Hybridized nanowires and cubes: A novel architecture of a heterojunctioned TiO_2/$SrTiO_3$ thin film for efficient water splitting. *Adv. Funct. Mater.* 20(24): 4287–94.

Ni, M., M. K. H. Leung, D. Y. C. Leung, and K. Sumathy. 2007. A review and recent developments in photocatalytic water-splitting using for hydrogen production. *Renew. Sust. Energy Rev.* 11: 401–25.

Nozik, A. J. 1975. Photoelectrolysis of water using semiconducting TiO$_2$ crystals. *Nature* 257: 383–86.

Ohnishi, T., Y Nakato, and H. Tsubumura 1975. Quantum yield of photolysis of water on titanium oxide. *Ber. Bunsenges. Phys. Chem.* 79: 523–25.

Ohno T., L. Bai, T. Hisatomi, K. Maeda, and K. Domen. 2012. Photocatalytic water splitting using modified GaN:ZnO solid solution under visible light: Long-time operation and regeneration of activity. *J. Am. Chem. Soc.* 134(19): 8254–59.

Ornelas, R., V. Baglio, F. Matteucci, et al. 2009. Solid polymer electrolyte water electrolyser based on Nafion-TiO$_2$ composite membrane for high temperature operation. *Fuel Cells* 9(3): 247–52.

Osterloh, F. E. 2016a. Nanoscale effects in water splitting photocatalysis. In: *Top Current Chemistry Series on Solar Energy for Fuels* (eds. H. Tüysüz and C. K. Chan). Switzerland: Springer, p. 104.

Osterloh, F. E. 2016b. Nanoscale effects in water splitting photocatalysis. *Top. Curr. Chem.* 371: 105–42.

Paracchino, A., N. Mathews, T. Hisatomi, M. Stefik, S. D. Tilley, and M. Gratzel. 2012. Ultrathin films on copper (I) oxide water splitting photocathodes: A study on performance and stability. *Energy Environ. Sci.* 5: 8673–81.

Peharz, G., F. Dimroth, and U. Wittstadt. 2007. Solar hydrogen production by water splitting with a conversion efficiency of 18%. *Int. J. Hydrogen Energy* 32(15): 3248–52.

Pendry, J. B., D. Schurig, J. J. Mock, et al. 2006. Metamaterial electromagnetic cloak at microwave frequencies. *Science* 314(5801): 977–80.

Phadke, S., J. Y. Lee, J. West, P. Peumans, and A. Salleo. 2011. Using alignment and 2D network simulations to study charge transport through doped ZnO nanowire thin film electrodes. *Adv. Funct. Mater.* 21: 4691–97.

Pratt, R. C., C. T. Lyons, E. C. Wasinger, and T. D. P. Stack. 2012. Electrochemical and spectroscopic effects of mixed substituents in bis (phenolate)–copper (II) galactose oxidase model complexes. *J. Am. Chem. Soc.* 134: 7367–77.

Puangpetch, T., T. Sreethawong, S. Yoshikawa, and S. Chavadej. 2009. Hydrogen production from photocatalytic water splitting over mesoporous-assembled SrTiO$_3$ nanocrystal-based photocatalysts. *J. Mol. Catal. A Chem.* 312(1–2): 97–100.

Radecka, M., M. Rekas, A. Trenczek-Zajac, and K. Zakrzewska. 2008. Importance of the band gap energy and flat band potential for application of modified TiO$_2$ photoanodes in water photolysis. *J. Power Sources* 181: 46–55.

Radecka M., A. Trenczek-Zajac, K. Zakrzewska and M. Rekas 2007. Effect of oxygen non-stoichiometry on photo-electrochemical properties of TiO$_2$–x, *J. Power Sources* 173: 816–821

Radecka M., M. Wierzbicka, S. Komornicki, and M. Rekas. 2004. Influence of Cr on photo-electrochemical properties of TiO$_2$ thin films. *Physica B* 348: 160–168.

Ramahi, O. M., T. S. Almoneef, M. AlShareef, and M. S. Boybay. 2012. Metamaterial particles for electromagnetic energy harvesting. *Appl. Phys. Lett.* 101: 173903.

Robatjazi, H., S. M. Bahauddin, C. Doiron and I. Thomann. 2015. Direct plasmon-driven photoelectrocatalysis. *Nano Lett.* 15(9): 6155–61.

Sabio E. M., R. L. Chamousis, N. D. Browning, and F. E. Osterloh. 2012. Photocatalytic water splitting with suspended calcium niobium oxides: Why nanoscale is better than bulk—A kinetic analysis. *J. Phys. Chem. C* 116(4): 3161–70.

Salvador, P. 1980. The influence of niobium doping on the efficiency of n-TiO$_2$ electrode in water photoelectrolysis. *Sol. Energy Mater.* 4: 413–21.

Schölin, R. M. Quintana, E. M. Johansson, et al. 2011. Preventing dye aggregation on ZnO by adding water in the dye-sensitization process. *J. Phys. Chem. C* 115: 19274–79.

Shalaev, V. M. 2007. Optical negative-index metamaterials. *Nature Photon.* 1: 41–48.

Shaner, M. R., K. T. Fountaine, S. Ardo, et al. 2014. Photoelectrochemistry of core-shell tandem junction n-p(+)-Si/n-WO_3 microwire array photoelectrodes. *Energy Environ. Sci.* 7: 779–790.

Sieber, K. D., C. Sanchez, J. E. Turner, and G. A. Somorjai. 1985. Preparation, electrical and photoelectrochemical properties of magnesium doped iron oxide sintered discs. *Mat. Res. Bull.* 20: 153–62.

Sivula, K., F. Le Formal, and M. Graetzel. 2011. Solar water splitting: Progress using hematite (α-Fe(2) O(3)) photoelectrodes. *ChemSusChem* 4: 432–49.

Smith, W., H. Fakhouri, J. Pulpytel, et al. 2012. Visible light water splitting via oxidized TiN thin films. *J. Phys. Chem. C* 116(30): 15855–66.

Srivastava, S., J. P. Thomas, M. A. Rahman, et al. 2014. Size-selected TiO_2 nanocluster catalysts for efficient photoelectrochemical water splitting. *ACS Nano* 8(11): 11891–98.

Standing, A., S. Assali, L. Gao, et al. 2015. Efficient water reduction with gallium phosphide nanowires. *Nat. Commun.* 6: 7824.

Staszak-Jirkovský, J., C. D. Malliakas, P. P. Lopes, et al. 2016. Design of active and stable Co–Mo–S_x chalcogels as pH-universal catalysts for the hydrogen evolution reaction. *Nat. Mater.* 15: 197–203.

Stubbe, J. and W. A. van der Donk. 1998. Protein radicals in enzyme catalysis. *Chem. Rev.* 98: 705–62.

Sun, L. W., H. Q., Shi, W. N. Li, et al. 2012. Lanthanum-doped ZnO quantum dots with greatly enhanced fluorescent quantum yield. *J. Mater. Chem.* 22: 8221–27.

Sun, Y., J. P. Bigi, N. A. Piro, et al. 2011. Molecular cobalt pentapyridine catalysts for generating hydrogen from water. *J. Am. Chem. Soc.* 133: 9212–15.

Sun, Y., J. Sun, J. R. Long, P. Yang, and C. J. Chang. 2013. Photocatalytic generation of hydrogen from water using a cobalt pentapyridine complex in combination with molecular and semiconductor nanowire photosensitizers. *Chem. Sci.* 4: 118–24.

Tao, H. A. C. Strikwerda, K. Fan, W. J. Padilla, X. Zhang, and R. D. Averitt. 2009. Reconfigurable terahertz metamaterials. *Phys. Rev. Lett.* 103: 147401.

Tian, Z. R., J. A. Voigt, J. Liu, et al. 2003. Complex and oriented ZnO nanostructures. *Nat. Mater.* 2: 821–26.

Tiedje, T. E., Yablonovitch, G. D. Cody and B. G. Brooks. 1984. Limiting efficiency of silicon solar cells. *IEEE Trans. Electron Dev.* 31: 711–716.

Tong, L., R. Zong, and R. P. Thummel. 2014. Visible light-driven hydrogen evolution from water catalyzed by a molecular cobalt complex. *J. Am. Chem. Soc.* 136: 4881–84.

Townsend, T. K., N. D. Browning, and F. E. Osterloh. 2012. Nanoscale strontium titanate photocatalysts for overall water splitting. *ASC Nano* 6(8): 7420–26.

Tumkur, G. Zhu, P. Black, Y. A. Barnakov, C. E. Bonner, and M. A. Noginov. 2011. Control of spontaneous emission in a volume of functionalized hyperbolic metamaterial. *Appl. Phys. Lett.* 99: 151115.

Varghese, O. K. and C. A. Grimes. 2008. Appropriate strategies for determining the photoconversion efficiency of water photo electrolysis cells: A review with examples using titania nanotube array photoanodes. *Sol. Energy Mater. Sol. C* 92(4): 374–84.

Verma, P., R. C. Pratt, T. Storr, E. C. Wasinger, and T. D. P. Stack. 2011. Sulfanyl stabilization of copper-bonded phenoxyls in model complexes and galactose oxidase. *Proc. Natl. Acad. Sci. USA* 108: 18600–605.

Vielstich, W., A. Lamm, and H. A. 2003. Gasteiger. *Handbook of Fuel Cells: Fundamentals, Technology, and Applications.* Chichester, England; New York: Wiley.

Vinke, I. C., J. Diepgrond, B. A. Boukamp, K. J. de Vries, and A. Burggraaf. 1992. Bulk and electrochemical properties of $BiVO_4$. *J. Solid State Ionics* 57: 83–89.

Walser, M. 2001. Electromagnetic metamaterials. In: *Proceedings of the SPIE 4467 Complex Mediums II: Beyond Linear Isotropic Dielectrics, Volume 1*. San Diego, CA, pp. 1–15.

Walter, M. G., E. L. Warren, J. R. McKone, et al. 2010. Solar water splitting cells. *Chem. Rev.* 110(11): 6446–73.

Wan, Y. and D. Zhao. 2007. On the controllable soft-templating approach to mesoporous silicates. *Chem. Rev.* 107: 2821–60.

Wang, C., T. Hisatomi, T. Minegishi, et al. 2016. Synthesis of nanostructured $BaTaO_2N$ thin films as photoanodes for solar water splitting. *J. Phys. Chem. C* 120(29): 15758–764.

Wang, H. and A. L. Rogach. 2014. Hierarchical SnO_2 nanostructures: Recent advances in design, synthesis, and applications. *Chem. Mater.* 26(1): 123–33.

Wang H., H. W. Lee, Y. Deng, et al. 2015. Bifunctional non-noble metal oxide nanoparticle electrocatalysts through lithium-induced conversion for overall water splitting. *Nat. Commun.* 6: 7261.

Wang, G., X. Yang, F. Qian, J. Z. Zhang, and Y. Li. 2010. Double-sided CdS and CdSe quantum dot co-sensitized ZnO nanowire arrays for photoelectrochemical hydrogen generation. *Nano Lett.* 10: 1088–92.

Wang, M., F. Ren, J. Zhou, et al. 2015. N doping to ZnO nanorods for photoelectrochemical water splitting under visible light: Engineered impurity distribution and terraced band structure. *Sci. Rep.* 5: 12925.

Wang, T., Z. Luo, C. Li, and J. Gon. 2014. Controllable fabrication of nanostructured materials for photoelectrochemical water splitting via atomic layer deposition. *Chem. Soc. Rev.* 43: 7469–84.

Warren, S. C. and E. Thimsen. 2012. Plasmonic solar water splitting. *Energy Environ. Sci.* 5: 5133–46.

Watts, C. M., X. Liu, and W. J. Padilla. 2012. Metamaterial electromagnetic wave absorbers. *Adv. Mater.* 24: OP98–120.

Whittaker, J. W. 2003. Free radical catalysis by galactose oxidase. *Chem. Rev.* 103: 2347.

Wong, M. S., H. P. Chou, and T. S.Yang. 2006. Reactively sputtered N-doped titanium oxide films as visible-light photocatalyst. *Thin Solid Films* 494: 244–49.

Woodhouse, M. and B. A. Parkinson. 2009. Combinatorial approaches for the identification and optimization of oxide semiconductors for efficient solar photoelectrolysis. *Chem. Soc. Rev.* 38(1): 197–210.

Wrighton, M. S., D. L. Morse, A. B. Ellis, D. S. Ginley, and H. B. Abrahamson. 1976a. Photoassisted electrolysis of alkaline water (12.0 M NaOH) by ultraviolet irradiation (313 nm) of an antimony doped stannic oxide electrode. *J. Am. Chem. Soc.* 98: 44–8.

Wrighton, M. S., A. B. Ellis, P. T. Wolczanski, D. L. Morse, H. B. Abrahamson, and D. S. Ginley. 1976b. Strontium titanate photoelectrodes. Efficient photoassisted electrolysis of water at zero applied potential. *J. Am. Chem. Soc.* 98: 2774–79.

Würfel, P., 2005. *Physics of Solar Cells*. Vol. I. Berlin, Germany: Wiley-VCH, p. 244.

Xu, L., L. Sang, C. Ma, et al. 2006. Preparation of mesoporous InVO4 photocatalyst and its photocatalytic performance for water splitting. *Chin. J. Catal.* 27(2): 100–02.

Xu, X., B. Peng, D. Li, J. Zhang, et al. 2011. Flexible visible–infrared metamaterials and their applications in highly sensitive chemical and biological sensing. *Nano Lett.* 11: 3232–38.

Yang, F., K. Sun, and S. R. Forrest. 2007. Efficient solar cells using all-organic nanocrystalline networks. *Adv. Mater.* 19: 4166–71.

Yang, J., C. Bao, T. Yu, et al. 2015. Enhanced performance of photoelectrochemical water splitting with ITO@α-Fe_2O_3 core–shell nanowire array as photoanode. *ACS Appl. Mater. Interfaces* 7(48): 26482–90.

Yang, J. L., S. J. An, W. I. Park, G. C. Yi, and W. Choi. 2004. Photocatalysis using ZnO thin films and nanoneedles grown by metal–organic chemical vapor deposition. *Adv. Mater.* 16:1661–64.

Yang, X., A. Wolcott, G. Wang, et al. 2009. Nitrogen-doped ZnO nanowire arrays for photo-electrochemical water splitting. *Nano Lett.* 9: 2331–36.

Yin, J., J. Ye, and Z. Zou. 2004. Enhanced photoelectrolysis of water with photoanode Nb:SrTiO$_3$. *Appl. Phys. Lett.* 85: 689–91.

Youn, D. H., S. Han, J. Y. Kim, et al. 2014. Highly active and stable hydrogen evolution electrocatalysts based on molybdenum compounds on carbon nanotube–graphene hybrid support. *ACS Nano* 8(5): 5164–73.

Zakrzewska, K., A. Brudnik, M. Radecka, and W. Posadowski. 1999. Reactively sputtered TiO$_{2-x}$ thin films with plasma-emission controlled departure from stoichiometry. *Thin Solid Films* 343: 152–55.

Zhang, J. and W. Que. 2010. Preparation and characterization of sol–gel Al-doped ZnO thin films and ZnO nanowire arrays grown on Al-doped ZnO seed layer by hydrothermal method. *Sol. Energy Mater. Sol. Cells* 94: 2181–86.

Zhang, X., J. Qin, Y. Xue, et al. 2014. Effect of aspect ratio and surface defects on the photocatalytic activity of ZnO nanorods. *Sci. Rep.* 4: 4596.

Zhao, J., E. Olide, and F. E. Osterloh. 2015. Enhancing majority carrier transport in WO$_3$ water oxidation photoanode via electrochemical doping. *J. Electrochem. Soc.* 162(1): H65–71.

Zou, Z. G. and H. Arakawa. 2003. Direct water splitting into H$_2$ and O$_2$ under visible light irradiation with a new series of mixed oxide semiconductor photocatalysts. *J. Photochem. Photobiol. A Chem.* 158(2): 145–62.

Zou, Z. G., J. H. Ye, K. Sayama, and H. Arakawa. 2001. Direct splitting of water under visible light irradiation with an oxide semiconductor photocatalyst. *Nature* 414(6864): 625–27.

Index

Printed and bound by CPI Group (UK) Ltd, Croydon, CR0 4YY

01/11/2024

01782614-0012